Conformations and Forces in Protein Folding

Conformations and Forces in Protein Folding

Edited by
Barry T. Nall and Ken A. Dill

AMERICAN ASSOCIATION FOR THE ADVANCEMENT OF SCIENCE

Library of Congress Cataloging-in-Publication Data

Nall, Barry T., and Dill, Ken A.
Conformations and Forces in Protein Folding / Barry T. Nall and Ken A. Dill
p. cm.
Includes bibliographic references.
ISBN 0-87168-394-6
1. Protein folding.
I. Nall, Barry T. II. Dill, Ken A.

QP551.C719 1991
547.7′5 – dc20 91-7369
CIP

Publication No. 91-05S

1333 H Street, N.W., Washington, D.C. 20005

Contents

Preface

The protein folding problem is deceptively simple to state. It is known that a globular protein adopts its three-dimensional equilibrium structure spontaneously under normal physiological conditions *(1)*. No cellular machinery is needed in this process; a linear polymer of amino acid residues in aqueous solvent will self-assemble automatically. Therefore, information that specifies structure must be encrypted within the amino acid sequence itself, functioning as an embedded stereochemical code. The challenge, then, is to crack the code. It is a problem of central importance in 20th century biology.

Impelled by the human genome initiative, the folding problem has taken on a new sense of urgency. Even a partial solution such as the successful prediction of α helix and β sheet would simplify enormously our ability to identify a sequence or an open reading frame as belonging to a functional class, because structure is far better conserved than sequence among related proteins. The folding problem and the genome initiative are often coupled for this reason. However, the two projects are different in kind. The genome initiative is a complex engineering effort, like putting a person on the moon. Debate ranges over such questions as, Who pays for it? Who works on it? and Which technologies are best? In contrast, protein folding remains an unsolved research problem. The relevant question is, How does it work?

Hard-hitting, box-office attractions lead inevitably to sequels. This authoritative volume is an outgrowth of the 1990 sequel to the now-familiar three-day seminar on protein folding sponsored by the American Association for the Advancement of Science (AAAS), which was organized by Barry T. Nall and Ken A. Dill. *Conformations and Forces in Protein Folding* is itself a welcome sequel to *Protein Folding: Deciphering the Second Half of the Genetic Code (2),* the companion volume resulting from an earlier AAAS seminar on folding.

This book is divided into four sections, each preceded by a short introduction. Part I, "Compact States, Electrostatics, and Folding," includes successful applications of theory to understanding the physical forces that determine conformation. Part II, "Relation of Amino Acid Sequence to Structure and Folding," presents the latest experimental approaches to complete structure determination and emphasizes nuclear magnetic resonance and x-ray crystallography. Part III, "Folding Mechanisms," is concerned with the *molten globule,* a key folding intermediate having native-like secondary structure with little or no specific tertiary structure. Finally, Part IV, "Auxiliary Factors and Folding: Membranes and Catalysis," turns to the crucial question of how proteins fold within a cell.

An especially valuable feature of this book is the sense of new direction that emerges from the juxtaposition of these disparate topics. Here is an example.

What does the molten globule, as studied by biophysical chemists, have to do with transport into the endoplasmic reticulum, the usual province of cell biologists? Most eukaryotic proteins are exported, and all export is gated through the endoplasmic reticulum (ER). From the cell's point of view, it is important to understand these molecular traffic signals because folding may be arrested at some intermediate state in those proteins destined for export. In this broadened perspective, *folding can be regarded as a regulatory process.*

Is there any connection between the molten globule and cellular transport? Auxiliary factors (Part IV) are involved in cellular regulation, and they appear to exert control by recognizing partially folded intermediates of exported proteins. If so, then the cell knows how to exploit general properties of protein folding. Like the cell, protein folders maintain a vital interest in partially folded intermediate states. The molten globule (Part III) is such a state. Those who follow this literature know that a molten globule-like state seems to be present in some proteins but conspicuously absent in others and that different investigators define the state differently. Reading this book, we are prompted to associate the following: (i) auxiliary factors can recognize general folding intermediates, and (ii) under suitable conditions, the molten globule state appears to be a general, on-pathway intermediate. Perhaps these ideas are connected.

This provocative volume is rich with such connections. It makes a valuable contribution to the protein folding literature.

George D. Rose

References

1. Anfinsen, C. B., *Science* **181,** 223 (1973).
2. Gierasch, L. M. and J. King, Eds. *Protein Folding: Deciphering the Second Half of the Genetic Code* (American Association for the Advancement of Science, Washington, D.C., 1990).

Contributors

Norma M. Allewell, Department of Molecular Biology and Biochemisty, Hall-Atwater-Shanklin Laboratories, Wesleyan University, Middleton, CT

Stephen Anderson, Center for Biotechnology, Rutgers University, Piscataway, NJ

Jean Baum, Department of Chemistry, Rutgers University, New Brunswick, NJ

Brandan A. Borgias, Cray Research Inc., San Ramon, CA

Linda J. Calciano, Department of Chemistry, University of California, Santa Cruz, CA

Hue Sun Chan, Department of Pharmaceutical Chemistry, Medical Sciences Building, University of California, San Francisco, CA

G. Marius Clore, Laboratory of Chemistry and Physics, National Institute of Diabetes and Digestive and Kidney Diseases, Bethesda, MD

Ken A. Dill, Department of Pharmaceutical Chemistry, Medical Science Building, University of California, San Francisco, CA

Christopher M. Dobson, Inorganic Chemistry Laboratory and Center for Molecular Sciences, University of Oxford, England

Philip A. Evans, Department of Biochemistry, University of Cambridge, Cambridge, England

Anthony L. Fink, Department of Chemistry, University of California, Santa Cruz, CA

Robert B. Freedman, Biological Laboratory, University of Kent, Canterbury, England

Lorrie D. Garvin, Department of Biochemistry, University of Texas Health Science Center, San Antonio, TX

Lila M. Gierasch, Department of Pharmacology, University of Texas Southwestern Medical Center, Dallas, TX

Michael K. Gilson, Department of Biochemistry and Molecular Biophysics, Columbia University, New York, NY

Mary P. Glackin, Department of Microbiology and Molecular Genetics, University of Vermont School of Medicine, Burlington, VT

Yuji Gotto, Department of Biology, Osaka University, Osaka, Japan

Angela M. Gronenborn, Laboratory of Chemistry and Physics, National Institute of Diabetes and Digestive and Kidney Diseases, Bethesda, MD

Claire Hanley, Inorganic Chemistry Laboratory and Center for Molecular Sciences, University of Oxford, England

Stephen C. Hardies, Department of Biochemistry, University of Texas Health Science Center, San Antonio, TX

Barry Honig, Department of Biochemistry, Columbia University, New York, NY

Paul M. Horowitz, Department of Biochemistry, University of Texas Health Science Center, San Antonio, TX

Thomas L. James, Department of Pharmaceutical Chemistry, University of San Francisco, San Francisco, CA

Thomas Kiefhaber, Laboratorium für Biochemie, Universität Bayreuth, West Germany

Irwin D. Kuntz, Department of Pharmaceutical Chemistry, University of San Francisco, San Francisco, CA

Kurt Lang, Laboratorium für Biochemie, Universität Bayreuth, West Germany

James B. Matthew, ICI Pharmaceuticals Group, ICI Americas Inc., Wilmington, DE

Sabine Mayer, Laboratorium für Biochemie, Universität Bayreuth, West Germany

Edith Wilson Miles, Laboratory of Biochemistry and Pharmacology, National Institutes of Health, Bethesda, MD

Hossein M. Naderi, Cray Research Inc., San Ramon, CA

Barry T. Nall, Department of Biochemistry, University of Texas Health Science Center, San Antonio, TX

C. N. Pace, Department of Biochemistry, Texas A & M University, College Station, TX

Daniel Palleros, Department of Chemistry, University of California, Santa Cruz, CA

Joseph H. B. Pease, Syntex Research, Palo Alto, CA

Oleg B. Ptitsyn, Institute of Protein Research, Academy of Sciences of the USSR, Moscow Region, USSR

Sheena E. Radford, Inorganic Chemistry Laboratory and Center for Molecular Sciences, University of Oxford, England

George D. Rose, Department of Biological Chemistry, Pennsylvania State University, Hershey Medical Center, Hershey, PA

E. Ralf Schönbrunner, Laboratorium für Biochemie, Universität Bayreuth, West Germany

Franz X. Schmid, Laboratorium für Bio chemie, Universität Bayreuth, West Germany

Gennady V. Semisotnov, Institute of Protein Research, Academy of Sciences of the USSR, Moscow Region, USSR

Dirk Stigter, Department of Pharmaceutical Chemistry, Medical Science Building, University of California, San Francisco, CA

Richard W. Storrs, Department of Biochemistry, University of Texas Health Science Center, San Antonio, TX

John F. Thomason, Cray Research Inc., San Ramon, CA

B. A. Wallace, Department of Chemistry and Center for Biophysics, Rensselaer Polytechnic Institute, Troy, NY

David E. Wemmer, Department of Chemistry, University of California, Berkeley, CA

Part I

Compact States, Electrostatics, and Folding

Ken A. Dill

Statistical mechanical models are playing an increasingly important role in understanding protein stability, allostery, binding, and solvation processes. This section includes four chapters that describe such statistical mechanical models and their applications to proteins. They are arranged in order from the most microscopically detailed model in chapter 1 to the least detailed in chapter 4. Microscopic detail has obvious advantages for specific predictions of properties of particular native structures, but exploration of protein stability and packing organization is presently possible only with lower resolution models.

Chapter 1 by Gilson and Honig reviews the development of an important new method to treat the electrostatic interactions in protein native states. It makes use of the finite difference Poisson-Boltzmann equation, which permits the treatment of the charge interactions within the protein and the distribution of mobile salt ions in the surrounding solution. The finite difference method is a way to solve the Poisson-Boltzmann equation for the complex geometric distributions of charge and dielectric heterogeneity within native proteins. The method has previously been successfully applied to several problems involving charge-charge interactions in proteins. In this chapter, Gilson and Honig describe a new application of this methodology to charge-solvent interactions. It appears that this algorithm can be combined with existing force-fields and may extend them in two important respects: to account for effects of ionic strength, and perhaps to offer a more proper treatment of the solvent screening of ionic interactions than existing empirical distance-dependent dielectric constants.

In chapter 2, Glackin, Matthew, and Allewell show the importance of electrostatic interactions for the binding and allosteric behavior of ATCase. These authors use Tanford-Kirkwood theory, modified to take into account the solvent exposure of the ionizable groups, to predict the pH dependence and the ionization of various groups within the protein, upon binding substrates, and in subunit assembly. The Tanford-Kirkwood theory treats the native protein as a sphere, and the surrounding mobile ion distribution follows the Poisson-Boltzmann distribution. The modified Tanford-Kirkwood theory considers each ion to have partial burial within the protein core in accordance with its estimated solvent exposure at the surface. This approach permits rapid calculation of pK values, which appear to be in reasonable agreement with those obtained in ATCase experiments.

Whereas the statistical mechanical models of electrostatic interactions in chapters 1 and 2 address issues of specific charge interactions in particular native proteins, in chapter 3 Stigter and Dill describe a treatment of electrostatic contributions to protein stability based on treating the denatured states on the same footing as the native states. This treatment is also based on the Poisson-Boltzmann equation: the native state is treated as a sphere with uniformly distributed charge, and the denatured state is treated as a sphere that is penetrated by a solution including the mobile ions. The theory shows that the radii of the denatured conformations decrease considerably with ionic strength, and thus, it appears that the complexities of electrostatic interactions cannot be entirely attributed to the complexity of the native state. Two of the main conclusions are that (i) the free energy of the denatured conformation is complex and dependent on ionic strength, and (ii) electrostatic interactions are not exclusively due to the energetics of charge-charge interactions, but also involve a considerable entropic component due to the freedom to distribute protons throughout the ionizable sites.

In chapter 4, Chan and Dill describe a lattice model for the exhaustive exploration of conformational spaces of short compact chains as a model for native proteins. It is found that the number of conformations accessible to a compact chain is strongly dependent on its shape. More notably, it is found that internal organization – helices and sheets – in polymers and proteins emerges as a simple consequence of compactness in single chains. Whereas chapter 4 describes studies of chains on two-dimensional square lattices, similar results have recently been obtained on three-dimensional simple cubic lattices *(1)*. These results suggest that perhaps a significant component of the structural organization within globular proteins is due to the compactness caused by the hydrophobic effect that drives proteins to native compactness.

Reference

1. Chan, H. S., and K. A. Dill, *Proc. Natl. Acad. Sci. U. S. A.* **87**, 6388 (1990).

1

Calculation of the Total Electrostatic Energy of a Macromolecular System

Solvation Energies, Binding Energies, and Conformational Analysis[1]

Michael K. Gilson, Barry Honig

Introduction

The total electrostatic energy of a macromolecular system can be partitioned into contributions from interactions between pairs of charged atoms on the macromolecule and from the interactions of individual charges with the solvent *(1)*. Charge-solvent interactions are important factors in the energetics of protein folding and in the binding of charged substrates to macromolecules. For example, the strong favorable interactions between ionized amino acids and the aqueous solvent account for the tendency of these groups to be found on or near the protein surface *(2, 3)*. In the past few years, a number of studies have demonstrated that finite difference solutions to the Poisson-Boltzmann (PB) equation (the FDPB method) provide a reliable means of obtaining charge-charge interactions in macromolecules *(4–9)*. In this study we show how the FDPB method can be extended to the calculation of charge-solvent interactions as well.

The simplest and best-studied example of electrostatic charge-solvent interactions is the hydration of monatomic ions. The Born model *(10, 11)* has been used for many years as a means of calculating enthalpies and free energies of hydration and has recently been shown to be quite accurate when consistently applied *(12)*. Although its original derivation was based on the nonintuitive physical process of charging and discharging an ion, as shown below and discussed previously *(13)*, the Born expression can be derived purely from reaction field considerations that have a clear physical basis. That is, the electrostatic contribution to the solvation free energy of an ion corresponds exactly to the interaction of the ion with the polarization it induces in the solvent. Unfortunately, the simplicity of the Born expression is lost for ions lacking spherical symmetry. However, this chapter describes how the FDPB method can be used to calculate the interaction of a charge with its induced polarization shape. This makes it possible, with only moderate computational cost, to obtain electrostatic solvation energies of molecules; electrostatic solvation contributions to the energies of free and bound sub-

1 Reprinted from *Proteins: Structure, Function, and Genetics* **4**, 7 (1988) by permission of Wiley-Liss, a division of John Wiley & Sons, Inc.

strates; and the total electrostatic energy of a macromolecule, explicitly accounting for solvation/desolvation effects.

Our implementation of the FDPB method has been discussed in a number of recent publications *(6, 8, 9, 14)*. The protein is described in terms of its three-dimensional structure with all atoms located at positions defined by the crystal structure or generated by model-building techniques. In order to account, in an average way, for the effects of electronic polarizability, atomic charges are considered to be embedded in a low dielectric medium. The surrounding solvent is assigned an appropriate dielectric constant, most frequently taken to be that of water, and may contain an electrolyte obeying the Poisson-Boltzmann equation [for example, see *(10)*].

As mentioned above, the FDPB method has provided a useful basis for the interpretation of experimental results on charge-charge interactions in a number of complex macromolecular systems. The method has not previously been applied to the calculation of charge-solvent interactions, in part because there has been no way to extract ion solvation energies from the electrostatic potential maps produced by the FDPB method. Rashin and Namboodiri *(13)* have recently used a boundary element technique to solve the Poisson equation for small polar molecules and have thus avoided the problem of extracting solvation energies from finite difference solutions. However, given the relative ease of application of the FDPB method to large molecules such as proteins and nucleic acids and the importance of accounting for ionic strength effects, which are not easily included in boundary element methods, we have preferred to extend the applicability of the FDPB method to the calculation of solvation effects.

This chapter describes the method that has been developed for this purpose. We begin by reviewing the definition of the total electrostatic energy of a molecule in classical electrostatics, emphasizing the separation into various charge-charge and charge-solvent interaction terms. The computational method is then described, and its accuracy is tested against analytic solutions to the linearized Poisson-Boltzmann equation.

The new methodology is applied first to the calculation of hydration energies of several small molecules. Good agreement with experimental hydration free energies is obtained using standard parameter sets. However, it should be possible to improve the accuracy of the results with some effort in developing parameters appropriate for solvation. Two applications to protein energetics are also presented: (i) an exploration of charge-solvent interaction energies in two proteins and of their dependence upon the depth of an atom from the protein's surface, and (ii) calculations of the charge-solvent interaction energy and total electrostatic energy of correctly and incorrectly folded conformations of two proteins. These studies are intended to demonstrate the wide applicability of the method and to illustrate the magnitude of charge-solvent interactions that are often neglected in the analysis of conformational and binding energies.

Materials and Methods

Theory

Model. In a previous work *(1)* we defined the total electrostatic energy of a protein in terms of the interactions of atoms which were treated as charged spheres. This approach made it possible to include solvation ("self-energy") terms in an expression for the total electrostatic energy which was based on continuum electrostatics. However, in this work, instead of using shells of charge, we use the standard molecular mechanics description of charged atoms as point charges centered at the nucleus. Each atom is treated as a low-dielectric sphere whose dielectric constant depends on the electronic polarizability of the atom. It should be emphasized that since with this description no charge is ever closer than a van der Waals radius from the surface of an atom, and hence from a dielectric boundary, discontinuities that arise in classical electrostatics from point charges at boundaries are never encountered. For any configuration of atoms, the total electrostatic energy becomes that of a set of point charges, q_i, each embedded in a low-dielectric cavity corresponding to its atomic radius. Taken together, the atomic radii define the low-dielectric cavity of the macromolecule, which was assigned a uniform dielectric constant ε_m. The atoms are surrounded by a solvent of dialectic constant ε_s, which may contain an ionic atmosphere.

Terms contributing to the total electrostatic energy. It is convenient to partition the total electrostatic energy into a contribution from the Coulombic interaction of the charges with each other, ΔG°_c; a contribution caused by the interaction of the charges with a polarizable solvent, ΔG°_p; and a contribution caused by the interaction of the charges with the ion atmosphere in the solvent, ΔG°_a *(1)*. Because the ion atmosphere introduces special complications if the nonlinear PB equation is used, in this work we will limit ourselves to the linearized form. The nonlinear PB equation will be considered in a future study.

For simplicity of notation, we combine the solvent polarization and ion atmosphere terms into a single charge-solvent interaction term, $\Delta G^\circ_s = \Delta G^\circ_p + \Delta G^\circ_a$. The total electrostatic energy can now be written as

$$\Delta G^\circ = \Delta G^\circ_c + \Delta G^\circ_s = \sum_i (\Delta G^\circ_{c,i} + \Delta G^\circ_{s,i}), \quad (1)$$

where the subscript i denotes the contribution of each charge. For the linearized PB equation, it is straightforward to show that

$$\Delta G^\circ = 1/2 \sum_i q_i (\varphi_{c,i} + \varphi_{s,i}), \quad (2)$$

where $\varphi_{c,i}$ and $\varphi_{s,i}$ denote, respectively, the Coulombic potential and the potential produced by the solvent at the location of charge i. The precise meaning of these terms can best be understood once an appropriate reference state has been introduced.

Reference state. It proves to be particularly convenient to define a reference state in which all charges are infinitely separated in a medium corresponding to the dielectric constant of the molecule, ε_m, and containing no ion atmosphere. All energies and potentials in this work are defined relative to this reference state. The advantage of using this state is that the Coulombic term, ΔG°_c, is zero since all charges are infinitely separated, while the solvent term, ΔG°_s, is zero because there are no dielectric boundaries in a homogeneous medium and the ionic strength has been set to zero.

The reference state is shown in the top panel of Fig. 1. Notice that the surface of the molecule is drawn in the figure but that this does not correspond to a dielectric boundary since the interior of the molecule and the surrounding medium have the same dielectric constant, ε_m. Figure 1 shows a pathway for assembling the charges from the reference state into their final configuration in the solvated macromolecule. The individual steps in this process are discussed in the following sections.

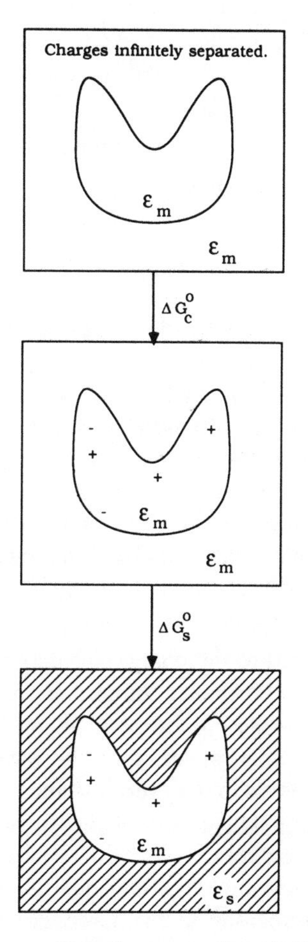

Fig. 1. Thermodynamic process for calculation of a molecule's total electrostatic energy.

The Coulombic term. In the first step of the process, the charges are brought together in the medium of dielectric constant ε_m. Note that the molecular surface still does not correspond to a dielectric boundary. Since this medium is homogenous, Coulomb's law is valid. The free energy change in this step, ΔG°_c is thus given by the Coulombic expression,

$$\Delta G^\circ_c = \sum_i \Delta G^\circ_{c,i} = 1/2 \sum_i \sum_{j \neq i} q_i q_j / \varepsilon_m r_{ij} \qquad (3)$$

The solvent term. After the charges have been assembled, the solvent environment is formed. ΔG°_s, the charge-solvent interaction energy of the system, corresponds to the work performed in bringing the solvent boundary from infinity to the position defined by the surface of the molecule. The total electrostatic free energy change of the system, ΔG° is then given by Eq. 1. The central contribution of this chapter is a new method for calculating ΔG°_s for complex systems.

It is instructive to decompose ΔG°_s into its individual contributions:

$$\Delta G^\circ_s = \sum_i (\Delta G^\circ_{s,ii} + \sum_{j \neq i} \Delta G^\circ_{s,ij}), \qquad (4)$$

where the first term describes the interaction of charge i with its own induced polarization (reaction field), while the second describes the interaction of charge j with the polarization induced by charge i.

The first term in Eq. 4 can be thought of as the sum of the solvation energies of each of the individual charged atoms in the configuration in which they are embedded in the macromolecule. Each element, $\Delta G^\circ_{s,ii}$, in the sum corresponds to the charge-solvent interaction energy of a particular charged atom assuming all other atoms to be neutral. It should be emphasized, however, that the other atoms still contribute to the overall shape of the macromolecule and hence to the solvent accessibility of atom i.

The second term in Eq. 4 accounts for solvent screening of the Coulombic interaction between charges i and j, as previously discussed *(1)*. Solvent screening is frequently incorporated into an effective dielectric constant, for example the distance-dependent dielectric constant used in molecular mechanics calculations. A comparison of the distance-dependent dielectric constant with the reductions of the present model has been presented *(9)*.

Special case of a single ion – the Born model. The derivation of an expression for the electrostatic solvation energy of a single spherical ion serves as a useful illustration of the theory developed above. Consider, for example, the transfer of an ion from vacuum ($\varepsilon = 1$) to water ($\varepsilon = 80$). For simplicity we assume zero ionic strength. To obtain the transfer energy, we first calculate the electrostatic energy of the ion in water and in vacuum, both relative to the reference state. The transfer energy is then just the difference between these two values.

The two energies can be obtained from the process depicted in Fig. 1. The reference state consists of a single point charge immersed in a continuous medium of dielectric constant ε_m. The molecular surface, in this case, is defined by a sphere whose radius is that of the ion. The Coulombic term, ΔG°_c, is zero, since the system contains only one charge. Similarly, there is no ij term (see Eq. 4) in ΔG°_s. Thus, the total electrostatic energy, ΔG°, will be given by $\Delta G^\circ_{s,ii}$.

As shown in Eq. 2, ΔG° is determined by the electrostatic potential at the center of the ion. In keeping with the present formalism, this potential for a single ion may be termed $\varphi_{s,ii}$. From classical electrostatic theory, the potential at charge, q_i, located at the center of a sphere of dielectric constant ε_m, which is embedded in a medium of dielectric constant ε_s, is given by

$$\varphi_{s,ii} = (332\, q_i/a_i)(1/\varepsilon_s - 1/\varepsilon_m) \qquad (5)$$

In Eq. 5, a_i is the ion's radius, and the constant 332 produces a result in kcal/mol. The electrostatic energy is thus given by

$$\Delta G^\circ = (166\, q_i^2/a_i)(1/\varepsilon_s - 1/\varepsilon_m). \qquad (6)$$

Substituting $\varepsilon_s = 80$ for water and $\varepsilon_s = 1$ for vacuum and taking the difference between the two resulting values for the ion's total electrostatic energy yields a transfer energy of $(166\ q_i^2 / a_i)(1/80 - 1)$, which will be recognized as the Born expression *(10, 11)*. Note that this expression has been obtained without invoking nonphysical charging and discharging processes. Rather, the electrostatic transfer energy results entirely from differences in charge-solvent interactions, which correspond to the second step in Fig. 1. It is worth pointing out that in this example, the transfer energy is independent of ε_m, but that this will be the case only for spherically symmetric boundaries and charge distribution.

It should be pointed out that in the more general instance where the object being transferred is not a single spherical ion, but rather is an entire molecule with a complex shape and charge distribution, the potentials cannot be obtained from Eq. 5 and must be calculated numerically (see below).

Conformational energies. The reference state used here is well defined for a given polypeptide sequence. The electrostatic energy of the unfolded state is poorly defined but is irrelevant when the energies of different folded conformations are being compared. As a consequence, the total electrostatic energies of different conformations of a given polypeptide may be compared in order to determine their relative electrostatic stabilities. However, the electrostatic energies of proteins having different sequences are not comparable in this way, because the reference states and the unfolded states of two different polypeptides are different.

Binding energies. The change in total electrostatic energy when two molecules form a complex may be calculated by means of a three-step thermodynamic process (Fig. 2) similar to that used in calculating the total energy of a single molecule. The initial state has both molecules fully solvated and infinitely separated from each other. In the first step, the solvent around each molecule is replaced by a medium having the dielectric constant of the molecular interior (ε_m) and zero ionic strength.

The work involved is $(\Delta G^\circ_{s,1} + \Delta G^\circ_{s,2})$, where 1 and 2 refer to the two separate

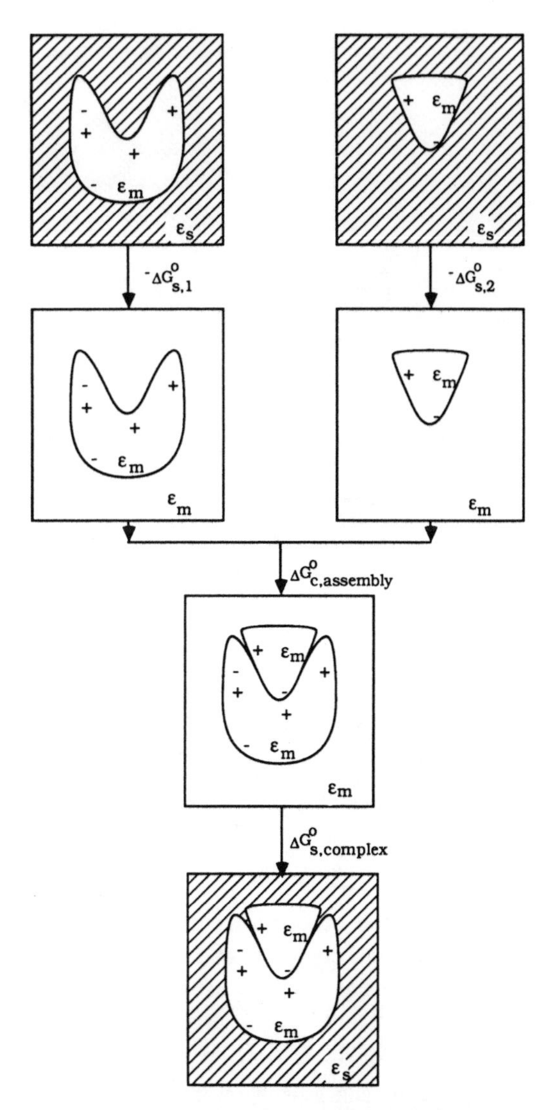

Fig. 2. Thermodynamic process for calculation of change in total electrostatic energy upon intermolecular binding.

molecules. Coulomb's law, with $\varepsilon = \varepsilon_m$, may now be used to compute the work of assembling the two molecules to form the complex (ΔG°_s assembly). Finally, the solvent is added back to yield the fully solvated complex. The work of performing this step is $\Delta G^\circ_{s,complex}$.

An alternate description of binding that may offer better insight in some circumstances is illustrated in Fig. 3, which shows a two-step thermodynamic process whose initial and final states are the same as those in the previous process. In step 1, each molecule is partially desolvated by removing the solvent from the

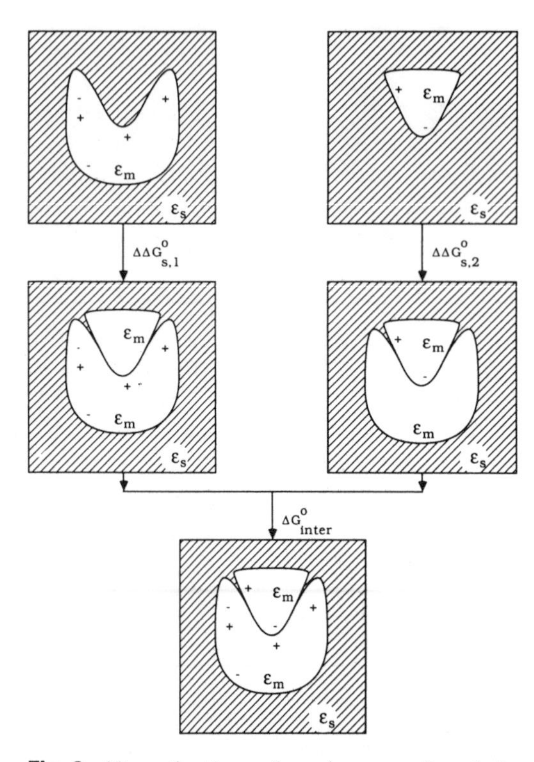

Fig. 3. Alternative thermodynamic process for calculation of change in total electrostatic energy upon intermolecular binding.

region that the other molecule will come to occupy in the complex and replacing the removed solvent by the usual medium of dielectric constant ε_m and zero ionic strength. The work involved ($\Delta\Delta G^\circ_{s,1} + \Delta\Delta G^\circ_{s,2}$) is the change in the charge-solvent interaction energies of the two species upon binding. Thus, if species 2 is a calcium ion, $\Delta\Delta G^\circ_{s,2}$ corresponds to the loss of solvation energy of the calcium upon binding to the protein. In the second step, the charges of molecule 1 are transferred to the low-dielectric space prepared next to molecule 2 in the previous step. The work in this step is the interaction energy between the charges of the two molecules, calculated with the complex surrounded by solvent (ΔG°_{inter}). (Actually, for this process to be properly defined, the charged and uncharged species are switched in the middle panels of Fig. 3. There is, of course, no electrostatic contribution from the uncharged species.)

Computational Method

The computational development that makes the present work possible is a procedure for calculating the change in solvent interaction of a molecule caused by an alteration in the dielectric constant of the solvent. For an arbitrary change in the solvent, let the initial solvent dielectric constant be ε_1 and the final value be equal to ε_2. Two finite difference calculations *(6, 14)* must be performed. In both, the charges are at their correct locations in the protein, and the dielectric constant of the protein region is set to its normal value, ε_m. However, in one calculation the region surrounding the protein is defined by ε_2. The difference in electrostatic potential at each charged atom for these two calculations must result entirely from the change in solvent parameters. It should be pointed out here that the finite difference method yields large, seemingly arbitrary potentials at the locations of charges. However, the differences in these potentials are meaningful and correspond to the change in the solvent polarization energy of the system upon altering the environment. This is given by

$$\Delta G^\circ_s = 1/2 \sum_i q_i \, \Delta\varphi_i, \qquad (7)$$

where $\Delta\varphi_i$ is the change in calculated potential at charge q_i upon altering the solvent. It should be noted that the calculations can be performed in a completely analogous fashion to obtain changes in electrostatic energy when the ionic atmosphere is changed.

In the applications described in the "Theory" section, where one is interested in determining the total electrostatic energy of a protein in a given conformation and solvent relative to the reference state, ε_1 must be equal to ε_m. Then Eq. 7 yields the work involved in the second step of the thermodynamic process described in Fig. 1. The energy of the first step, assembly of all charges in a medium of ε_m, is given by straightforward application of Coulomb's law as discussed above and is carried out analytically.

The finite difference calculations described here were carried out using the software package DelPhi *(9, 14)* which solves the Poisson-Boltzmann equation. Two techniques may be used to reduce the errors that result from replacing continuous functions by functions on the finite difference grid. In the "focusing" technique *(14, 15)* two or more finite difference calculations are performed, in which each calculation uses the same number of grid vertices, but the grid is made finer with each run by reducing the spatial extent of the lattice. The boundary conditions for each calculation (except the first) are obtained from the results of the preceding coarser calculation, and the series of runs thus "focuses in" on the molecule under study. This approach provides a relatively fine grid, together with accurate boundary conditions. The second technique, rotational averaging *(14)*, involves performing a series of finite difference calculations, where each calculation has the physical system under study rotated by a different angle with respect to the finite difference grid. The electrostatic potential fields from the runs are then averaged to yield results of improved accuracy. The focusing and rotational averaging techniques may be combined with each other *(14)*.

Since it has been shown that accurate solutions to the linearized Poisson-Boltzmann equation can be obtained for charges as close as ~1 Å to a dielectric boundary or from another charge *(14)*, it appears likely that the potential difference at a charge produced by altering the solvent can be accurately calculated if the solvent remains at least 1 Å from the charge. This condition should be easy to meet, as the solvent cannot approach closer than the van der Waals radius of an atom, and the smallest van der Waals radius, that of hydrogen, is about 1 Å.

Atomic and Molecular Parameters

A set of atomic radii and a solvent probe radius are required in order to define the dielectric boundary between a molecule and the solvent. As discussed below, it would be desirable to develop a parameter set that reproduces the experimentally observed solvation energies of biologically important ionic species. However, since such a parameter set is not yet available, standard van der Waals radii are used in this work.

The study of charge-solvent interactions as a function of charge depth below the surface is based on the same set of "united atom" van der Waals radii *(16)* that were used in recent studies on subtilisin *(8, 9)*. The other calculations in this chapter treat polar hydrogens bound to nitrogens and oxygens explicitly. These hydrogens are given a radius of 1.2 Å and the nitrogens and oxygens are given radii of 1.5 Å *(17)* and 1.6 Å *(16)*, respectively.

Unless otherwise specified, the polar hydrogen parameter set 19 from the molecular mechanics program CHARMM *(18)* provided the atomic charges used in this work. The CHARMM charges used for methyl ammonium, acetate, and methanol were obtained from the amino acid side chains lysine, aspartate (or glutamate), and serine, respectively. The charge set for ammonium was derived from that for the lysine side chain by transferring 0.1 proton charges from the nitrogen to the hydrogen replacing the methylene group, thus producing a symmetrical charge distribution.

Protein structures

The Brookhaven Protein Data Bank *(19)* provided the atomic coordinates of rhodanese (1RHD) *(20)*, crambin (1CRN) *(21)*, *Themiste dyscritum* met-hemerythrin (chain A of 1HMQ) *(22)*, and the VL domain of an Fab immunoglobulin fragment (residues 1–113 of 1MCP) *(23)*. The energy-minimized and misfolded structures of hemerythrin and VL domain, which were very generously provided to us by Drs. J. Novotny and R. Bruccoleri, have been fully described and analyzed by a number of techniques *(24)* and are described briefly in "Results and Discussion" below. The misfolded conformations used here were generated by a torsional search over side-chain conformations followed by energy minimization, using the program CONGEN *(25)*.

A problem that arose in the conformational analysis of the hemerythrin and VL domain structures was the placement of hydrogen atoms, since the coordinates of hydrogens are not determined by x-ray crystallography. (Hydrogen bonding is treated in CHARMM and CONGEN as at least partly electrostatic.) On the other hand, hydrogen coordinates were included in some of the computationally generated structures that were provided to us. In order to place all of the structures on an equal footing in this regard, all were stripped of any hydrogens, and polar hydrogens were added back using the CHARMM command HBUILD to generate a first set of structures. A second set of structures was obtained by energy-minimizing the first set of structures with only the hydrogens free to move. The minimizations were performed in 200 conjugate gradient steps, using CHARMM's default energy parameters and updating the nonbonded lists every ten steps. The RMS changes in all hydrogen positions resulting from minimization were roughly 0.2 Å. The two sets of structures are discussed further in "Results and Discussion."

The electrostatic calculations on hemerythrin and VL domain assumed the chain termini (including the artificial C-terminus at residue 113 of the VL domain) to be charged; all histidines to be neutral; and all aspartates, glutamates, lysines, and arginines to be fully ionized. In accord with the work of Novotny and co-workers *(24, 26)*, the iron atoms in hemerythrin have been omitted from all calculations. No distance cutoff was used in applying Coulomb's law. The Coulombic interactions reported in this study exclude all interactions between atoms connected by one or two covalent bonds (1–2 and 1–3 interactions) but include all interactions between atoms connected by three or more bonds (1–4 interactions, etc.). The purpose of the exclusions is to avoid large artifactual changes in interaction energies associated with slight variations in bond lengths and angles, which should be accounted for completely by bonded energy terms. 1–4 interactions were included because distances between atoms joined by three covalent bonds may change significantly with dihedral angle. For example, the distance between two main-chain nitrogen atoms changes from 2.9 Å to 3.8 Å when the intervening ψ angle varies from 0° to 180°.

Results and Discussion

Accuracy tests

The numerical technique for calculating charge-solvent interaction energies was tested by comparisons with the results of analytic solutions for a low-dielectric sphere surrounded by a high-dielectric solvent containing an electrolyte *(1, 27)*. A number of cases were tested. In each case, the sphere's internal dielectric constant was set to 2; the final external dielectric constant and ionic strength were 80 and 0.15 M, respectively, while the initial external dielectric constant and ionic strength were 2 and 0.0 M and correspond to the state illustrated in the second panel of Fig. 1. The thickness of the Stern layer *(10, 14)* was set to zero.

Table 1 presents the results of these tests. The first four have a single charge 1 Å below the surface of a 30 Å sphere, while the last test has five charges arranged in a 4 Å sphere, forming a small "ion." As shown in Table 1, it is possible to calculate charge-solvent interaction energies accurate to within about 5% for each of these systems. Note that the single-charge cases have charges very close to the sphere's surface, taxing the accuracy of the numerical method *(14)*. The results for deeper charges are equivalent or better (results not shown).

Based on the results for the model "ion," the calculations of hydration energies of small molecules (presented below) were performed using no focusing and no rotational averaging (see "Materials and Methods" for descriptions of these techniques) but a small grid size. The other charge-solvent calculations presented in this chapter also used no focusing, but they did use rotational averaging over six

Table 1. Accuracy tests.[a]

b	q	Foc	Rot	Final grid size (Å)	Numerical results (kcal/mol)	Tanford-Kirkwood results (kcal/mol)	Error (%)
30	(1,29,0,0)	N	Y	1.36	−36.0	−40.6	13
30	(1,12.9,24.9,7.5)	N	Y	1.36	−38.0	−40.6	5
30	(1,12.9,24.9,7.5)	Y	Y	1.00	−42.5	−40.6	4
30	(1,29,0,0)	Y	Y	1.00	−38.9	−40.6	4
4	(1,0,0,0)	N	N	0.18	−56.9	−60.0	5
	(1,0,2.5)						
	(1,0,−2.5)						
	(−1,2.5,0,0)						
	(−1,−2.5,0,0)						

[a]Comparisons between charge-solvent interaction energies calculated using Tanford-Kirkwood theory and using the numerical method described in this chapter. b = radius of low-dielectric sphere centered at origin; q = charges, where first value is charge in atomic units, and remaining three values are Cartesian coordinates; Foc = Y if two-step focusing used, N if not; Rot = Y if six-fold rotational averaging used, N if not; Final grid size = distance between vertices of finite difference grid. (If focusing used, this value corresponds to the most detailed grid.)

angles. Although focusing would have improved the accuracy somewhat, this improvement did not appear to be worth the additional computer time given the qualitative emphasis of this work.

Comparison with experiment: Hydration energies of small molecules

As a means of testing the physical model, we have used the method described above to calculate the electrostatic component of the charge-solvent interaction energies of four polar compounds whose hydration energies are known from experiment: methyl ammonium, ammonium, acetate, and methanol.

Two finite difference calculations were performed for each molecule: one with an external dielectric constant of 80 (water) and one with an external dielectric constant of 1 (vacuum). The internal dielectric constants were taken to be 2. In order to speed convergence, physiological ionic strength was used together with the external dielectric constant of 80. This appears to have a negligible effect on the calculated hydration energies, since redoing one of the acetate calculations using zero ionic strength changed the hydration energy by only 0.1%. As a further test, we repeated the continuum hydration enthalpy calculations of Rashin and Namboodiri *(13)* for acetate, obtaining results that agree within 3%.

The calculated electrostatic free energies of hydration for methyl ammonium, ammonium, acetate, and methanol are presented in Table 2, along with experimentally determined hydration free energies. The comparison between calculated and experimental values is complicated by the fact that there are no direct experimental measurements of individual ions. Rather, the experimental values that are reported depend upon the assumptions used to obtain the hydration free energy of a single reference ion, ΔG°_{s}, where "ion" refers to either H^+ or K^+ [see *(28, 29)*].

These single-ion hydration energies are subject to considerable uncertainty and tend to vary by approximately 10 kcal/mol. Table 2 emphasizes the experimental results associated with $\Delta G^{\circ}_{s}(H^+) = -262.5$ kcal/mol, which appears to be the most generally accepted value (see footnote to Table 2 for references) and yields better agreement with our calculations. However, experimental results obtained for $\Delta G^{\circ}_{s}(H^+) = -254.3$ kcal/mol *(29)* are also included.

The calculated values differ from the experimental values calculated using

Table 2. Calculated electrostatic component of hydration free energies, compared with experimental hydration free energies, for three small molecules.[a]

	Hydration free energy (kcal/mol)
Methyl ammonium	
Experimental[b]	–71. (–63.)
Calculated	–77.
Ammonium	
Experimental[b]	–79. (–71.)
Calculated	–86.
Acetate	
Experimental[c]	–79. (–87.)
Calculated	–70.
Methanol	
Experimental[d]	– 5.1
Calculated	– 4.3

[a]All values correspond to hypothetical 1 M reference states in vapor and aqueous phases. For the ions, the first values correspond to a proton hydration free energy ($\Delta G^{\circ}_{s}(H^{+})$ of –262.5 kcal/mol *(28, 29, 36–38)* and the parenthesized values correspond to $\Delta G^{\circ}_{s}(H^{+} = -254.3)$ kcal/mol, suggested by Marcus *(29)*. The gas phase data used in calculating the ionic hydration energies have uncertainties of about ± 3 kcal/mol. [b]*(39)*. [c]Calculated using thermodynamic cycle described and referenced by Scheraga and co-workers *(33)* except that the hydration free energy of neutral acetic acid was obtained from *(40)*. [d]*(41)*.

$\Delta G^{\circ}_{s} = -262.5$ kcal/mol by 8–19%. (With $\Delta G^{\circ}_{s}(H^{+}) = -254.3$ kcal/mol, the maximum error rises to 24%, for acetate.) These results are encouraging, especially in light of the fact that the atomic parameters – standard van der Waals radii, and the CHARMM charge set – have not been optimized for the present method of calculating hydration free energies. On the other hand, the calculations yield only the electrostatic contribution to the energies, so a comparison with experimental results is not entirely valid. Dispersion forces and the hydrophobic effect, for example, are likely to make contributions of the same order of magnitude as the differences between the calculated and experimental values.

As previously noted by Rashin and Namboodiri *(13)* in a study of hydration enthalpies, the continuum electrostatic model appears capable of yielding results whose quality equals that of more computationally demanding molecular simulations. The present results are similar to those obtained in free energy simulations of amino acids in water *(30)*. Direct comparison of results is not possible, however, since somewhat different molecules were studied. Ultimately, it should be possible to improve the electrostatic free energies provided by the continuum model by finding atomic charges and radii that reproduce the experimental values for hydration free energies of amino acid side chains, metal ions, and other groups of biological interest. However, it appears that even without further paramaterization, the FDPB approach introduced in this work may be expected to yield reasonable estimates of solvent polarization energies. In the following sections we use the method to determine the magnitude of these terms in a number of different systems.

Applications to proteins

Charge-solvent interactions as a function of charge depth. Although burying a charged atom within a protein incurs a substantial penalty in solvation energy, a buried charge may still interact significantly with the solvent. Conversely, an atom whose surface is accessible to solvent may yet be partially desolvated. In this section the magnitude of charge-solvent interaction energies are calculated for a number of atoms, and the relationship between these energies and the distances between the charges and the solvent is examined.

The solvent interaction energies of 45 charged atoms in the large protein rhodanese (293 residues) and ten charged atoms in the small protein crambin (46 residues) were calculated under the assumption that all other atoms in the protein were neutral. The results, therefore, yield the interaction of each atom with its own induced polarization in the solvent. In order to facilitate comparison, the atoms examined were considered to bear unit charges. (Their solvent-interaction energies may be corrected for a nonunit charge q by multiplying the values provided here by q^2.) Each atom's depth from the surface was also calculated by determining its distance from the closest point of a Connolly dot surface *(31)* generated using the same parameters as used in defining the protein-solvent dielectric boundary (see "Materials and Methods").

The energies were calculated using an internal dielectric constant of 2, an initial external dielectric constant of 2, an initial ionic strength of 0.0 M, a final external dielectric constant of 80, and a final external ionic strength of 0.15. Thus the energies are determined relative to that of a charge in the reference state, which here is assumed to have $\varepsilon_m = 2$. These energies may also be regarded as electrostatic solvation energies of the charges as they are found in the protein, relative to the solvation energies they would have if they were completely buried in an infinitely large protein. Repeating the present calculations with an initial ionic strength of 0.0 M has very little effect (~1%) on the energies. No Stern layer was used, so mobile ions in the solvent approach to the protein's dielectric boundary.

The selected atoms for rhodanese are the main chain nitrogens of residues 1, 29, 39, 54, 86, 111, 113, 116, 129, 249; all 14 aspartate OD1 atoms; all 20 arginine NH1 atoms, and CD1 of isoleucine 270. The isoleucine atom was selected because it is the deepest (7.6 Å) atom in the protein when a 1.4 Å probe is used to generate the surface. The crambin atoms examined are the main chain nitrogens of residues 1, 7, 13, 17, 23, 27, 31, 35, 40, and 46.

As shown in Fig. 4, the solvent-interaction energies fall between 0 and −50 kcal/mol. The less negative values correspond to atoms that are more deeply buried and so are further from the solvent. The fact that atoms at the surface can have solvent-interaction energies ranging from ~−50 kcal/mol to ~−25 kcal/mol (Fig. 4) results from the present definition of depth, which does not distinquish between atoms almost completely surrounded by solvent and atoms with only a small solvent-exposed patch.

The atoms with the strongest solvent interactions tend to be those belonging to the ionizable side chains −i.e., OD1 of Asp and NH1 of Arg − because these atoms are usually well solvated. The main chain nitrogen with a very negative energy belongs to the N-terminus, which is exposed to solvent; and the OD1 and NH1 atoms with the weakest interactions with solvent (~−9 kcal/mol) form an unusual buried salt bridge *(20)*. The results for atoms not in direct contact with the solvent should be insensitive to their assumed atomic radii because these atoms do not contribute to the van der Waals surface of the protein. On the other hand, the fact that an atom is not in direct contact with the solvent does not mean that its interaction with the solvent is negligible. For example, the main nitrogen of residue 113 in rhodanese is 4.8 Å below the surface, but has an interaction with the solvent of −7.5 kcal/mol; and CD1 of isoleucine 270, 7.6 Å deep, retains − 6.7 kcal/mol of interaction with the solvent. These results suggest that in the sulfate-binding protein of *S. phimurium,* whose structure was solved by Pflugrath and Quiocho *(32),* the sulfate, although buried, retains a substantial interaction with the solvent.

Figure 4 also demonstrates that the sol-

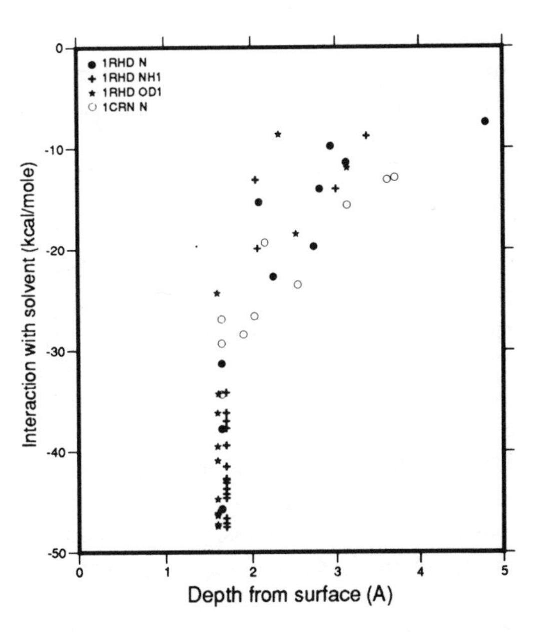

Fig. 4. Charge-solvent interaction energies as a function of atomic depth, for proteins rhodanese (1RHD) and crambin (1CRN). N, main-chain nitrogen; NH1,Nη1 of arginine; OD1,Oδ1 of aspartate.

vent interaction energies for crambin tend to be more favorable than those for rhodanese for a given charge depth. This must result from crambin's small size, which causes each atom to be nearer the solvent than an atom of similar nominal depth in a larger protein.

The fact that buried charges may retain substantial electrostatic interactions with the solvent suggests that desolvation costs will tend to be overestimated by free energy functions *(33, 34)*, which determine the work of dehydrating a charged atom based on measures of the atom's accessibility to solvent. A method of treating electrostatic charge-solvent interactions that would account for their long range and yet be computationally simple might serve as a useful correction to such free energy functions.

Conformational energy minimization and misfolded proteins. The total conformational energy of a protein is known to be the sum of many large contributions such as hydrophobic interactions, chain entropy, electrostatics, etc. In attempting to distinguish between correctly and incorrectly folded polypeptide conformations, it is important to have as accurate a treatment as possible for each of these terms. The methodology introduced in this work provides a realistic treatment of electrostatic interactions and would thus be useful in conformational analysis. In fact, Moult and James *(35)* have found that the electrostatic energy alone can be used to select stable conformations of loops in a trypsin-like enzyme from *Streptomyces griseus.*

They accounted for solvent screening of pairwise interactions using image charges (thus approximating the second term in Eq. 4), but did not account for the solvation of individual charged atoms (the first term in Eq. 4). In this section we illustrate the application of the FDPB method to the calculation of electrostatic contributions to conformational energies of entire proteins. We also address the question of whether relative electrostatic energies can be used to identify stable protein conformations.

The proteins studied are hemerythrin and the VL domain of the immunoglobulin Fab fragment MC/PC 603 (see "Materials and Methods"). For each protein, the conformations analyzed were *(24)* the crystal structure; a conformation generated from the crystal structure by energy minimization using CHARMM; and a severely misfolded conformation, generated using the program CONGEN. As described by Novotny *et al. (24)*, the misfolded conformation of hemerythrin was produced by assigning to its main-chain atoms the coordinates of the main-chain atoms of corresponding residues of the VL domain and then using the iterative torsional search program CONGEN *(25)* to optimize the side-chain conformations This procedure is made possible by the fact that hemerythrin and the VL domain have the same number of residues (113). The misfolded form of the VL domain was produced *(24)* by the converse progress of giving its main chain the hemerythrin fold and optimizing its side chains in this new structure.

We used the methods described above to calculate each structure's total electrostatic energy, which consists of a solvent-interaction term and a Coulombic term. The calculations use an interior dielectric constant of 2, a solvent dielectric constant of 80, a solvent ionic strength of 0.15 M (physiological), and no Stern layer. As noted in "Materials and Methods," all 1–2 and 1–3 interactions were omitted in calculating the Coulombic terms.

Table 3 presents the electrostatic energies calculated for each conformation of the two proteins. The first column contains the interactions with solvent, relative to the interactions with a medium of dielectric constant 2; the second column contains the Coulombic energies of assembling the charges from the reference state. The third column contains the sums of the first two columns and represents total electrostatic energies relative to the reference state defined in "Materials and Methods." The absolute energies are given, as are the energies relative to the corresponding values for the crystal structure of each protein. The entries are averages of results from two sets of structures of each protein: one set with hydrogens added using the CHARMM command HBUILD, and the other with the hydrogens subsequently energy-minimized (see "Materials and Methods"). Although these two sets of structures yield electrostatic energies that differ by up to tens of kcal/mol, the signs and rankings of the entries in Table 3 do not depend on whether the hydrogens have been energy-minimized.

The calculations are subject to a number of errors. The charge-solvent interaction energies could be in error by on the order of 20%, based on the comparisons between calculated and experimentally determined hydration energies of small molecules, presented above. Further errors may result from the assumptions made concerning the ionization states of titratable residues, particularly in the case of histidines, which have been assumed neutral, and of which there are seven in hemerythrin and one in the VL domain; from the neglect of hemerythrin's ions; and, as discussed below, from the treatment of hydrogen bonding.

In considering the results in Table 3, we presume that for each protein, the most "correct" structure is the crystal structure; the next most correct is the energy-minimized crystal structure; and the misfolded structures are completely incorrect. It is striking that for both proteins, the solvent interaction energies rank the structures in order of correctness: the crystal structures show the strongest interaction with solvent; the energy-minimized structures have solvent interactions of intermediate strength; and the misfolded structures have the least favorable interactions with solvent. It is interesting in this regard that Novotny *et al.* *(24)* found that misfolded hemerythrin had nine buried ionizable groups, while the misfolded VL domain had only four. The significant loss of solvent polarization energy in misfolded hemerythrin is likely to be due to the large number of buried ionizable groups.

In contrast to the solvent interaction energies, the Coulombic energies cannot be simply related to the "correctness" of a particular structure. Indeed for the VL domain, they are inversely correlated with the expected order of correctness. As a consequence, the total electrostatic energies are also ranked incorrectly. This would appear to suggest, in contrast to the conclusion of Moult and James *(35)*, that electrostatic energies alone cannot be used to identify stable protein conformations. However, the following considerations indicate that Table 3 should not be interpreted in this way.

First, the α-helical conformations (crystal and energy-minimized hemerythrin and misfolded VL domain) contain more hydrogen bonds than the β-sheet conformations (misfolded hemerythrin and crystal and energy-minimized VL domain). Since the Coulombic energies were calculated for all charge-bearing atoms, including those participating in hydrogen bonds, it is not surprising that the α-helical conformations have the lowest Coulombic energies. However, the electrostatic component of hydrogen bonds is not well defined, and the use of partial charges to obtain hydrogen bonding potentials is, to a

Table 3. Electrostatic energies of hemerythrin and VL domain for crystal structures, energy-minimized structures, and misfolded structures.[a]

	Solvent		Coulombic		Total	
	Absolute	Relative	Absolute	Relative	Absolute	Relative
Hemerythrin						
Crystal	–764.	0.	–741.	0.	–1,505.	0.
Minimized	–691.	+ 73.	–918.	–177.	–1,609.	– 104.
Misfolded	–618.	+146.	–873.	–132.	–1,491.	+ 14.
VL domain						
Crystal	–569.	0.	–566.	0.	–1,135.	0.
Minimized	–512.	+ 57.	–688.	–122.	–1,200.	– 65.
Misfolded	–496.	+ 73.	–773.	–207.	–1,269.	– 134.

[a]Energies in kcal/mol.

great extent, a matter of convenience. For this reason, it would be of interest to evaluate the total electrostatic energy in the absence of hydrogen bonding contributions.

A second problem is that it is not really meaningful to compare the energies of the crystal and energy-minimized conformations, even if they contain identical secondary structures. Energy minimizations tend to produce overly compact structures since they neglect conformational entropy. These structures will have excessively negative Coulomb energies because, for example, hydrogen bond energies will be optimized by the minimization procedure. Thus, although the total electrostatic energies of the minimized structures are lower than those of the presumably more correct crystal structures, total electrostatic energies may still be useful in comparing the stabilities of different computer-generated conformations, all of which, for example, might be energy minimized.

Given the various complications, our results do not at this stage allow us to draw conclusions concerning the suitability of the total electrostatic energy as a criterion for identifying the relative stability of different protein conformations. However, our calculations do suggest that solvent interaction energies can help identify unstable conformations. This result, taken together with the loop conformation study of Moult and James *(35)*, suggest that total electrostatic energy may provide a useful measure of stability, if the complications associated with different types of secondary structure can be resolved. On the other hand, it should be emphasized here that there is no a priori reason to use electrostatic energy as a measure of stability since the total free energy of a protein is the sum of many contributions [see the report by Novotny *et al.* *(24)* for a discussion of other factors relating to hemerythrin and the VL domain]. In any case, the methods introduced in this work now make it possible to account in a consistent fashion for both solvent screening of pair-wise interactions and solvation energies in the analysis of protein conformations.

Conclusions

In this chapter we have shown that the FDPB method can be extended to account for the effects of electrostatic charge-solvent interactions. In previous work we have shown how the method can be used to calculate solvent-screened charge-charge interactions. Thus, it is now possible to calculate the total electrostatic energy of a given macromolecular system. This work describes how the methodology can be applied to the calculation of the free energies of substrate binding and protein folding. Other possible applications include the study of molecular recognition and the improvement of energy functions used in molecular mechanics.

The numerical results presented in this work are intended to demonstrate the magnitude of various solvent polarization effects in a qualitative sense and to illustrate how the FDPB method can be used in different applications. However, if truly quantitative agreement with experiment is to be obtained, it will be necessary to develop a set of atomic parameters that will yield appropriate hydration energies for compounds representative of the chemical groups found in proteins such as carboxylates, ammoniums, and alcohols. In the meantime, the method makes it possible to obtain reasonable estimates of the rather large solvent polarization terms that are frequently neglected in conformational analysis molecular mechanics simulations.

Acknowledgments

We are very grateful to Dr. J. Novotny and Dr. R. Bruccoleri for providing us with the energy-minimized and misfolded protein structures. We also thank Drs. R. Bruccoleri, R. Fine, H. Rodman Gilson, J. Novotny, A. Rashin, K. Sharp, and P. Youkharibache for stimulating discussions and suggestions. This work was supported by grants from the NIH (GM-30518) and ONR (N00014-86-K-0483).

References

1. Gilson, M. K., A. Rashin, R. Fine, B. Honig, *J. Mol. Biol.* **183,** 503 (1985).
2. Paul, C. H., *J. Mol. Biol.* **155,** 53 (1982).
3. Rashin, A. A., and B. H. Honig, *J. Mol. Biol.* **173,** 515 (1984).
4. Warwicker, J., and H. C. Watson, *J. Mol. Biol.* **157,** 671 (1982).
5. Rogers, N. K., G. R. Moore, M. J. E. Sternberg, *J. Mol. Biol.* **182,** 613 (1985).
6. Klapper, I., R. Hagstrom, R. Fine, K. Sharp, B. Honig, *Proteins* **1,** 47 (1986).
7. Warwicker, J., *J. Theor. Biol.* **121,** 199 (1986).
8. Gilson, M. K., and B. H. Honig, *Nature* **330,** 84 (1987).
9. Gilson, M. K., and B. H. Honig, *Proteins* **3,** 32 (1988).
10. Bockris, J. M., and A. K. N. Reddy, *Modern Electrochemistry,* vol. 1 (Plenum Press, New York, 1977).
11. Born, M., *Z. Physik* **1,** 45 (1920).
12. Rashin, A. A., and B. Honig, *J. Chem. Phys.* **89,** 5588 (1985).
13. Rashin, A. A., and K. Namboodiri, *J. Phys. Chem.* **91,** 6003 (1987).
14. Gilson, M. K, K. A. Sharp, B. H. Honig, *J. Comput. Chem.* **9,** 327 (1988).
15. McAllister, D., and J. R. Smith, *Computer Modeling in Electrostatics* (Research Studies Press, Wiley, New York, 1985).
16. McCammon, J. A., P. G. Wolynes, M. Karplus, *Biochemistry* **18,** 927 (1979).
17. Weast, R. C., Ed., *CRC Handbook of Chemistry and Physics,* 56th ed. (CRC Press, Cleveland, 1975).
18. Brooks, B. R., *et al., J. Comput. Chem.* **4,** 187 (1983).
19. Bernstein, F. C., *et al., J. Mol. Biol.* **112,** 535 (1977).
20. Ploegman, J. H., G. Drent, K. H. Kalk, W. G. J. Hol, *J. Mol. Biol.* **123,** 557 (1978).
21. Hendrickson, W. A., and M. M. Teeter, *Nature* **290,** 107 (1981).
22. Stenkamp, R. E., L. C. Sieker, L. H. Jensen, *J. Am. Chem. Soc.* **106,** 618 (1984).
23. Satow, Y., G. H. Cohen, E. A. Padlan, D. R. Davies, *J. Mol. Biol.* **190,** 593 (1986).
24. Novotny, J. A. Rashin, R. E. Bruccoleri, *Proteins* **4,** 19 (1988).
25. Bruccoleri, R. E., and M. Karplus, *Biopolymers* **26,** 137 (1987).
26. Novotny, J., R. Bruccoleri, M. Karplus, *J. Mol. Biol.* **177,** 787 (1984).
27. Tanford, C., and J. G. Kirkwood, *J. Am. Chem. Soc.* **79,** 5333 (1957).
28. Desnoyers, J. E., and C. Jolicoeur, *Mod. Aspects Electrochem.* **5,** 1 (1969).
29. Marcus, Y., *Ion Solvation* (John Wiley & Sons, Ltd., New York, 1985).
30. Bash, P., U. C. Singh, R. Langridge, P. Kollman, *Science* **236,** 564 (1987).
31. Connolly, M. L., *Science* **221,** 709 (1983).
32. Pflugrath, J. W., and F. A. Quiocho, *Nature* **314,** 257 (1987).
33. Kang, Y. K., G. Nemethy, H. A. Scheraga, *J. Phys. Chem.* **91,** 4118 (1987).
34. Eisenberg, D., and A. D. McLachlan, *Nature* **319,** 199 (1986).
35. Moult, J., and M. James, *Proteins* **1,** 146 (1986).
36. Friedman, H. L., and C. V. Krishnan, in *Water: A Comprehensive Treatise,* F. Franks, Ed. (Plenum Press, New York, 1973), vol. 3, chap. 1.
37. Randles, J. E. B., *Trans. Farad. Soc.* **52,** 1573 (1956).
38. Rosseinsky, D. R., *Chem. Rev.* **65,** 467 (1965).
39. Taft, R. W., *Prog. Org. Chem.* **14,** 247 (1983).
40. Wolfenden, R., L. Andersson, P. M. Cullis, *Biochemistry* **20,** 849 (1981).
41. Hine, J., and P. K. Mookerjee, *J. Org. Chem.* **40,** 292 (1975).

2

Electrostatic Effects and Allosteric Regulation in E. coli *Aspartate Transcarbamylase*

Mary P. Glackin, James B. Matthew, Norma M. Allewell

Introduction

Because cellular functions are controlled via the regulation of macromolecular assemblies, the principles that govern the operation of these assemblies are of great importance, albeit as yet poorly understood. One of the archetypal systems for investigating molecular mechanisms of recognition, communication, and regulation in macromolecular assemblies is aspartate transcarbamylase (ATCase) from *E. coli*, a large (310 kD), highly regulated, multisubunit protein [reviewed in *(1–7)*]. We discuss here our current understanding of the role of electrostatic effects in its assembly and regulation.

Function

ATCase catalyzes the first committed step in pyrimidine biosynthesis, transfer of a carbamoyl group from carbamoyl phosphate to the α-amino group of L-aspartate to generate the skeleton of the pyrimidine ring (Fig. 1) *(8, 9)*. Enzymatic activity is regulated by binding of substrates, nucleotides, and protons. The substrate L-aspartate binds cooperatively *(10)*: while at nonsaturating concentrations of L-aspartate, adenosine triphosphate (ATP) is an activator *(10)*, cytosine triphosphate (CTP) is an inhibitor *(10)*, and uridine triphosphate (UTP) potentiates inhibition by CTP *(11)* (Fig. 2). These modes of regulation serve to maintain appropriate ratios of purines to pyrimidines within the cell. In addition, both catalytic and regulatory functions are highly sensitive to pH *(12–14)*, implying regulation by proton binding.

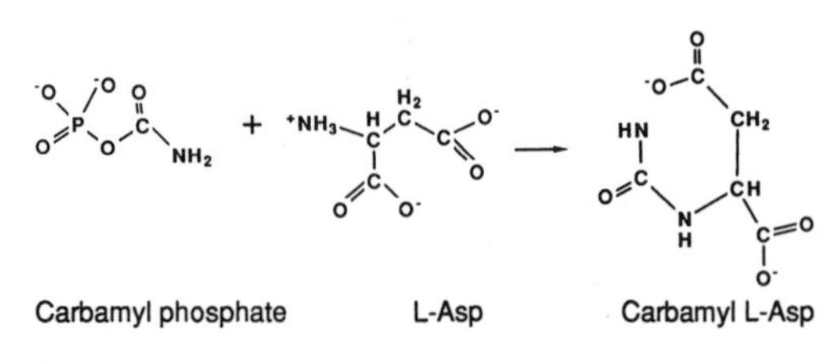

Fig. 1. Reaction catalyzed by ATCase.

Structure

ATCase is comprised of twelve polypeptide chains: six catalytic chains organized as two catalytic (c_3) subunits and six regulatory chains organized as three regulatory (r_2) subunits *(15, 16)*. The isolated molecule has D_3 symmetry; however, symmetry about the molecular two-fold axes is lost in the crystal as a result of crystal contacts. Studies with isolated catalytic and regulatory subunits indi-

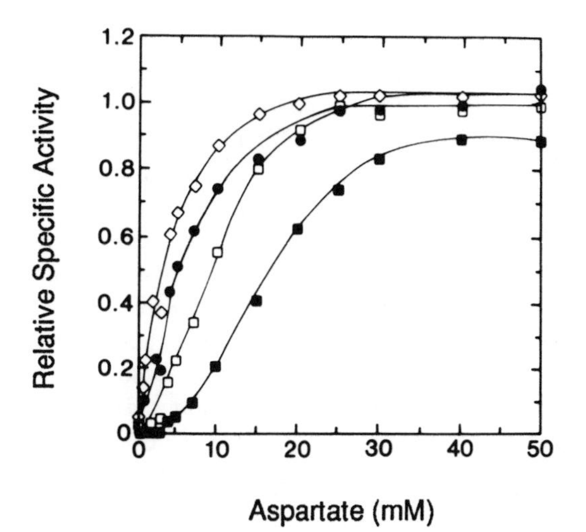

Fig. 2. Effects of nuceloside triphosphates on the velocity of the reaction catalyzed by ATCase at pH 7, 28°C. ●, control; ◇, 2 mM ATP; ☐ 2 mM CTP; ■ 2mM CTP + 2 mM UTP. Reprinted from *(11)*.

cate that only catalytic subunits have enzymatic activity *(15)*. The active site is located between catalytic chains *(17, 18)*. Both catalytic and regulatory subunits are able to bind nucleoside triphosphates *(19)*; however, feedback regulation of enzymatic activity by nucleoside triphosphates depends upon binding to the regulatory subunits of the holoenzyme.

ATCase undergoes a dramatic change in molecular architecture when L-aspartate or one of its competitive inhibitors binds in the presence of carbamoyl phosphate or when the bisubstrate analog PALA (N-(phosphonacetyl)-L-aspartate) binds *(20)*. Structures of the protein determined by x-ray crystallography in the absence and presence of these ligands indicate that in this structural reorganization, the enzyme expands by 12 Å along its threefold axis, while catalytic and regulatory subunits rotate by approximately 10° and 15°, respectively, about their symmetry axes *(21–26)*. The trigger for this reorganization appears to be movement of two loops in the catalytic chain, the 240s loop containing residues 231–246 and the 80s loop containing residues 77–86. The 240s loop swings into the active site to allow nearby residues to interact with functional groups of L-aspartate, while the 80s loop swings into the adjacent active site to interact with carbamoyl phosphate (Fig. 3) *(24)*. The resulting change in quaternary structure disrupts one set of noncovalent interactions between chains or domains and facilitates the formation of a second set (Fig. 4). Interactions of the types C1-C4 between

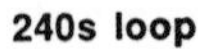

Fig. 3. Superimposed α carbon traces of a catalytic chain in the unliganded and PALA-liganded structures *(23)* illustrating shifts of the 80s and 240s loops. Heavy line: Unliganded structure; light line: PALA-liganded structure.

catalytic chains in the two catalytic subunits, C1-R4 between catalytic and regulatory chains, and R1-R1 between domains of regulatory chains are eliminated, while new interactions of the types C1-C2 between catalytic chains within catalytic subunits, C1-R1 between catalytic and regulatory chains, and C1-C1 between domains of catalytic chains develop.

Although nucleoside triphosphates have pronounced effects on enzymatic activity, the structural changes induced by their binding appear smaller, both in solution and in crystals. The change in sedimentation coefficient is −0.5%, compared to a change of −3.5% when PALA binds *(27)*. In the crystal, binding of CTP results in small shifts in the backbone in the vicinity of the nucleoside triphosphate binding site, at the interdomain interface in the regulatory chain, and at the interface between catalytic and regulatory chains *(28)*.

Electrostatic effects

Because carbamoyl phosphate, L-aspartate, and all of the nucleoside triphosphates bear several negative charges, ATCase might be expected to have evolved so as to utilize electrostatic effects in catalysis and regulation. The pronounced pH dependence of the apparent K_m for binding L-aspartate *(12)*, V_{max} *(12)*, the Hill coefficient *(12)*, and the extent of inhibition by CTP *(13)* and activation by ATP *(14)* indicate that this is, in fact, the case. Site-directed mutagenesis experiments also implicate several ionizable residues in both the cooperative binding of substrates and regulation by nucleoside triphosphates [reviewed in *(3, 5, 7)*]. Attempts have been made to derive pK values for some of these groups by fitting several types of experimental data to appropriate models *(29–37)*. Results obtained with other proteins indicate, however, that pK values of many residues change when large conformational changes occur. This was also shown to be the case for ATCase when electrostatic effects accompanying assembly of the holoenzyme from its subunits were modeled *(38)*. These considerations indicate a need for further modeling studies, since all of the groups involved cannot be resolved experimentally or identified directly from stuctural studies.

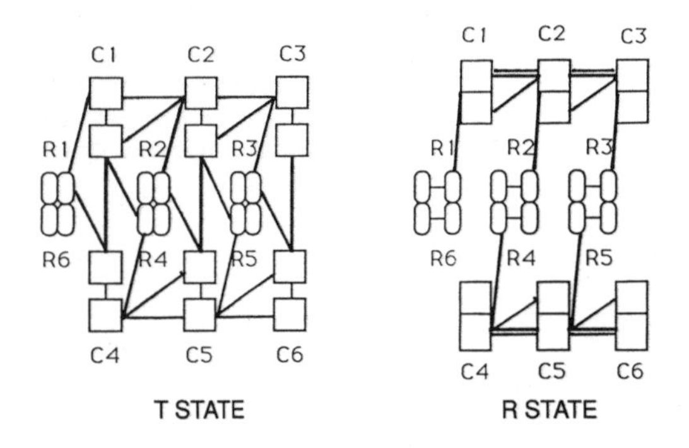

Fig. 4. Changes in interdomain and interchain contacts associated with the T-R transition induced by binding of PALA or dicarboxylic acids. Each square represents a domain of a catalytic chain; each ellipsoid a domain of a regulatory chain. Chains are numbered as defined in *(56)*. Catalytic chains within catalytic subunits and regulatory subunits are related by a three-fold symmetry axis. The molecular two-fold axes relating the two catalytic subunits and regulatory chains within regulatory subunits are suppressed by molecular contacts in the crystal, making the upper and lower halves of the molecule nonequivalent.

The Electrostatic Model

This study uses the modified Tanford-Kirkwood model *(39)*. Tanford-Kirkwood theory provides analytical expressions for the electrostatic free energy of a set of discrete point charges on the surface of a low dielectric sphere, which is surrounded by an ion exclusion shell of defined thickness, imbedded in a high dielectric medium *(40)*. Laplace's equation is applied to the low dielectric region and ion exclusion shell, while the Poisson-Boltzmann equation, with the same assumptions that are inherent in Debye-Huckel theory, is applied to the high dielectric mobile ion region.

Modified Tanford-Kirkwood theory incorporates a burial factor, as defined by Lee and Richards *(41)*, for each pair of ionizable groups. This procedure eliminates the adjustable burial parameters required in order to obtain agreement with experiments when the original theory is applied to proteins and allows for the possibility that the dielectric constants experienced by various ionizable groups will differ. The approximations of this model have been the subject of considerable discussion *(42–50)*; however, pK values and other parameters calculated with this model generally agree well with experimental results. Furthermore, Matthew has recently shown that the changes in pK values that result from internal motions of proteins are substantially larger than the differences in the results obtained with various electrostatic models *(51)*.

Coordinates for the unliganded structure were those derived from the most recent refinement *(54)*, for the PALA-liganded holoenzyme those of Ke *et al.* *(24)*, and for the CTP-liganded holoenzyme those of Kim *et al.* *(28)*, all at pH 5.8 or 5.9. Coordinate sets for subunits were constructed by abstracting subunit coordinates from the coordinates of the holoenzyme. These procedures do not take into account structural changes resulting from changes in pH or dissociation of the holoenzyme. Values of the model parameters used in the calculations are given in Table 1.

Table 1. Values of model parameters.

Temperature	298K
Ionic strength	0.02 M
Ion exclusion shell (Å)	2
Radii of low dielectric sphere (Å)	
Unliganded ATCase	69
CTP-liganded ATCase	69
PALA-liganded ATCase	68
Catalytic subunit	
C1C2C3	45
C4C5C6	47
Regulatory subunit	34
Low dielectric constant	4
High dielectric constant	78

Results

Agreement with experiment

Because every model involves approximations, it is important to compare the results of calculations with experimental results. Although individual pK values are frequently used to test the validity of predictions from models, only one experimentally determined pK value is available for ATCase at this time. Kleanthous *et al.* *(52)* have estimated the pK of His 134 in the catalytic chain near the active site at <6; we calculate a value of 5.9.

The most extensive experimental data with which the calculations can be compared at this time are measurements of the stoichiometries of linked proton binding for assembly of the holoenzyme, binding of substrate analogs to both the catalytic subunit and the holoenzyme, and binding of nucleotides to catalytic and regulatory subunits and the holoenzyme as a function of pH *(33–36)*. Representative comparisons of calculated and experimental results are shown in Fig. 5. Although there are 810 ionizable groups in ATCase and 261 in the catalytic subunit, calculated and experimental results agree in every case to within two protons. In the case of binding of PALA to the catalytic subunit and CTP binding to ATCase, the calculations also reproduce the pH dependence of the experimental results.

Neverthess, agreement is not as good as in small, more thoroughly analyzed systems such as hemoglobin where agreement within a few tenths of a proton can be achieved with these methods. Agreement is weakest for assembly, perhaps because the coordinates for the subunits were abstracted from the structure of the holoenzyme, and there are likely to be significant differences in the structures of the subunits both in isolation and incorporated into the holoenyzme. Even here, however, calculated and experimental values agree to within two protons over the pH range examined.

Electrostatic stabilization

The calculated contributions of electrostatic effects to the free energies of folding unliganded, PALA-liganded, and CTP-liganded ATCase are shown in Fig. 6. Also shown are analogous calculations based upon the protein coordinates for the PALA- and CTP-liganded structures, with the atoms of the ligand deleted. As shown, the free energies of electrostatic stablization of the unliganded and PALA-liganded structures are very similar and significantly more negative than that of the CTP-liganded structure over the pH range for which the calculations were carried out. The calculations for PALA indicate that the PALA-liganded structure is stabilized by electrostatic interactions at the binding sites; the free energy of this structure increases by 40–50 kcal/m^{-1} when the PALA atoms are deleted. Since the total electrostatic free energies of the PALA-liganded and unliganded structures are very similar, these results for PALA suggest that energetically favorable electrostatic interactions at the active site of the PALA-liganded structure may be "paid for" by the elimination of favorable electrostatic interactions within the protein upon PALA binding. In contast, the free energy of the CTP-liganded structure decreases by 10–15 kcal/m^{-1} when the CTP atoms are deleted.

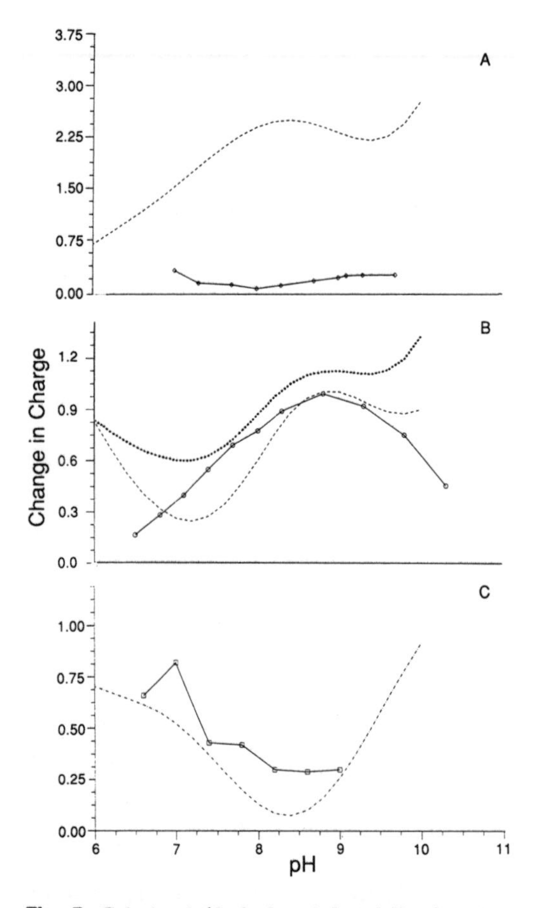

Fig. 5. Calculated (**dashed** and **dotted lines**) and experimental (**solid lines**) values of the change in charge resulting from (**a**) Assembly of ATCase from its catalytic and regulatory subunits, (**b**) binding of PALA to the catalytic subunit, and (**c**) binding of CTP to ATCase. The dashed and dotted curves in (**b**) correspond to binding to C1C2C3 chains and C4C5C6 chains, respectively. Experimental results from (**a**) *(36)*; (b) *(34)*; (c) *(35)*.

This is an unexpected result, since interactions of the phosphate groups of CTP with the protein are known to stabilize the complex. It may reflect the fact that the structure used in this calculation was determined at pH 5.9, where the phosphate groups are not fully ionized and do not interact optimally with the protein and where the pattern of ionic bonding is influenced by intermolecular contacts in the crystal *(28)*. Our calculations indicate that the γ phosphate of CTP is not fully ionized even at pH 8 in the protein:CTP complex, although free CTP, in the conformation it assumes in the crystal, ionizes normally.

Table 2 compares results from these calculations with values of the total free energies of interactions determined experimentally. Electrostatic effects make the largest contribution to the free energies of assembly of the holoenzyme and to PALA binding. Their contribution to the formation of r:r interfaces is small, and their contribution to CTP binding at pH 5.9 where the crystal structure was determined appears to be energetically unfavorable, as noted above. Under these condi-

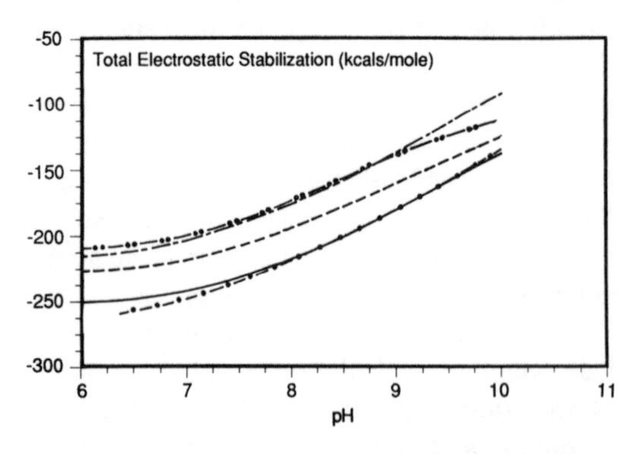

Fig. 6. Electrostatic component of the free energies of folding unliganded ATCase (—), PALA-liganded ATCase (–•–•), CTP-liganded ATCase (-—-), T-state ATCase (CTP-liganded ATCase with CTP atoms eliminated from the calculations) (— —), R-state ATCase (PALA-liganded ATCase without the PALA atoms) (••–••–).

Table 2. Comparison of experimentally determined free energies and calculated values of their electrostatic component at pH 8, 298 K, and [I] = 0.02 M.

	Experimental ΔG (kcal-mol^{-1})	Calculated ΔG_{el} (kcal-mol^{-1})
Assembly		
c:r in c_6r_6	−7.3[a] −10[b,c]	−5.9
r:r in r_2	−9 to −14[d]	−1.4
PALA binding		
c_3	−9.7[e]	−4.6 (C1C2C3) −5.2 (C4C5C6)
CTP binding		
c_6r_6	−7.9[f]	+3.5
r_2	−7.9[f]	+3.8

[a]*(57)* [b]*(58)* [c]*(59)* [d]*(60)* [e]*(33)* [f]*(35)*.

tions, binding of CTP is apparently driven by other kinds of forces such as hydrogen bonds and van der Waals interactions.

Potential surfaces

The potential surfaces of PALA-liganded and CTP-liganded ATCase are compared in Color Plate I. As shown, the potential surfaces around both structures are extremely asymmetric with large regions of both positive and negative potential. Total volumes of the regions of positive and negative potential are very similar for the two structures: approximately 5800 A^3 for regions of positive potential and 17,500 A^3 for regions of negative potential at a contour interval of 2kT. Corresponding values at a contour interval of 5kT are 1400 A^3 and 2000 A^3, respectively. The potential surface around most of the protein is negative, except in the vicinity of the active and nucleoside triphosphate binding sites. This distribution will favor productive binding at these sites and minimize nonproductive binding at other sites. The major effect of binding both PALA and CTP is to reduce the volumes of regions of positive potential. These potential surfaces will be analyzed in more detail in a subsequent publication.

Changes in pK values

The ability to calculate pK values of all ionizable groups within a protein rapidly and accurately is a major strength of modified Tanford-Kirkwood theory. Hence this method provides the information that is most needed to correlate predicted and experimental results and to obtain mechanistic insights. We have exploited this feature of the method to predict the changes in pK values that result from assembly and binding of PALA and CTP by comparing the values for isolated subunits and holoenzyme and holoenzyme with and without ligands bound.

All three processes produce large and widespread changes in pK values. The locations of residues whose pK values are predicted to change by more than 0.2 pH units in these processes are shown in Fig. 7. All three processes alter pK values of many residues in both the carbamoyl phosphate and L-aspartate binding domains of the catalytic chain, the interface between catalytic and regulatory chains, and the interdomain interface and nucleotide binding site of the regulatory chain. Because the coordinates for the subunits used in the assembly calculations were identical to the subunit coordinates for the holoenyzme, the predicted pK changes for assembly reflect simply the effects of all of the interchain interactions shown in Fig. 4. pK changes associated with ligand binding include the effects of ligand-induced conformational changes, which are largest at ligand binding sites and interdomain and interchain interfaces.

Because allosteric proteins are widely believed to function as two state systems [cf. *(53)*], it is of interest to compare the sets of residues in ATCase that undergo large pK changes as a result of assembly and ligand binding. The more closely ATCase conforms to a two-state system, the more similar these sets are expected to be. Residues whose pK values change by one or more pH units in any one of these processes are listed in Table 3. Residues that undergo pK changes greater than 0.5 pH units in all three processes are starred and plotted in Fig. 8. These residues

Table 3. Calculated pK changes > 1 induced by assembly of ATCase, PALA binding, and CTP binding, pH 8, 298 K, [I] = 0.02 M. Residues that are predicted to undergo pK shifts > 0.5 in all three processes are marked with an asterisk.

c chain

	Assembly		PALA binding		CTP binding	
	c1	c6	c1	c6	c1	c6
Asp 14*	0.39	0.63	0.32	0.58	< 0.1	1.07
Arg 17	0.29	0.14	–0.41	0.15	–0.70	1.15
Arg 54	0.33	0.40	2.26	2.33	0.92	–0.12
Asp 75	0.18	0.22	0.56	1.05	0.37	0.93
Lys 84	0.30	0.43	0.87	0.14	0.15	–0.54
Asp 100	< 0.1	< 0.1	1.44	< 0.1	0.95	< 0.1
Asp 129	0.86	1.04	–0.15	< 0.1	< 0.1	0.32
His 134	0.26	0.34	1.93	1.81	0.29	0.59
Asp 162	0.36	0.42	0.97	1.04	0.18	0.13
Lys 164*	1.93	2.04	< 0.49	–0.18	< 0.1	–0.64
Tyr 165*	1.24	0.97	1.60	1.90	–0.26	1.72
Arg 167*	0.44	0.52	0.99	1.55	0.23	0.67
Arg 226	0.15	0.19	1.01	0.87	0.22	0.24
Arg 229	0.36	0.28	2.57	2.32	1.41	–0.12
Lys 232*	0.52	1.09	0.69	–0.39	0.94	–0.94
Arg 234*	1.05	1.12	0.86	1.24	0.41	0.58
Asp 236	0.17	0.28	0.87	1.21	0.39	0.77
Glu 239	–0.29	–0.26	1.12	1.56	0.78	0.74
His 255	0.36	0.28	–1.24	–0.18	–0.72	–0.14
His 265	0.41	0.44	1.19	1.62	0.60	< 0.1

r chain

	Assembly		PALA binding		CTP binding	
	r1	r6	r1	r6	r1	r6
Glu 10	< 0.1	< 0.1	1.14	0.83	0.77	0.53
Arg 14	0.14	< 0.1	0.95	2.02	0.13	1.08
Asp 19	< 0.1	< 0.1	0.35	0.42	1.81	1.34
His 20	0.20	0.15	0.38	1.06	1.08	1.11
Arg 55	0.26	0.16	1.12	0.74	–0.48	1.26
Lys 56	0.20	0.23	0.11	< 0.1	0.45	1.94
Asp 87	< 0.1	< 0.1	–0.71	–1.38	0.41	0.40
Lys 94	0.11	0.12	0.13	0.32	1.59	2.08
Arg 128*	0.90	1.12	1.42	1.59	–1.02	–0.60
Arg 130*	0.24	1.26	1.92	1.95	< 0.1	0.65
Lys 139*	1.86	0.97	–1.04	0.68	–0.67	0.87
Tyr 140	1.73	1.79	< 0.1	< 0.1	0.32	–0.60
Cys 141	0.99	1.29	0.36	< 0.1	0.36	< 0.1
Lys 143*	2.77	2.99	–0.98	–1.10	–0.57	–0.53

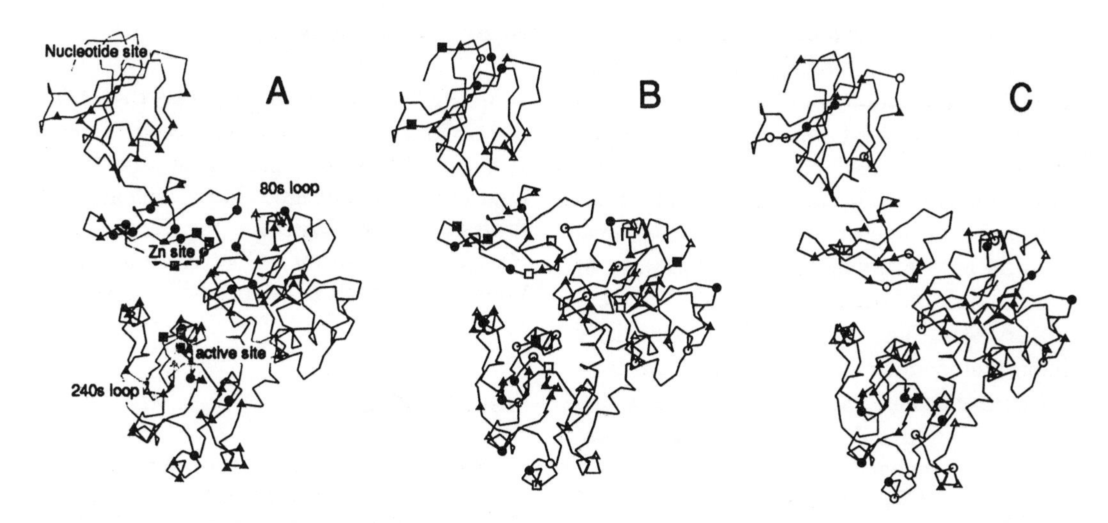

Fig. 7. Residues in ATCase whose calculated pK values change by more than 0.2 pH units as a result of (**a**) assembly of ATCase from c_3 and r_2 subunits, (**b**) binding of PALA, and (**c**) binding of CTP. Open symbols, pK change < 0; closed symbols, pK change > 0. Squares, pK change > 1; circles, pK change > 0.5; triangles, pK change > 0.2 pH units.

are localized in the 80s and 240s loops of the catalytic chain and the c:r interface of the regulatory chain, regions that both the structural studies and site-directed mutagenesis have implicated in the allosteric mechanism. This distribution provides additional evidence for the central role of these regions in molecular function. However, the fact that these residues represent a relatively small subset of the complete set of perturbed residues suggests that a complete description of the allosteric mechanism will require more than a two-state model.

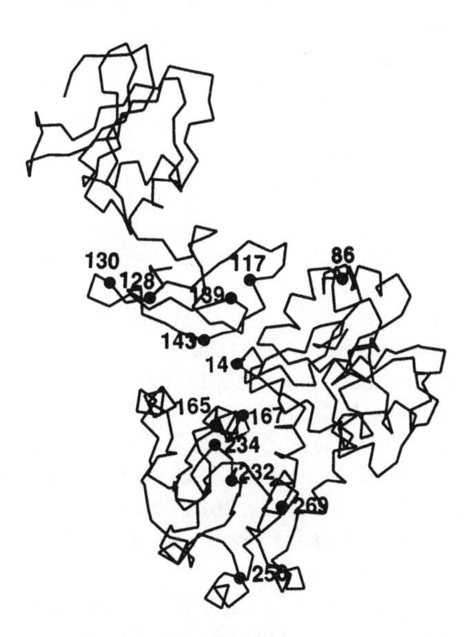

Fig. 8. Residues whose pK values shift by 0.5 pH units as a result of assembly, PALA binding and CTP binding.

Discussion

Allosteric regulation of ATCase and other systems requires a redistribution of energy within the molecule. In this study, we have used pK values as probes of the energies and environments of ionizable residues. Changes in pK values may arise as a result of changes in both charge configuration and solvent accessibilities; they result in a change in energy only when the group is charged.

Our results indicate that each of the three major processes required for enzyme function (assembly, binding of substrates, and binding of nucleotides) results in large and widespread changes in pK values. pK changes predicted for assembly can arise only from protein:protein interactions and associated changes in solvent accessibility, since conformational changes were not built into the calculations. pK changes associated with binding of both PALA and CTP reflect, in addition, the conformational changes associated with ligand binding, which are large for PALA and small for CTP.

Many of the residues that are affected by assembly or ligand binding are remote from the site of interaction, for example, residues in the regulatory chain that are perturbed by binding of PALA and residues in the catalytic chain perturbed by binding of CTP. Site-directed mutagenesis of these residues may provide new information about pathways of communication between catalytic and regulatory sites. Such studies will be particularly important in understanding the mechanism of action of nucleotides, since the crystal structures of CTP- and ATP-liganded ATCase are remarkably similar *(55)*.

Values of the electrostatic component of the free energy of stabilization must be interpreted cautiously, since these values are very sensitive to the degree of refinement of the structure. For example, the apparent electrostatic stabilization of the unliganded structure increased by 60 kcal-m^{-1} as a result of the most recent refinement. In addition, they represent only the terms in the protein energy balance that are sensitive to pH and ionic strength. Estimates of the contribution of local electrostatic interactions at a binding site, derived by comparing results obtained for the same protein structure with and without the ligand, have a higher degree of reliability. However, even these results require further exploration, since the result that CTP binding is associated with energetically unfavorable electrostatic interactions is unexpected. As noted previously, this may arise because the crystal structure upon which the calculation was based was determined at pH 5.9 where the phosphate groups of CTP are not fully ionized and crystal contacts influence ionic bonding. Analogous calculations with the structures that have now been determined at pH 7 will resolve this point in the future.

These calculations have clearly established the importance of electrostatic effects in mediating and regulating the biological function of ATCase. They provide evidence for long-range electrostatic effects and lay the groundwork for investigating the mechanisms by which electrostatic effects propagate through the structure.

Acknowledgments

Supported by NIH grant DK17335 (to NMA) and a Connecticut High Technology Fellowship (to MPG). We thank professor W. N. Lipscomb and colleagues for coordinate sets, Professor James Wild for Fig. 2, and Paul Harkins, Himanshu Oberoi, and John Wareham for their assistance with figures.

References

1. Kantrowitz, E. R., S. C. Pastra-Landis, W. N. Lipscomb, *Trends Biochem. Sci.* **5**, 124 (1980).
2. Kantrowitz, E. R., S. C. Pastra-Landis, W. N. Lipscomb, *Trends Biochem. Sci.* **5**, 150 (1980).
3. Kantrowitz, E. R., and W. N. Lipscomb, *Science* **241**, 669 (1988).
4. Schachman, H. K., *J. Biol. Chem.* **263**, 18583 (1988).
5. Allewell, N. J., *Ann. Rev. Biophys. Biophys. Chem.* **18**, 71 (1989).
6. Hervé, G., in *Allosteric Enzymes*, G. Hervé, Ed. (CRC Press, Boca Raton, FL, 1989), pp. 61–79.
7. Kantrowitz, E. R., and W.N. Lipscomb, *Trends Biochem. Sci.* **15**, 53 (1990).
8. Jones, M. E., L. Spector, F. Lipman, *J. Am. Chem. Soc.* **77**, 819 (1955).
9. Reichard, P., and G. Hanshoff, *Acta Chem. Scand.* **10**, 548 (1956).
10. Gerhart, J. C., and A. B. Pardee, *J. Biol. Chem.* **237**, 891 (1962).
11. Wild, J. R., S. J. Loughrey-Chen, T. S. Corder, *Proc. Natl. Acad. Sci. U.S.A.* **86**, 46 (1989).
12 Pastra-Landis, S. C., D. R. Evans, E. R.Kantrowitz, *J. Biol. Chem.* **253**, 4624 (1978).
13. Kerbiriou, D., and G. Hervé, *J. Mol. Biol.* **78**, 687 (1973).
14. Thiry, L., and G. Hervé, *J. Mol. Biol.* **125**, 515 (1978).
15. Gerhart, J. C., and H. K. Schachman, *Biochemistry* **4**, 1054 (1965).
16. Weber, K., *Nature* **218**, 1116 (1968).
17. Monaco, H. L., J. L. Crawford, W. N. Lipscomb, *Proc. Natl. Acad. Sci. U.S.A.* **75**, 5276 (1978).
18. Robey, E. A., and H. K. Schachman, *Proc. Natl. Acad. Sci. U.S.A.* **82**, 361 (1985).
19. Suter, P., and J. P. Rosenbusch, *J. Biol. Chem.* **252**, 8136 (1977).
20. Gerhart, J. C., and H. K. Schachman, *Biochemistry* **7**, 538 (1968).
21. Ladner, J. E., *et al., Proc. Natl. Acad. Sci. U.S.A.* **79**, 3125 (1982).

22. Krause, K. L., K. W. Volz, W. N. Lipscomb, *Proc. Natl. Acad. Sci. U.S.A.* **82,** 1643 (1985).
23. Krause, K. L., K. W. Volz, W. N. Lipscomb, *J. Mol. Biol.* 193, **527** (1987).
24. Ke, H.-M., W. N. Lipscomb, Y. Cho, R. B. Honzatko, *J. Mol. Biol.* **204,** 725 (1988).
25. Gouaux, J. L., and W. N. Lipscomb, *Proc. Natl. Acad. Sci. U.S.A.* **85,** 4205 (1988).
26. Gouaux, J. E., and W. N. Lipscomb, *Biochemistry* **29,** 389 (1990).
27. Howlett, G. J., and H. K. Schachman, *Biochemistry* **16,** 5077 (1977).
28. Kim, K. H., Z. Pan, R. B. Honzatko, H.-M. Ke, W. N. Lipscomb, *J. Mol. Biol.* **196,** 853 (1987).
29. Beard, C. B., and P. D. Schmidt, *Biochemistry* **12,** 2255 (1973).
30. Ireland, C. B., and P. D. Schmidt, *J. Biol. Chem.* **252,** 2262 (1977).
31. Mosberg, H. I., C. B. Beard, P. D. Schmidt, *Biophys. Chem.* **6,** 1 (1977).
32. Roberts, M. F., *et al.* , *J. Biol. Chem.* **251,** 5976 (1976).
33. Knier, B. L., and N. M. Allewell, *Biochemistry* **17,** 784 (1978).
34. Allewell, N. M., G. E. Hofmann, A. Zaug, M. Lennick, *Biochemistry* **18,** 3008 (1979).
35. Burz, D. S., and N. M. Allewell, *Biochemistry* **21,** 6647 (1982).
36. McCarthy, M. P., and N. M. Allewell, *Proc. Natl. Acad. Sci. U.S.A.* **80,** 6824 (1983).
37. Léger, C., and G. Hervé, *Biochemistry* **27,** 4293 (1988).
38. Glackin, M. P., M. P. McCarthy, D. Mallikarachchi, J. B. Matthew, N. M. Allewell, *Proteins: Struct. Funct. & Genet.* **5,** 66 (1989).
39. Shire, S. J., G. I. H. Hanania, F. R. N. Gurd, *Biochemistry* **13,** 2967 (1974); Shire, S. J., G. I. H. Hanania, F. R. H. Gurd, *Biochemistry* **13,** 2974 (1974); Shire, S. J., G. I. H. Hanania, F. R. N. Gurd, *Biochemistry* **14,** 1352 (1975); Matthew, J. B., G. I. H. Hanania, F. R. N. Gurd, *Biochemistry* **18,** 1919 (1979); Matthew, J. B., G. I. H. Hanania, F. R. N. Gurd, *Biochemistry* **20,** 571 (1979).
40. Tanford, C., J. G. Kirkwood, *J. Am. Chem. Soc.* **79,** 5333 (1957).
41. Lee, B., and F. M. Richards, *J. Mol. Biol.* **55,** 379 (1971).
42. Warshel, A., S. T. Russell, *Quart. Rev. Biophys.* **17,** 283 (1984).
43. Zauhar, R. J., and R. S. Morgan, *J. Mol. Biol.* **186,** 815 (1985).
44. Matthew, J. B., *Ann. Rev. Biophys. Biophys. Chem.* **14,** 387 (1985).
45. Honig, B., W. Hubbell, R. Flewelling, *Ann. Rev. Biophs. Biophys. Chem.* **15,** 163 (1986).
46. Lohman, T., *CRC Crit. Rev. Biochem.* **19,** 191 (1986).
47. Rogers, N. K., *Prog. Biophys. Mol. Biol.* **48,** 37 (1986).
48. Wodak, S. J., M. DeCrombrugge, J. Janin, *Prog. Biophys. Mol. Biol.* **49,** 29 (1987).
49. Harvey, S. C., *Proteins: Struct. Funct. & Genet.* **5,** 78 (1989).
50. Sharp, K., and B. Honig, *Ann. Rev. Biophys. Biophys. Chem.*, **19,** 301 (1991)
51. Wendoloski, J. J., J. B. Matthew, *Proteins: Struct. Funct. & Genet.* **5,** 313 (1989).
52. Kleanthous, C., D. E. Wemmer, H. K. Schachman, *J. Biol. Chem.* **263,** 13062 (1988).
53. Perutz, M. F., *Quart. Rev. Biophys.* **22,** 139 (1989).
54. Stevens, R., and W. N. Lipscomb, personal communication.
55. Lipscomb, W. N., personal communication.
56. Honzatko, R. B., *et al., J. Mol. Biol.* **160,** 219 (1982).
57. Bothwell, M. A., and H. K. Schachman, *J. Biol. Chem.* **255,** 1962 (1980).
58. Chan, W. W.-C., *FEBS Lett.* **44** 178 (1974).
59. Chan, W. W.-C., *J. Biol. Chem.* **250,** 668 (1975).
60. Cohlberg, J. A., *et al., Biochemistry* **11,** 3396 (1972).
61. Wild, J. R., *et al., Proc. Natl. Acad. Sci. U.S.A.* **86,** 46 (1989).
62. Honzatko, R. B., *et al., J. Mol. Biol.* **160,** 219 (1982).

3

Charge Effects on Folded and Unfolded Proteins

Dirk Stigter, Ken A. Dill

The stability of a protein is generally dependent upon pH. The denaturation temperature of a protein is often observed to have a broad maximum around the isoelectric pH of the molecule *(1–3)*. This suggests that at the pH extremes where there is net charge on the molecule, the electrical free energy of the protein can be reduced by unfolding, a process that increases the average separation of like charges. The electrical driving force for unfolding can also be affected by ionic strength, since small ions are able to shield intrachain charge repulsions. Our purpose in this chapter is to develop a theory for these electrical contributions to the stability of globular proteins.

The effects of charge are treated by using the following model. We consider the unfolded molecule to have charges randomly distributed throughout a "porous" sphere, which is otherwise filled by the solvent, a solution containing monovalent salt. The folding process is considered to follow a two-part fictitious thermodynamic pathway, chosen for mathematical convenience and previously described for modeling the thermal stabilities of proteins *(4, 5)*. (i) The protein molecule undergoes a density increase, whereby the ionizable groups become more concentrated and the salt solution is expelled, until the chain reaches a maximum compactness. Within the sphere, the ionizable groups remain distributed randomly. Along the radius, the degrees of ionization vary with the electrical driving forces. (ii) Then the molecule rearranges its residues to put its charges on the surface of the protein, a sphere of low dielectric constant. Both the compact spherical native protein and the porous spherical unfolded molecule are taken to be in ionization equilibrium with an aqueous solution of given pH and ionic strength, with the potential distribution given by the Poisson-Boltzmann equation. The model and the potential calculation are described in detail here.

Our models for the electrostatic free energies of porous and nonporous spheres are variants of models used earlier for the titration of proteins *(6)* and of polyelectrolytes *(7)*. The problem of the titration equilibrium is closely related to that of how electrostatic interactions affect stability *(8)*.[1] The present approach is developed to treat this combined problem. Various aspects of the theory are illustrated by using experimental titration data on both the folded and unfolded states of myoglobin, a well-studied example of a one-domain protein.

Reprinted from *Biochemistry* **29**, 1262 (1990) with permission.

Characterization of the Protein Model

The unfolded protein is represented by a sphere in which protein residues and solution are mixed randomly, in ionization equilibrium with the surrounding solution of given pH and ionic strength. Along the contraction path, the radius R of this sphere decreases from R^* in the unfolded state to R_p in the randomly condensed state, as indicated in Fig. 1. At the same time the protein density of the sphere, $\rho = (R_p/R)^3$, increases from the initial value $\rho^* = (R_p/R^*)^3$ to $\rho = 1$ for the fully condensed state which excludes any solution. In keeping with the earlier lattice treatment *(4, 5)*, the chain is divided into n segments which occupy n lattice sites, each with volume $v = (4\pi/3n)R_p^{\ 3}$. In general, then, a sphere with density ρ has n/ρ lattice sites, n of which are filled with protein and the rest, $n\,(1/\rho - 1)$ sites, with salt solution. With an arrangement of such sites in concentric shells, when r is the radial coordinate, all sites in the outer shell, between $r = R$ and $r = R_{core} = R - v^{1/3}$, are counted as exterior or surface sites. This constitutes a fraction $f_e = 1 - (R_{core}/R)^3$ of the total number n/ρ of sites in the sphere. In the final state, which is globular and compact ($\rho = 1$), the $(1 - f_e)n$ interior or core sites are occupied predominantly by a fraction Φ of the hydrophobic residues in the chain.

The ionic charge on the protein derives from t different types of acid and basic groups, n_i groups of type $i = 1, 2, ..., t$ per protein molecule, each with intrinsic ionization constant k_i for proton binding. A fraction f_s of each type of these ionic groups is in surface sites, in the outer shell between $r = R_{core}$ and $r = R$; the charge deriving from these groups is treated as a uniform surface charge at $r = R$. The remaining ionizable groups, the

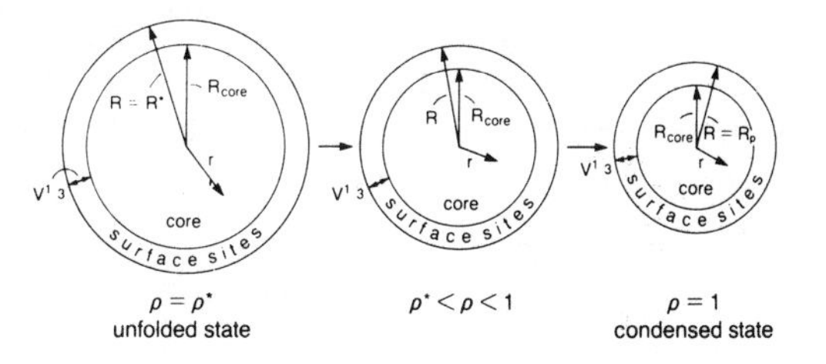

Fig. 1. Along the contraction path, the radius R of this sphere decreases from R^* in the unfolded state to R_p in the randomly condensed state.

fraction $1 - f_s$, are distributed uniformly in the core, in the volume $r < R_{core}$. Along the condensation path, the random distribution always gives $f_s = f_e$; hence, f_s is a function of the variable density ρ. Along the reconfiguration path, where $\rho = 1$, f_s is a function of the composition, Φ, and of the reorder variable θ, which is defined as the fraction of surface sites occupied by hydrophobic segments *(4, 5)*. The relation is

$$f_s = [(1 - \theta)/(1 - \Phi)]f_e \qquad \text{for } \rho = 1 \qquad (1)$$

This completes the description of the folding pathway. We now turn to the electrostatic problem.

Self-consistent Charge Potential Treatment

The electrostatic free energy of a system is determined by the potential field experienced by each charge, but this potential, in turn, is generated by the charge distribution. Hence, the charge distribution and the potential field must be determined self-consistently. In the present model the potential field, ψ, is spherically symmetric and thus depends only on the radial distance r from the center of the molecule. If this field $\psi(r)$ is known, then the

1 In this chapter the electrical free energy g_{el} is derived in terms of the net charge number Z and the interaction factor w, taken empirically from titration curves as $g_{el} = kTwZ^2$ for both native and denatured protein. This approach implies that w is independent of the protein charge number Z in agreement with the solid sphere model of the globular protein in Eq. 23. We note that the factor w is not exactly constant for our model of the unfolded protein. It follows from Eq. 30 that for the porous sphere model w depends on Z because the change of the α_i with Z depends on r and, hence, the shape of the charge distribution changes with Z. However, the resulting variations of w with Z are small for low densities, and for the ρ^* values in Fig. 5, they are well within the errors inherent in the determination of w from experimental titration curves.

degree of ionization, α_i, for each type i of ionic group can be calculated as a function of the bulk proton concentration, $H = 10^{-pH}$, by using

$$q_i \ln \frac{\alpha_i(r)}{1-\alpha_i(r)} = \ln H + \ln k_i - \frac{e\psi(r)}{kT} \qquad i = 1, 2, ..., t \tag{2}$$

where $q_i = +1$ for basic and $q_i = -1$ for acid groups, e is the protonic charge, and kT is Boltzmann's constant multiplied by absolute temperature. The first two terms on the right-hand side of Eq. 2 simply describe the ordinary binding equilibrium for ligands with intrinsic binding constant k_i, with independent binding sites, i.e., Langmuir-type binding. Electrostatic interactions between the ionized groups bias the binding and are taken into account in the last term on the right-hand side of Eq. 2, through the potential field $\psi(r)$ as noted above. This correction has been commonly used to treat the titration of proteins [for a review, see *(9)*] and polyelectrolytes *(10)*. An alterative way to look at this potential correction is to recognize that whereas H is the bulk proton concentration, $H \exp(-e\psi/kT)$ is the proton concentration at the local binding site in the presence of the field of the other charges.

Our purpose is to calculate the α_i values. They are determined by the self-consistent theory described below. Then, when the α_i values are known, the surface charge density of the sphere is given in terms of them by

$$\sigma = \frac{f_s e}{4\pi R^2} \sum_{i=1}^{t} q_i n_i \alpha_i(R) \qquad \text{at } r = R \tag{3}$$

and the fixed charge density in the core as a function of r is

$$\rho_{el}(r) = \frac{(1-f_s)e}{(4\pi/3)R_{core}^3} \sum_{i=1}^{t} q_i n_i \alpha_i(r) \qquad \text{for } 0 < r < R_{core} \tag{4}$$

with Eq. 1 for f_s.

The fixed charges of the protein are neutralized by mobile small ions surrounding the spherical folded molecule and, along the condensation path, also inside the porous spherical unfolded states of the molecule. The distributions of these charges and of the electrostatic potential are related by the Poisson-Boltzmann (PB) equation. Although this equation is not exact and has a controversial history, Monte Carlo computations have shown that it is a good approximation for distributions of small monovalent ions *(11–15)*. For fields generated by typical proteins in aqueous salt solutions, the linearized form of the PB equation has been found to be a good approximation *(16)*. Outside the sphere we thus have

$$\frac{d^2\psi}{dr^2} + \frac{2}{r}\frac{d\psi}{dr} = \kappa_w^2 \psi \qquad r > R \tag{5}$$

The Debye length outside the sphere, $1/\kappa_w$, is given by

$$\kappa_w^2 = \frac{e^2}{\varepsilon_0 \varepsilon_w kT} \sum_s n_s z_s^2 \qquad r > R \tag{6}$$

n_s is the concentration (number density) of ions of type s with valency z_s in the bulk solution, ε_0 is the permittivity of vacuum, and ε_w is the relative permittivity or dielectric constant of the solvent around the sphere.

We make the assumption that the potential vanishes in the bulk solution, that is

$$\psi = 0 \qquad \text{at } r = \infty \tag{7}$$

As long as the distribution of the charge fixed to the sphere has spherical symmetry, its details play no role is determining the field in the outside solution. Only the total net charge Z_{tot} of the sphere, which includes the mobile small ions inside it, is relevant. Then the outside field, satisfying Eqs. 5 and 7, is the well-known solution *(17)*

$$\psi(r) = \psi(R) \frac{Re^{-\kappa_w(r-R)}}{r} \qquad r > R \tag{8}$$

where $\psi(R)$ is the surface potential of the sphere

$$\psi(R) = \frac{Z_{tot}\, e}{4\pi\varepsilon_0\varepsilon_w R(1+\kappa_w R)} \tag{9}$$

The potential field inside the porous sphere arises from an averaging of the interactions among the many fixed and mobile charges. Therefore, we take for the dielectric constant inside the sphere, ε_{sphere}, the volume average of the solution value ε_w and the protein value ε_p

$$\varepsilon_{sphere} = (1-\rho)\varepsilon_w + \rho\varepsilon_p \qquad r < R \quad (10)$$

Similarly, we use for ion concentration inside the sphere the average values $(1-\rho)n_s$. Therefore, the Debye length inside the sphere, $1/\kappa_{sphere}$, is given by

$$\kappa^2_{sphere} = \frac{(1-\rho)\varepsilon_w}{(1-\rho)\varepsilon_w + \rho\varepsilon_p}\kappa^2_w \quad (11)$$

which follows from replacement of n_s/ε_w by $(1-\rho)n_s/\varepsilon_{sphere}$ in Eq. 6. In the outer shell the fixed ionic charges, treated as a surface charge, are taken into account through the boundary conditions below. Therefore, the PB equation for the outer shell of the sphere is with Eq. 11 for κ^2_{sphere}

$$\frac{d^2\psi}{dr^2} + \frac{2}{r}\frac{d\psi}{dr} = \kappa^2_{sphere}\psi$$

$$R_{core} < r < R \quad (12)$$

For the core this equation is modified by the fixed charge density $\rho_{el}(r)$ of Eq. 4.

$$\frac{d^2\psi}{dr^2} + \frac{2}{r}\frac{d\psi}{dr} = \kappa^2_{sphere}\psi - \frac{\rho_{el}}{\varepsilon_0\varepsilon_{sphere}}$$

$$0 < r < R_{core} \quad (13)$$

We now discuss the boundary conditions and the solution of Eqs. 12 and 13. Equation 8 gives for the potential gradient at the outside surface of the sphere

$$\left(\frac{d\psi}{dr}\right)_R = -\frac{1+\kappa_w R}{R}\psi(R) \qquad r = R^+ \quad (14)$$

At $r = R$, ψ is continuous. Gauss's law combined with Eq. 14, with the surface charge from Eq. 3, and with the change of dielectric constant from ε_w to ε_{sphere} from Eq. 10 yields the gradient at the surface inside the sphere

$$\left(\frac{d\psi}{dr}\right)_R = -\frac{\varepsilon_w(1+\kappa_w R)}{\varepsilon_{sphere}R}\psi(R) + \frac{\sigma}{\varepsilon_0\varepsilon_{sphere}}$$

$$r = R^- \quad (15)$$

Since at $r = R_{core}$ there is no surface charge and the dielectric constant is uniform, ψ and $d\psi/dr$ are continuous at this boundary. Finally, without a point charge at $r = 0$, symmetry requires that

$$d\psi/dr = 0 \quad \text{for } r \to 0 \quad (16)$$

Equation 8 gives the outside field, except for the initially unknown surface potential $\psi(R)$. The potential field inside the sphere is obtained by the successive stepwise numerical integration of Eqs. 12 and 13 by means of a fourth-order Runge-Kutta integration routine *(18)*. This computer routine is designed for the simultaneous integration of a set of coupled first-order ordinary differential equations such as

$$dy_1/dr = f_1(r, y_1, y_2)$$

$$dy_2/dr = f_2(r, y_1, y_2) \quad (17)$$

The technique is used for our second-order differential equations by defining $y_1 = \psi$ and $y_2 = d\psi/dr$. This converts Eq. 12 into the following set of coupled equations

$$dy_1/dr = y_2$$

$$\text{for } R_{core} < r < R \quad (18)$$

$$dy_2/dr = \kappa^2_{sphere}y_1 - (2/r)y_2$$

and Eq. 13 becomes

$$\frac{dy_1}{dr} = y_2$$

$$\text{for } r < R_{core} \quad (19)$$

$$\frac{dy_2}{dr} = \kappa^2_{sphere}y_1 - \frac{\rho_{el}(y_1)}{\varepsilon_0\varepsilon_{sphere}} - \frac{2}{r}y_2$$

For each given potential $y_1 = \psi$, the charge density ρ_{el} in Eq. 19 is obtained from Eqs. 2 and 4.

For given values of R, R_{core}, f_s, n_i, q_i, H, and k_i, the self-consistent surface potential $\psi(R)$ is determined in an iteration procedure aimed at satisfying the final boundary condition given by Eq. 16. An integration is started at the surface of the sphere with a choice of $y_1 = \psi(R)$ and the corresponding value $y_2 = (d\psi/dr)_{R^-}$ from Eq. 15 using Eqs. 3 and 2 for σ. With such initial values $[y_1(R), y_2(R)]$ the integration of Eq. 18 now proceeds stepwise inward from $r = R$ to $r = R_{core}$. At this point the current values $[y_1(R_{core}), y_2(R_{core})]$ are used to initialize the integration of Eq. 19 which is terminated near $r = 0$ with a test of Eq. 16. The latter boundary condition is satisfied only for the exact, but initially unknown, starting potential $\psi(R)$. The final slope, $y_2 = (d\psi/dr)_{r\to 0}$, is positive or negative, depending on whether the integration was started with a surface potential which is too high or too low. This is used as a criterion in a bisection iteration method *(18)* to approach the value of $\psi(R)$ that satisfies Eq. 16 and thus determine the self-consistent potential and charge distributions in the sphere.

We describe below tests of the model against experimental data. Then in the section following, we develop a theory for the effects of charge on protein stability.

Protein Titrations

Globular proteins

Linderstrøm-Lang *(6)* was the first to apply the Debye-Hückel theory *(17)* to protein titrations. The Linderstrøm-Lang theory assumes that the folded protein is a nonporous spherical Z-valent ion with radius R which excludes (the center of) small ions to a radius R_{ex}. The treatment leads to the prediction of activity coefficients. However, it also predicts an interaction factor w which more readily can be determined from experiments. The factor w is related to the electrical free energy g_{el} for charging the native protein [see, e.g.,Tanford and Kirkwood *(34)*] and is defined by

$$\mathrm{pH} + q_i \log\frac{\alpha_i}{1-\alpha_i} = -\mathrm{p}K_i - 0.868wZ$$

$$= -\mathrm{p}K_i - \frac{0.434}{kT}\frac{\partial g_{el}}{\partial Z} \qquad (20)$$

where $\mathrm{p}K_i = -\log k_i$ and the change from natural log introduces $\log e = 0.434$. For the Linderstrøm-Lang model we have

$$g_{el} = \int_0^Z e\psi(R, Z')\,dZ' \qquad (21)$$

The Debye-Hückel theory *(17)* gives for the surface potential of the protein ion with uniform surface charge Ze

$$\psi(R, Z) = \frac{Ze}{4\pi\varepsilon_0\varepsilon_w R} - \frac{Ze}{4\pi\varepsilon_0\varepsilon_w}\left(\frac{\kappa_w}{1+\kappa_w R_{ex}}\right) \qquad (22)$$

The first term in Eq. 22 is the Coulomb potential of the charge Ze at the surface of the protein ion. The second term is the potential of the ionic atmosphere with charge $-Ze$. Substitution of Eq. 22 into Eqs. 21 and 20 yields the well-known expression for w

$$w = \frac{1}{2ZkT}\frac{\partial}{\partial Z}\int_0^Z e\psi(R, Z')\,dZ' = \frac{e\psi(R, Z)}{2ZkT}$$

$$= \frac{e^2}{8\pi\varepsilon_0\varepsilon_w kT}\left(\frac{1}{R} - \frac{\kappa_w}{1+\kappa_w R_{ex}}\right) \qquad (23)$$

Predictions of Eq. 23, with $R_{ex} = R + 2.5$ Å, are often somewhat too high; nevertheless, agreement with experiments is better than 20% in a number of cases *(9, 19–22)*.

Our model for the globular state of proteins differs from the Linderstrøm-Lang model in one principal respect, Our model is simpler because we assume $R_{ex} = R$; i.e., we do not assume a charge-free shell of thickness $R_{ex} - R$ that is impenetrable by small ions. Since, as a consequence, in our model the countercharge may approach the protein up to the surface of the fixed charges, the predicted surface potentials and w factors are somewhat lower than the predictions of the Linderstrøm-Lang model. We believe our approach is better justified since the surfaces of globular proteins are not smooth. The charged surface groups may extend somewhat further into the solution to reduce their Born energy, allowing small counterions to penetrate into the high-potential regions between the fixed charges, similar to the situation at the surface of detergent micelles *(23)*. Therefore, our simpler model should be a better approximation for globular proteins than the Linderstrøm-Lang model. In any case, the differences in w are small.

Comparison of theoretical predictions of w with experiments requires choice of radius $R_p = R = R_{ex}$ of the globular sphere in Eq. 23. In Table 1 we compare experimental interaction values, w_{exp}, for several globular proteins with two model results obtained as follows. Assuming a specific volume of 0.73 mL/g for globular protein with molecular weight M, a radius R_0 is calculated from $(4/3)\pi R_0^3 = 0.73\,M/N_{av}$. Table 1 shows values of w for model spheres with $R_p = R_0$ and, in parentheses, for $R_p = R_0 + 3Å$. The data show that there is satisfactory agreement between theory and experiment for an assumed radius R_p of the model sphere which is not much greater than R_0.

Tables 2 and 3 show the experimental results on metmyoglobin reported by Breslow and Gurd *(19)*. The experimental w values in Table 3 were derived from the titration of histidines in the globular protein and from the titration of carboxyls in the unfolded state. For the globular protein, wexp agrees quite well with the model calculation, with R = Rex = 18 Å in Eq. 23, as shown by the comparison in Table 3. The results in this section show that the model gives satisfactory agreement with experiments on the titration of representative one-domain globular proteins.

Denatured protein

In this section, we apply the present theory to the prediction of w for unfolded proteins as a function of the molecular density ρ of the porous sphere. This is then compared with titration experiments on unfolded myoglobin. For globular proteins the application of Eq. 20 is simple because all charged groups are at the surface and, in our model, at the same potential $\psi(R)$. For the globular molecules the interaction correction $2wZ$ in Eq. 20 can be evaluated directly as the potential $e\psi/kT$ at the charged groups, as indicated in Eqs. 2 and 23. This is not the case for the unfolded protein where the experimentally determined factor w refers to an average value for the whole protein molecule, with its uneven distribution of charge and potential. Now w is derived most conveniently from the charge and the electrical free energy as follows.

From Eqs. 3 and 4, the number of fixed charges Z of the porous sphere is

$$Z = \sum_{i=1}^{t} \left[f_s q_i n_i \alpha_i(R) + \frac{(1-f_s) q_i n_i}{(4\pi/3) R_{core}^3} \int_0^{R_{core}} \alpha_i(r) 4\pi r^2 \, dr \right] \tag{24}$$

where the first term in brackets gives the fixed surface charge and the second term gives the fixed charge distributed in the core. The electrical free energy, g_{el} is evaluated as the work done reversibly to charge up the porous sphere to Z charges. This work does not depend on the particular choice of charging process, but only on the final state, that is, on the distribution of the fixed charge as given in Eq. 24, subject to the self-consistent potential distribution. In the Debye-Hückel linear approximation the work of charging equals the sum, or integral, of $(1/2)\psi\ d(Ze)$ over all fixed

Table 1. Comparison of experimental interaction factors w_{exp} with theoretical results from Eq. 23 ,with $R_p = R_{ex} = R_0$ and, in parentheses, with $R_p = R_{ex} = R_0 + 3$ Å.

Protein	R_0 (Å)	M_{salt}	w_{exp}	w_{calc}
Myoglobin[a]	17.3	0.06	0.085	0.086 (0.067)
		0.16	0.050	0.063 (0.048)
Bovine serum albumin[b]	26.8	0.01	0.054	0.071 (0.061)
		0.03	0.036	0.053 (0.044)
		0.08	0.028	0.038 (0.032)
		0.15	0.024	0.030 (0.025)
Ribonuclease[c]	15.8	0.01	0.112	0.149 (0.117)
		0.03	0.093	0.119 (0.092)
		0.15	0.061	0.075 (0.056)
Conalbumin[d]	28.1	0.01	0.046	0.066 (0.057)
		0.03	0.035	0.049 (0.041)
		0.10	0.025	0.032 (0.027)

[a] *(19)* [b] *(21)* [c] *(22)* [d] *(20)*.

Table 2. Ionizable groups of metmyoglobin following Breslow and Gurd *(19)* at 25° C.

Group	n_i	pK_i	Group	n_i	pK_i
ε-amino	19	10.60	Histine (unfolded)	12	6.48
α-amino	1	7.80	Hemic acid	1	8.90
Histidine (globular)	6	6.62	Carboxyl	23	4.40

Table 3. Titrations of metmyoglobulin.

	Globular		Unfolded	
M_{salt}	w_{exp}	w_{calc}	w_{exp}	ρ^*
0.16	0.050	0.059	0.034	0.51
0.06	0.085	0.081	0.044	0.39

charge Ze where at the surface $\psi = \psi(R)$, and in the core $\psi = \psi(r)$ varies with r. In view of Eq. 24 the electrical free energy of the porous sphere becomes

$$g_{el} = \sum_{i=1}^{t} \left[\frac{f_s q_i n_i \alpha_i(R) e\psi(R)}{2} + \frac{(1-f_s) q_i n_i}{(4\pi/3) R_{core}^3} \int_0^{R_{core}} \frac{\alpha_i(r) e\psi(r)}{2} 4\pi r^2 \, dr \right] \qquad (25)$$

After the determination of the self-consistent charge-potential distribution of the model sphere for the desired parameters by the method outlined in the previous sections, the information then is available for the evaluation of Eqs. 24 and 25. Since the protein charge varies with pH, we may compute w using the difference equation

$$w = \frac{1}{2ZkT} \frac{\partial g_{el}}{\partial Z} = \frac{1}{2ZkT} \frac{\Delta g_{el}/\Delta \text{pH}}{\Delta Z/\Delta \text{pH}} \qquad (26)$$

Using the above procedure, with $\Delta\text{pH} = 0.02$ in Eq. 26, we have evaluated the interaction factor w for the myoglobin model along the condensation path, with the parameters defined in Table 2, and taking for the globular molecule radius $R_p = 18$ Å, $n = 110$ lattice sites (corresponding to 153 amino acids), and fraction of hydrophobic residues $\Phi = 0.45$. The dielectric constant of globular proteins is not well-known. On the basis of recent theoretical work by Gilson and Honig *(24)*, we choose $\varepsilon_p = 3.5$. The results for w as a function of the chain density ρ are plotted (Fig. 2) for the two values of the ionic strength in Table 3. The curves show how sensitive w is to the salt concentration and to the density ρ.

The circles in Fig. 2 indicate the experimental values for the unfolded state, w_{exp} in Table 3. When these experimental results for w and the known salt concentrations are substituted in the theory, Fig. 2 shows that the predicted density of the unfolded state is $\rho^* = 0.51$ and $\rho^* = 0.39$ in solutions of 0.16 and 0.06 M ionic strength, respectively. This prediction, that the unfolded state has extremely high density, is consistent with predictions from a very different theoretical treat-

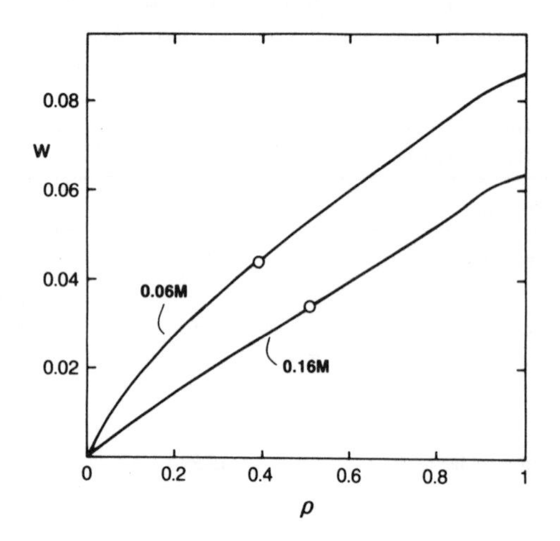

Fig. 2. Electrostatic interaction factor w from Eq. 26 for titration of carboxyl groups of metmyoglobin along hypothetical contracting path, as function of density ρ of model sphere, at 25°C and pH 3.5, ionic strength as indicated in figure. Circles indicate experimental w values from Table 3.

ment *(4, 5)*. This predicted density for the unfolded state is much greater than that a polypeptide chain would have in a θ solvent, in which the chain would obey random flight statistics. For a random flight chain, we may calculate the density as follows. The length of the fully stretched myoglobin chain of 153 residues is about $L = 153 \times 3.8 = 581$ Å. With a persistence length $P = 19$ Å for polypeptide chains *(25)*, one obtains a radius of gyration of the coil $s = (LP/3)^{1/2} = 61$ Å and an equivalent sphere radius $R = s\,(5/3)^{1/2} = 78$ Å, giving a protein density $\rho = (R_p/R)^3 = (18/78)^3 = 0.012$ in the sphere. This is much less than the values for the unfolded state, $\rho^* = 0.39$ and $\rho^* = 0.51$, obtained above.

Moreover, the experimental interaction factors, w_{exp}, are considerably larger than those which would be predicted for the titration of a polyelectrolyte molecule in solution, further supporting the view that the unfolded protein is relatively dense. Titration curves of synthetic linear polyelectrolytes such as poly(acrylic acid) and poly(methacrylic acid) are well explained with a model of a uniformly charged cylinder *(26, 27)*. Neglecting end effects for a cylinder with length L, radius a, and linear charge density Ze/L, the Debye-Hückel approximation for the interaction factor of w of such a model is, from Eq. 23,

$$w = \frac{1}{2Z}\frac{e\psi_s}{kT} = \frac{e^2}{4\pi\varepsilon_0\varepsilon_w kTL}\left(\frac{K_0(\kappa_w a)}{\kappa_w a K_1(\kappa_w a)}\right) \quad (27)$$

where ψ_s is the surface potential of the cylinder *(28)* and K_0 and K_1 are modified Bessel functions. With a radius of $a = 3.65$ Å for the unfolded myoglobin cylinder and L = 581 Å as before, Eq.27 predicts $w = 0.014$ in 0.16 M and $w = 0.018$ in 0.06 M ionic strength solutions. These interaction factors, predicted for a polypeptide in a good solvent, are much less, by a factor of nearly 3, than the experimental values, w_{exp} in Table 3, for the unfolded state. We conclude that the values of w_{exp} obtained by experiments on unfolded myoglobin are consistent only with a high unfolded-state density, similar in some regards to polymers in poor solvents.

A further test of the theory can be made

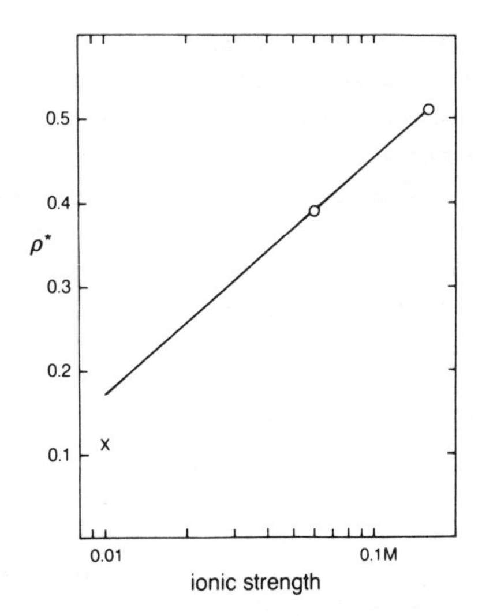

Fig. 3. Density of unfolded metmyoglobin at 25°C as a function of ionic strength. **(Circles)** From titration data, w_{exp} in Fig. 2. **(Cross)** From viscosity data by Privalov *et al.* *(3)*.

by comparison with viscosity measurements of the unfolded state density. In Fig. 3, we make a linear extrapolation to predict the density the unfolded molecules would have in a solution at 0.01 M ionic strength, to compare with viscosity measurements which have been made at that ionic strength by Privalov *(3)*. They have reported intrinsic viscosity data for unfolded metmyoglobin in 10 mM sodium acetate buffer solutions at various temperatures and pH values. In the pH range where metmyoglobin unfolds at 25°C, the intrinsic viscosity increases from $[\eta] = 3.2$ cm^3/g at pH 4.80 to $[\eta] = 20.5$ cm^3 /g at pH 3.40. Neglecting drainage of the unfolded coil, a high estimate of its density is given by the ratio of $[\eta]$ values, $\rho^* = 3.2/20.5 = 0.156$.

Partial drainage may be introduced by polymer viscosity theory, on the basis of the model of a porous sphere *(29, 30)*. This lowers the density of unfolded metmyoglobin with $[\eta] = 20.5$ cm^3/g to $\rho^* = 0.125$ if we use a high estimate of the friction factor of the fully drained sphere and to $\rho^* = 0.096$ if we use a low estimate of this fraction factor, which is an uncertain parameter of the hydrodynamic model. We take the average, $\rho^* = 0.11$, as the experimental density of the unfolded state in 0.01 M salt solutions, indicated by a cross in Fig. 3.

We compare the latter value with the above results for ρ^* from titration data. In Fig. 2 the present model predicts a lower ρ^*, that is, an increase in coil size, for decreasing ionic strength I, in agreement with polyelectrolyte theory *(7, 31, 32)*. The theory *(32)* suggests that a $\rho^* - \log I$ plot is slightly convex. Therefore, results of the linear extrapolation in Fig. 3 of the ρ^* values in Fig. 2 to $\rho^* \simeq 0.17$ in 0.01 M ionic strength solution may be somewhat too high. Additional experiments would be helpful. In summary, the titration and viscosity data yield three approximate ρ^* values of unfolded myoglobin in different ionic strength solutions, to be compared later with predictions of the protein folding theory.

The model for folded and unfolded states is quite simple, but its surprisingly good performance suggests that whatever errors it has may compensate each other. Corrections can be made for (i) the self-energy of the smeared charges representing discrete ionized groups, (ii) a nonrandom distribution of ionic groups at the surface of the protein sphere, and (iii) a spread of intrinsic pK values in each class of ionizable groups. Some of these considerations have been addressed in more recent work reviewed, e.g., by Matthew *(33)*, including the relation between charge site geometry and dielectric value *(34)*, and the relation between solvent accessibility and effective pK values *(35)*. For most proteins it is found that only a few percent of the ionizable groups exhibit abnormal pK values because they are inaccessible to solvent *(36)*. Of particular interest is the recent work by Gilson and Honig *(37)*, who, for arbitrary geometry and distributions of discrete fixed charges, solve the Poisson-Boltzmann equation by a finite difference method.

It follows from the comparison of theory and experiment above, for the titration behavior of folded and unfolded proteins, that the present approach should also provide a satisfactory foundation for treatment of charge effects on protein stability, undertaken below. The prediction of protein titration behavior is probably more demanding of an electrostatics theory than the prediction of protein stability. Titration theory can be tested on native and unfolded protein separately, but it is only differences in free energy that enter the theory of protein stability. These differences may be considerably averaged, because of the wide variation in the many different environments of the ionizable groups in the ensemble of unfolded states. This further suggests that the refinements indicated above may be of less importance for stability than for titrations.

Free Energy of Charged Proteins

In the remainder of the chapter, we extend the theory above to treatment of the electrostatic contributions to protein stability, the free energy of the folded state minus the free energy of the unfolded state. In an earlier paper *(16)*, we compared two ways to evaluate the free energy of charged colloids, by a thermodynamic charging process and by using the binding polynomial. We developed both methods for colloidal particles on which all fixed charges are at the same electrostatic potential. This uniformity of the electrostatic potential applies to our present model of the folded protein, since it is characterized here as a nonporous sphere with a uniform surface charge. We found that the results of both methods, although in different analytical form, are numerically not significantly different. It is difficult to apply the binding polynomial method rigorously to systems in which the potential of the fixed charges is nonuniform. On the other hand, application of the thermodynamic charging method remains straightforward. Therefore, we use here the charging process for the porous sphere model along the folding pathway.

We consider a solution of colloid spheres which each carry t types of acid and basic groups, n_i groups of type $i = 1, 2, ..., t$. Using the charging process, we derived Δg_b, the proton binding free energy per molecular sphere, relative to its un-ionized state. When all ionic groups are at the surface, $r = R$, the earlier treatment *(16)* yields

$$\Delta g_b = kT \sum_{i=1}^{t} n_i \ln(1 - \alpha_i) + g_{dl} \quad (28)$$

where the ionization degrees α_i satisfy the relevant Eqs. 2 for $r = R$. The *total* free energy of the electrical double layer in Eq. 20, g_{dl}, and the *electrical* free energy of the electrical double layer in Eq. 20, E_{el}, are related by *(38)*

$$g_{dl} = - \int_0^{\psi(R)} Z(\psi')ed\psi'(R)$$

$$= -Ze\psi(R,Z) + \int_0^z e\psi(R,Z')dZ'$$

$$= -Ze\psi(R,Z) + g_{el} \quad (29)$$

In the Debye-Hückel approximation, when Z and $\psi(R)$ are linearly related, we have

$$g_{dl} = -g_{el} = -1/2Ze\psi(Z,R)$$

$$= -\frac{e\psi(R)}{2} \sum_{i=1}^{t} q_i n_i \alpha_i(R) \quad (30)$$

In a study of the charge effects on the stability of ribonuclease, Hermans and Scheraga *(8)* derived Eq. 28 for a single class i of ionizable groups on the protein.

We now turn to the porous sphere model, with a fraction f_s of the ionic groups at the surface and the remainder distributed in the core. Similar to Eq. 28, we have contributions from the surface and now also from each spherical shell with thickness dr in the core. Integrating over the core, we now obtain for all groups of the porous sphere

$$\Delta g_b = g_\alpha + g_{dl} \quad (31)$$

where

$$g_\alpha = kT \sum_{i=1}^{t} \left[f_s n_i \ln(1 - \alpha_i(R)) + \frac{(1 - f_s)n_i}{(4\pi/3)R_{core}^3} \int_0^{R_{core}} \ln(1 - \alpha_i(r)) 4\pi r^2 \, dr \right] \quad (32)$$

which largely arises from the translational entropy of the protons on the ionizable groups and, using the Debye-Hückel approximation as in Eq. 30, $g_{dl} = -g_{el}$ is now given by Eq. 25 for the porous sphere.

In Fig. 4 the proton binding free energy of metmyoglobin in solutions of 0.01 M ionic strength at 25°C is plotted for three pH values along the hypothetical folding pathway. Panel A shows $\Delta g_b/kT$ along the condensation path, when the proton density of the model sphere increases from $\rho = 0.1$ to $\rho = 1$. In panel B, for the reconfiguration path at $\rho = 1$, the reorder parameter θ decreases from $\theta = \Phi = 0.45$ for random distribution to its minimum value, reached for $f_s = 1$ in Eq. 1, when all ionic groups are at the surface of the sphere.

Figure 4 shows that at pH 7, very near the isoelectric point of myoglobin, Δg_b is essentially identical for the folded and unfolded states. This is because $\psi \approx 0$ in the model sphere along the entire folding path and, hence, the α_i values in Eqs. 2, 25, and 32 are independent of configuration. Although there is no electrostatic interaction, $g_{dl} = -g_{el} = 0$, $\Delta g_b = g_\alpha$ has a large negative value relative to the hypothetical reference state in which $\alpha_i = 0$ for all acid and basic groups. For pH 4 and 10, the free energy becomes large for the collapsed disordered "intermediate," $\rho = 1$ and $\theta = 0.45$. In this hypothetical intermediate state the average distance between the ionic charges is at a minimum; hence, their repulsion is maximal and, as a result, the Δg_b curves exhibit a maximum.

We now turn to the dependence of Δg_b on

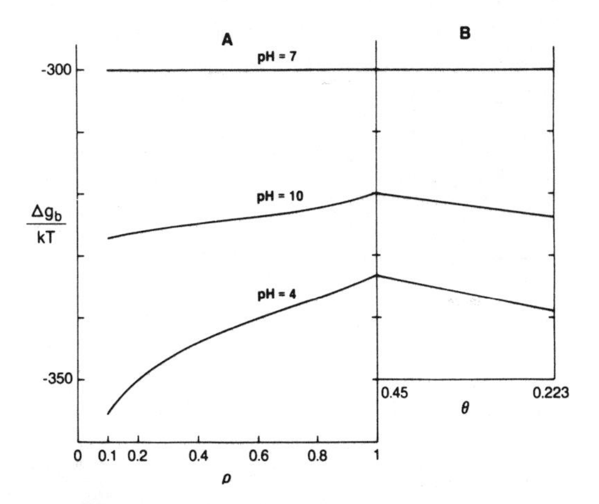

Fig. 4. Proton binding free energy of metmyoglobin with unfolded ionic groups of Table 2 along hypothetical folding pathway in aqueous solutions of 0.01 M ionic strength at 25°C for pH values as indicated. **(A)** Contracting path with variable density ρ. **(B)** Reconfiguration path with variable reorder parameter θ.

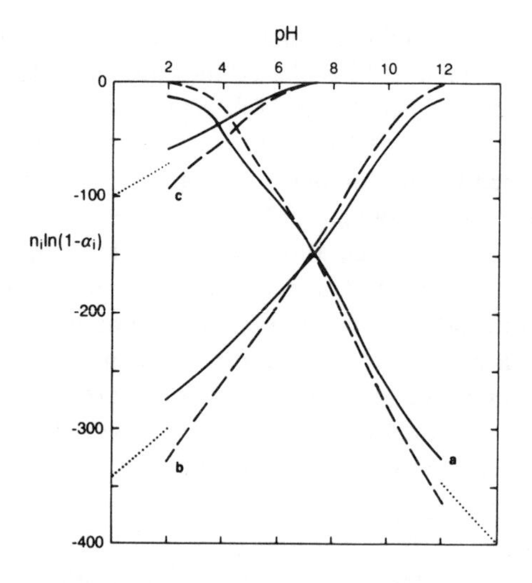

Fig. 5. Proton binding entropy of various groups in metmyoglobin with unfolded groups of Table 2 as function of pH in aqueous solutions of 0.01 M ionic strength at 25°C. **(Solid curves)** Globular state. **(Dashed curves)** Unfolded state with $\rho = 0.1$. **(Dotted lines)** Limiting slope at extreme pH. **(Curves a)** Carboxylic acid, $n_i = 23$. **(Curves b)** ε-amino, $n_i = 19$. **(Curves c)** Histidine, $n_i = 12$.

pH, already discernible in Fig. 4. In Fig. 5 the separate contributions for the different values of i of $n_i \ln(1 - \alpha_i)$ to g_α/kT are shown as a function of pH for the three main types of ionic groups of myoglobin, ε-amino, imidazole, and carboxylic acid, as solid curves for the globular molecule and as dashed curves for the unfolded state with density $\rho = 0.1$, in solutions of 0.01 M ionic strength. The general shape of the curves is as expected from ionization behavior. For example, at low pH the acid groups tend to un-ionized, $\alpha_i \rightarrow 0$, and the basic groups tend to full ionization, $\alpha_i \rightarrow 1$. Un-ionized groups do not contribute to g_α since $\ln(1 - \alpha_i) \rightarrow \ln 1 = 0$. For nearly full ionization, the protein charge is constant and, hence, the potential is also constant with further change in pH. In that case, at the pH extremes $\alpha_i \rightarrow 1$ and

$$\ln(1 - \alpha_i) = \ln \alpha_i - q_i(\ln H + \ln k_i - (e\psi/kT))$$

$$= (q_i \ln 10)\,\mathrm{pH} + \text{constant} \qquad (33)$$

Equation 33 shows that at low pH $n_i \ln(1-\alpha_i)$ for the basic groups, with $q_i = 1$, have a positive limiting slope $2.303n_i$ (see Fig. 5, curves b and c). Similarly, curves a for the acid groups, with $q_i = -1$, have a negative limiting slope, $-2.303\,n_i$, at high pH. The limiting slopes, indicated in Fig. 5 by dotted lines, are outside the pH range of the calculations for ionic strength 0.01 M.

The shape of the curves in Fig. 5 is influenced by the interaction between the groups; this is measured by the electrostatic potential ψ in Eq. 2. At low pH, ψ is on average more positive for the globular than for the unfolded protein since there is higher charge density in the globular state. Therefore, at low pH, the acid groups are more ionized and the basic groups are less ionized in the globular than in the unfolded state. The opposite holds at high pH, where ψ is more negative for the folded than for the unfolded state. In between, at the IEP near pH 7, there are no interactions because $\psi = 0$ everywhere and, hence, the two sets of curves cross at the IEP.

Figure 6 shows g_α and g_{dl} as a function of pH, for the globular myoglobin as solid curves and as dashed curves along the unfolding pathway, for $\rho = 0.9$ and 0.1. The general shape of the curves is typical for mixed basic and acid polyelectrolytes, with a maximum at the IEP. At this maximum the free energy of the electrical double layer, g_{dl}, vanishes because the molecule has no net charge. For charged proteins, on either side of the IEP, g_{dl} is negative because formation of the electrical double layer is a spontaneous process (compare also Eq. 29). At the IEP the acid groups are still partly ionized, as are the basic groups. Therefore, at the IEP, the entropic contribution to the binding free energy, g_α, is large and negative with respect to the hypothetical standard state of the protein in which all acid and basic groups are un-ionized (compare Fig. 5). The net change of g_α upon unfolding is small due to near cancellation of the large contributions. The same is true for the change of g_{dl} along the unfolding path. It is interesting that the variations with pH of g_α and g_{dl} are of the same order of magnitude, as are the changes of g_α and g_{dl} along the unfolding path, although here the changes may be in opposite directions, resulting in smaller change of the sum Δg_b.

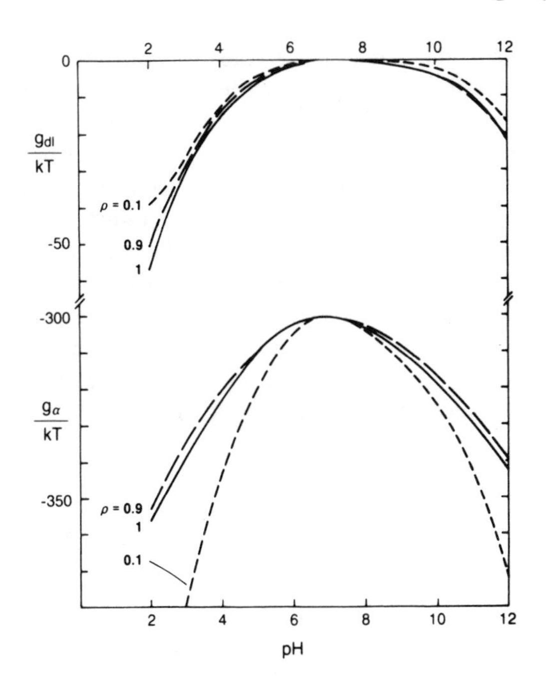

Fig. 6. Proton binding entropy g_α and free energy of electrical double layer g_{dl} of metmyoglobin with unfolded charges of Table 2 as a function of pH in aqueous solutions of ionic strength 0.01 M at 25°C. **(Solid curves)** Globular state. **(Dashed curves)** Unfolded states with density ρ as indicated.

Electrostatic Contribution to Free Energy of Folding

Protein stability is a balance of various free energy contributions. If, in addition, ligand molecules are present which can bind to the protein, they will lead to another contribution to the free energy. The ligands of interest in this case are protons. If the free energy of binding ligand to one state of the protein differs from the free energy of the binding to another state, then the ligand binding will affect the relative stabilities to the same degree. That is, the free energy differences in stability of the folded state relative to any particular unfolded state (ρ,θ) due to proton binding is

$$\delta g_f(\rho,\theta) = \Delta g_b(\text{folded}) - \Delta g_b(\rho,\theta) \quad (34)$$

Results of $\delta g_f / kT$ versus pH for the system under discussion are shown (Fig. 7) for various ρ values along the contracting path. Whereas, at the extremes of pH, the predominant driving forces for binding are due to the binding entropies in g_α (as noted above and see Fig. 6), the binding entropies nearly cancel in taking the difference to compute the relative stability (Eq. 34). From the thermodynamic relation [*(8)*; see also *(16)*, Eqs. 4 and 6]

$$(\partial \Delta g_b / kT) / \partial \ln H = -Z \quad (35)$$

it follows that the slope of the curves in Fig. 7 is related to the charge difference δZ between the folded and the unfolded (ρ,θ) state according to

$$(\partial \delta g_f / kT) / \partial \text{pH} = (\ln 10)\delta Z \quad (36)$$

In a test of the self-consistency and accuracy of our computations we found that for the systems of Fig. 7, where δZ ranges about from -7 to $+7$ protonic charges, Eq. 36 was always satisfied to better than 0.01 charge in δZ.

In many cases ionic groups are buried inside the structure of a native protein. For example, in metmyoglobin, 6 of the 12 histidines are not titrated in the native form (see Table 2). Consequently, these six un-ionized histidines do not contribute to the Δg_b of the folded state. This is readily taken into account in the reference state of δg_f. Figure 8 shows these results for δg_f on the basis of a reference globular state in Eq. 34 with 6 instead of 12 histidines. Comparison with Fig. 7 reveals that, as expected, the change of reference state does not modify δg_f at high pH where all histidines are un-ionized even in contact with

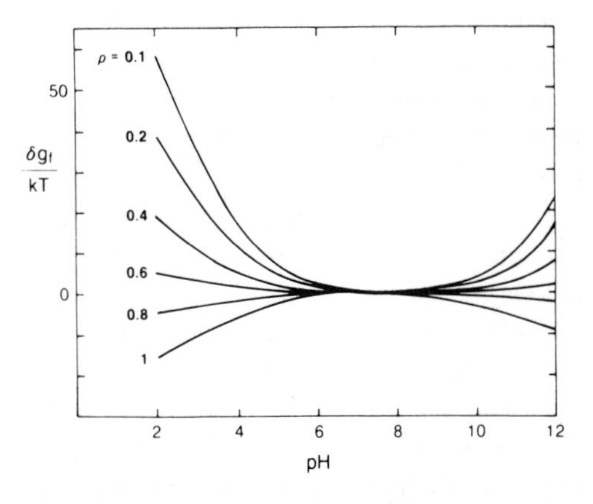

Fig. 7. Proton binding contribution to folding free energy of metmyoglobin as function of pH along contracting path, with density ρ as indicated, in aqueous solutions of 0.01 M ionic strength at 25°C. Unfolded ionic groups of Table 2 also for globular reference state.

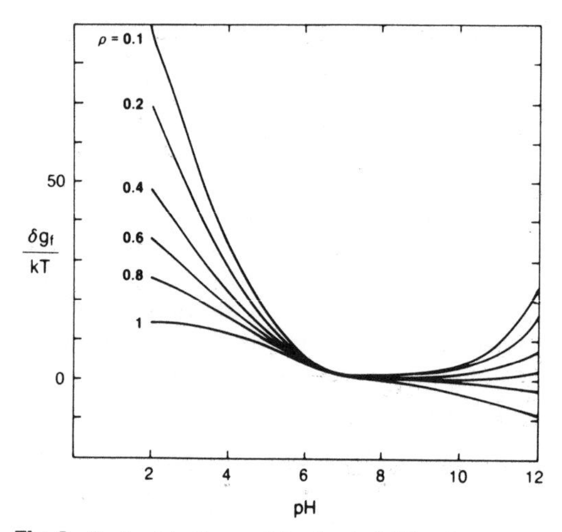

Fig. 8. Proton binding contribution to folding free energy for systems of Fig.7 with 6, instead of 12, titratable histidines in globular reference state (see Table 2).

the solution. However, the change at low pH is substantial. Here the increased proton binding capacity, relative to the folded state, increases δg_f by $2.303kT$ per pH unit for every extra proton per molecule bound by the unfolded protein [compare *(16)*]. The overall influence of ionic groups buried in native proteins is that the stability versus pH curve of the protein becomes more asymmetric.

We now consider the dependence of the charge effects in protein folding on ionic strength for the systems of Figs. 7 and 8. The influence of ionic strength on the binding free energy Δg_b along the folding path, and of the globular state, is quite large. However, the difference δg_f changes little with ionic strength. In Fig. 9 δg_f is plotted versus ionic strength in solutions of pH 3, for various densities ρ of myoglobin with the unfolded set of ionic groups of Table 2. A major factor in this dependence is the average distance between the protein charges in the two states, relative to the Debye length.

We now account for the buried ionic groups in globular myoglobin. Figure 10 is for the same system as Fig. 9, but now with 6, instead of 12, histidines in the folded reference state of δg_f in Eq. 34. The main effect is a shift of all curves to more positive values. In general, the δg_f curves in Fig. 10 rise with increasing ionic strength. It is inappropriate to interpret this as implying that the folded protein becomes less stable upon the addition of salt. As discussed earlier, the equilibrium density ρ^* of the unfolded state increases when salt is added. Such ρ^* values, taken from the straight line in Fig. 3 without consideration of the possible influence of pH differences, are marked as stars in Fig. 10. The connecting curve decreases with increasing ionic strength, indicating that the folded state is favored by the addition of salt. This illustrates the importance of proper accounting for environmental changes on ρ^* and the unfolded states.

Born Energy and Other Factors

We assume, in Eq. 10, that our spherical protein has a lower dielectric constant than the surrounding aqueous solution. It is, therefore, appropriate to consider a correction which accounts for the fact that the Born energy of an ionized group in the core of the model sphere is increased relative to the energy of the same ionized group at the surface of the sphere or in the surrounding solution. Such a correction changes the intrinsic binding constant k_i in Eq. 2 for the ionic groups in the core. Therefore, we have repeated all of the above calculations with a model that now also includes an additional term to account approximately for the Born energy. We find that predictions for the folded state, and unfolded states with density $\rho^* \leq 0.85$ are little affected by this additional term. Born free energies are, of course, maximal for ionic groups buried in the native, globular state *(36)*. The consequence is that in most cases they are not ionized at all. Therefore, such buried groups can be dealt with satisfactorily by simply discounting them in the proton binding calculation of the globular state (see Fig. 8). In summary, since we believe this approach is less arbitrary than using simplified models of the Born energy, we have preferred it for the protein stability calculations.

Consistent with the omission of the Born energy of the buried groups, we have used the same intrinsic pK values in the folded and unfolded state for the remaining ionic groups. This is not an essential feature of the model

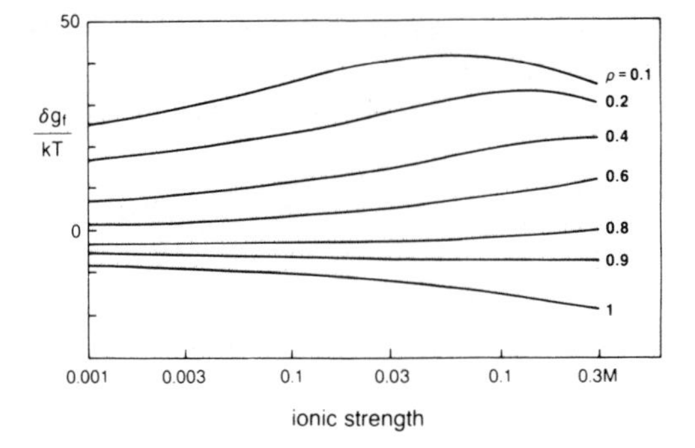

Fig. 9. Proton binding contribution to folding free energy of metmyoglobin as function of ionic strength along contracting path, with density ρ as indicated, in aqueous solutions of pH 3 at 25°C. Unfolded ionic groups of Table 2 also for globular reference state.

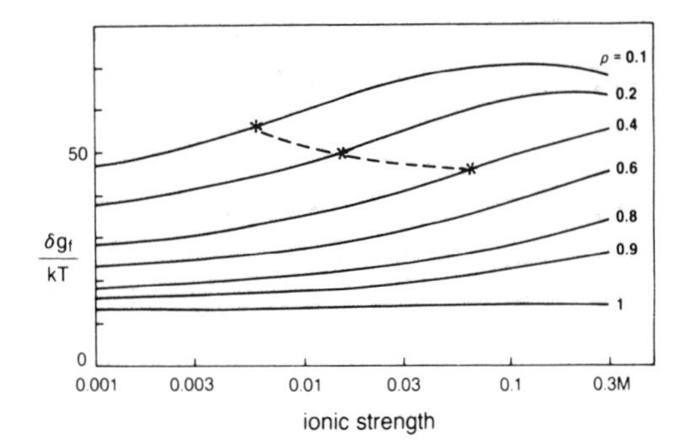

Fig. 10. Proton binding contribution to folding free energy for systems (Fig. 9) with 6, instead of 12, titratable histidines in globular reference state (see Table 2). Stars indicate ρ^* values from Fig. 3.

and can be readily modified.

The present model only attempts to treat the nonspecific effects of charge repulsion on protein stability. The smeared-charge treatment we have used does not address effects of specific interactions, such as salt bridges. It may, however, serve as a base-line model, to which electrostatic free energies of specific interactions might be added, by summing interactions of charges at specific sites in the folded protein [by use of methods described by Friend and Gurd *(39)*, for example]. A general type of interaction, neglected here, is the attraction between ions of opposite charge, which will lead to a negative electrostatic contribution to the free energy. Calculations *(39)* have shown that these interactions might lead to considerable stabilization.

Conclusions

We have presented a model for the effects of proton binding equilibria on folded and unfolded states of globular protein molecules. Our purpose has been to provide a framework in which folded and unfolded states can be treated on equal footing. Although several methods currently exist for treating specific electrostatic interactions in folded proteins *(37, 39, 40)*, no corresponding method has been available to predict effects of electrostatics on the unfolded state. Therefore, theory has been unavailable for calculating electrostatic effects on protein stability, the difference of free energies between folded and unfolded states. The present treatment was developed to address that need. It is a first approximation which accounts for the nonspecific effects of charge repulsions. Our folded-state model is a simplification of the 1924 model of Linderstrøm-Lang. The folded molecule is considered to be a sphere with a uniform surface charge surrounded by salt solution, as in the earlier model, but without an exclusion shell for the charge of the small ions, which is now allowed up to the charged protein surface. The charge effects are modeled by using the Poisson-Boltzmann equation. The model gives somewhat better predictions than the Linderstrøm-Lang model for the pH titration curves of globular proteins. The unfolded state of the protein is modeled as a porous sphere of low dielectric constant, penetrated by salt solution. We study the range of possible densities of the unfolded molecules. Using this model for the unfolded state to interpret several different experimental results leads to the conclusion that the density of the unfolded state is highly dependent upon ionic strength and is often quite high. Finally, we consider the effects of the proton binding equilibria and ionic strength on protein stability. Because there is no net charge at the isoelectric point, the predicted electrostatic contribution to stability is negligible at that pH. However, at the extremes of pH, charge effects can lead to destabilization of the folded state, relative to the unfolded molecule, by several tens of *kT*. Moreover, when there are buried non-titratable ionic groups, then the electrostatic contribution to destabilization becomes asymmetric, and the maximum stability of the protein is not at the isoelectric pH.

Acknowledgment

We are indebted to the Department of Materials Science and Mineral Engineering, University of California, Berkeley, for hospitality to D.S.

References

1. Acampora, G., and J. Hermans, Jr., *J. Am. Chem. Soc.* **89,** 1543 (1967).
2. Privalov, P. L., and N. N. Khechinashvili, *J. Mol. Biol.* **86,** 665 (1974).
3. Privalov, P. L., Yu. V. Griko, S. Yu. Venyaminov, V. P. Kutyshenko, *J. Mol. Biol.* **190,** 487 (1986).
4. Dill, K. A., *Biochemistry* **24,** 1501 (1985).
5. Dill, K. A., D. O. V. Alonso, K. Hutchinson, *Biochemistry* **28,** 5439 (1989).
6. Linderstrøm-Lang, K., *C. R. Trav. Lab. Carlsberg* **15,** No.7 (1924).
7. Hermans, J. J., and J. Th. G. Overbeek, *Recl. Trav. Chim. Pays-Bas* **67,** 761 (1948).
8. Hermans, J., Jr., and H. A. Scheraga, *J. Am. Chem. Soc.* **83,** 3283 (1961).
9. Steinhardt, J., and S. Beychok, *Proteins (2nd Ed.),* Chapter 8 (1964).
10. Katchalsky, A., and J. Gillis, *Recl. Trav. Chim. Pays-Bas* **68,** 879 (1949).
11. Torrie, G. M., and P. Valleau, *Chem. Phys. Lett.* **65,** 343 (1979).
12. Snook, I., and W. van Megen, *J. Chem. Phys.* **75,** 4104 (1981).
13. Linse, P., G. Gunnarson, B. Jonsson, *J. Phys. Chem.* **86,** 413 (1982).
14. LeBret, M., and B. H. Zimm, *Biopolymers* **23,** 271 (1984).
15. Mills, P., C. F. Anderson, M. T. Record, Jr., *J. Phys. Chem.* **89,** 3984 (1985).
16. Stigter, D., and K. A. Dill, *J. Phys. Chem.* **93,** 6737 (1989).
17. Debye, P., and E. Hückel, *Phys. Z.* **24,** 185 (1923).
18. Rice, J. R., *Numerical Methods, Software and Analysis* (McGraw-Hill, New York, 1983).
19. Breslow, E., and F. R. N. Gurd, *J. Biol. Chem.* **237,** 371 (1962).
20. Wishnia, A., I. Weber, R. C. Warner, *J. Chem. Soc.* **83,** 2071 (1961).
21. Tanford, C., S. A. Swanson, W. S. Shore, *J. Am. Chem. Soc.* **77,** 6414 (1956).
22. Tanford, C., J. D. Hauenstein, D.G. Rands, *J. Am. Chem. Soc.* **77,** 6409 (1956).
23. Stigter, D., and K. J. Mysels, *J. Phys. Chem.* **59,** 45 (1955).
24. Gilson, M. K., and B. H. Honig, *Biopolymers* **25,** 2097 (1986).
25. Flory, P., *Statistical Mechanics of Chain Molecules* (Interscience, New York, 1969), pp. 42, 111.
26. Kotin, L., and N. Nagasawa, *J. Chem. Phys.* **36,** 873 (1962).
27. Sugai, S., and K. Nitta, *Biopolymers* **12,** 1363 (1973).
28. Stigter, D., *J. Colloid Interface Sci.* **53,** 296 (1975).
29. Debye, P., and A. M. Bueche, *J. Chem. Phys.* **16,** 573 (1948).
30. Brinkman, H. C., *Proc. Acad. Sci., Amsterdam* **50,** 618, 821 (1948).
31. Stigter, D., *Macromolecules* **15,** 635 (1982).
32. Stigter, D., *Macromolecules* **18,** 1619 (1985).
33. Matthew, J. B., *Annu. Rev. Biophys. Chem.* **14,** 387 (1985).
34. Tanford, C., and J. G. Kirkwood, *J. Am. Chem. Soc.* **79,** 5333 (1957).
35. Matthew, J. B., *et al., Biochem. Biophys. Res. Commun.* **81,** 416 (1978).
36. Rashin, A., and B. Honig, *J. Mol. Biol.* **173,** 515 (1984).
37. Gilson, M. K., and B. Honig, *Proteins: Struc. Funct., Genetics* **4,** 7 (1988).
38. Verwey, E. J. W., J. Th. G. Overbeek, *Theory of the Stability of Lyophobic Colloids* (Elsevier, New York, 1948).
39. Friend, S. H., and F. R. N. Gurd, *Biochemistry* **18,** 4612 (1979).
40. Weiner, S. J., *et al., J. Am. Chem. Soc.* **106,** 765 (1984).

4

Compact Polymers[1]

Hue Sun Chan, Ken A. Dill

Introduction

The compact conformations of a chain molecule comprise a very small but important subset of all the physically accessible conformations. Their importance derives from the fact that the native conformations of all globular proteins are compact. By "compact," we refer to those configurations of single chain molecules that are tightly packed, i.e., fully contained within a volume of space (a box) with the minimal (or near minimal) surface/volume ratio. In contrast, by "open," we refer to the complete superset of all accessible conformations, including those that are compact. Relatively little attention has previously been directed toward the set of compact conformations of polymers. Considerably more effort in polymer science has focused on the more open conformations of chains, because they are far greater in number and because they are the predominant conformations of chains in solution or in the bulk.

Our purpose in this chapter is to explore in some detail the nature of the compact conformations and to show how they differ from the larger superset of all possible conformations. Our purpose here is served by exhaustive simulation of every possible conformation of short chains on two-dimensional square lattices. There are three advantages of studies in two dimensions for our present purposes. First, certain predictions can be compared with a significant literature dating back to the work of Orr *(1)* in 1947 on exhaustive lattice simulations of open conformations, most of which have been in two dimensions. Second, we can explore greater chain lengths for a given amount of computer time. And third, the surface/volume ratio, a principal determinant of the driving force for a protein to collapse *(2)*, for long chains in three dimensions is more closely approximated by short chains in two dimensions than by equally short chains in three dimensions. Also, it has been shown recently that a two-dimensional square lattice model of short chains predicts certain general features of protein behavior *(3)*: copolymers of sequences of H (nonpolar) and P (other) residues, subject to excluded volume, and with increasing HH attraction energy, will collapse to a relatively small number (often one) of maximally compact conformations in which there is a core of H residues.

In addition, one principal motivation of the present study is to explore the hypothesis *(4)* that excluded volume forces may be responsible for the formation of certain specific forms of internal chain organization in compact polymers. For open chains subject to excluded volume, we have recently found that the most probable configurations of chains containing two intrachain contacts are helices and antiparallel sheets, the principal form of internal chain organization (secondary structures) in globular proteins *(4)*. Here we ex-

---

1 Reprinted from *Macromolecules* **22**, 4559 (1989) with permission.

plore in addition the hypothesis that excluded volume may also be a principal driving force for the formation of secondary structures in more compact chains, and thus, perhaps a driving force for the formation of secondary structures in globular proteins. Although the present study is limited to two dimensions, other work *(5)* shows similar behavior in three dimensions.

The Model

We begin with a summary of some definitions and terminology of the model used in our analysis. For brevity, only the essentials are included here, since further details can be found elsewhere *(4)*. In the two-dimensional square lattice with coordination number $z = 4$, chains are comprised of a sequence of monomers or residues, each occupying one lattice site. Residues are numbered sequentially from one end of the chain, starting with 1. Coordinates of the ith residue are given by the vector $\mathbf{r}_i$; bond lengths are normalized to unity, such that $\mathbf{r}_i = (n, m)$, where n, m are integers. Bond angles are limited to ±90° and 180°. Excluded volume is taken into account by forbidding two different residues from occupying the same lattice site, i.e., $\mathbf{r}_i \neq \mathbf{r}_j$ for $i \neq j$. For a chain with $N + 1$ residues, the chain length is defined to be the number of bonds in the chain, which equals N.

Two residues i and j are taken to be in contact when they are nearest neighbors on the lattice, $|\mathbf{r}_i - \mathbf{r}_j| = 1$. There are two types of contacts: the *connected contacts* are those between residues i and $i + 1$, implied by the connectivity of the chain, whereas *topological contacts* are made between residues not adjacent in the sequence. A topological contact is represented as an ordered pair $(i, j), j \neq i - 1, i, i + 1$. The *order* k of a topological contact is the number of bonds along the chain between the two contacting residues, given by $k = |j - i| > 1$.

A convenient way to represent contacts is in terms of the *contact map,* a two-dimensional matrix of size $(N + 1) \times (N + 1)$, in which a dot at the ith row and the jth column denotes the topological contact (i, j). Clearly the contact map is symmetric under row-column interchange, hence only the upper triangular half needs to be shown; see Fig. 1. The shaded main diagonal corresponds to the connected contacts, which are always implicit and thus are not represented with dots. Any line parallel to the main diagonal is termed a *diagonal.* Contacts of the same order lie along the same diagonal. The contact map characterizes the *topology* of a chain conformation, in contrast to the list of coordinates which characterizes the *geometry* of a chain conformation.

One of the principal virtues of the contact-map representation is its direct representation of secondary structures, which are defined by their topologies rather than by their geometries. The principal secondary structures observed in globular proteins — helices, parallel and antiparallel sheets, and turns — are all represented by very simple regular patterns on the contact map; see Fig. 1. The patterns representing secondary structures are strings of dots *(6)*. Clearly, at least two contacts are needed to form a string; accordingly, the dotted boxes in parts a–c of Fig. 1 encircle "minimal units" that must be present in order to be qualified as helices or sheets. Each of these minimal units consists of two dots on the contact map. In terms of chain geometry, the minimal units for helices have six residues, while those for sheets have four. There is no upper limit to the length of the strings of contacts for secondary structures, since secondary structures can be extended indefinitely. Hence, the number of residues participating in a helix may be 6, 8, 10, ..., while the corresponding numbers of residues in a sheet are 4, 6, 8, 10, ... As indicated by the dotted box in Fig. 1d, there are always two residues at a turn, and they must be connected to an antiparallel sheet.

It is worthwhile to comment on the distinction between topological and geometric descriptions of secondary structures. In principle and in practice, helices can be readily represented either by their geometries or topologies. The topological representation entails specification of a particular pattern of spatial-neighboring, but nonconnected, res-

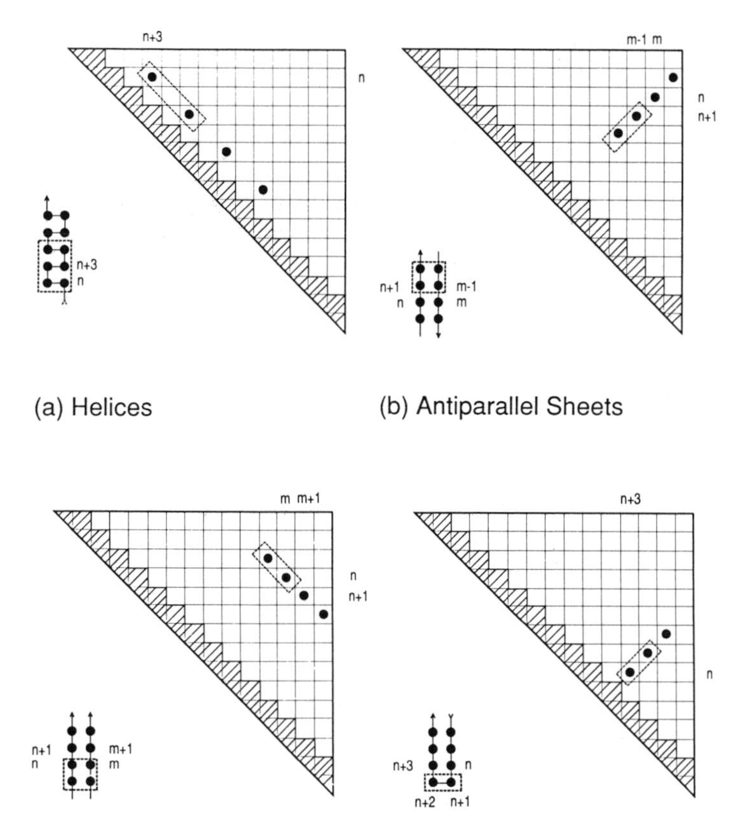

(c) Parallel Sheets

(d) Turns

Fig. 1. Secondary structures as contact patterns: The dotted boxes encircle minimal units that must be present to be qualified as secondary structures. **(a)** Contacts for helices are along the order 3 diagonal. **(b)** and **(d)** Contacts for antiparallel sheets and turns form strings that are perpendicular to the main diagonal. **(c)** Contacts for parallel sheets form strings that are parallel to the main diagonal.

idues along the chain. The geometric representation, on the other hand, entails specification of a particular region of acceptable ϕ–ψ angles among the connected backbone neighboring monomers. Either the geometric or topological representation can be suitably used to identify helices. However, it is clearly much less sensible to represent sheets by specifying the geometries, the ϕ–ψ angles among connected neighboring residues, for the following reasons. Although ϕ–ψ angles can distinguish sheets from helices, they cannot so readily distinguish antiparallel and parallel sheets. This distinction is immediately clear, however, in the topological representation. Moreover, for long sheets of either type, specification in terms of local bond angles would require extreme precision, and in so doing could not capture the essence of small twists and bends. It is our conviction therefore that what is meant by "sheet" must be more akin to the "train-track" nature of organization of strands of chain, rather than some small class of bond angles among connected residues. Hence, for our present purposes, we adopt the view that secondary structures are defined in terms of spatially localized units of intrachain contacts, rather than by precise details of the geometries of bond angles among connected neighbors.

It follows that lattice models can be expected to adequately represent secondary structures, even though they simplify the representation of bond angle geometries into a small number of discrete states. Even within geometric representations of secondary structures, quite wide latitude is generally taken in defining broad regions of ϕ–ψ angles of peptide backbones for classifying secondary structures *(7–10)*. The discretized conformations of lattice models probably do no worse than this. Moreover, topological properties are relatively independent of lattice type and spatial dimensionality. The intrinsic α-helical contact pattern represented by $(i, i+3)$, for example, can be readily represented on several different types of lattices in two or three dimensions, including the two-dimensional square lattice studied here. In three-dimensional models, we obtain predictions that are qualitatively similar to those presented below for the two-dimensional square lattice *(5)*.

Enumeration of the Chain Conformations

In this chapter, we carry out exhaustive explorations of every accessible conformation of a polymer chain molecule on the two-dimensional square lattice, as a function of the chain "compactness." In later sections, we study the properties of these conformations; in the present section, we simply enumerate how many conformations there are at each compactness, as a function of the chain length. We first define variables and then summarize related past efforts on exhaustive explorations of chain conformations on lattices. We then devote considerable attention to the enumeration of conformations that are maximally compact.

A fundamental measure of conformational freedom is the total number of conformations, or states, that are accessible. For open chains with $N+1$ residues, modeled here as unconstrained self-avoiding walks of N steps, this is given by $\Omega_0(N)$, which is the number of all such walks. Among these, some conformations will involve a certain number of topological self-contacts, hence $\Omega_0(N)$ may be written as the sum

$$\Omega_0(N) = \sum_{t=0}^{t_{\max}} \Omega^{(t)}(N) \qquad (1)$$

where $\Omega^{(t)}(N)$ is the number of conformations with exactly t contacts, and $t_{\max}$ is the maximum number of topological contacts that can be present in chains with $N+1$ residues. $\Omega^{(t)}(N)$ measures conformational freedom as a function of number of intrachain contacts. The compactness of the molecule is proportional to t, as described in more detail below, and is clearly related to the density of the molecule, the number of chain segments per unit volume.

In the two-dimensional square lattice, the *perimeter* of an isolated lattice site is 4, hence the *maximum perimeter*, $P_{\max}(N)$, of a chain that occupies $N+1$ sites is $2(N+2)$, when there are no intrachain contacts among the $N+1$ residues. In general, the number of topological contacts t is related to the perimeter $P(N)$ of such chains by the equation

$$t = \frac{1}{2}[P_{\max}(N) - P(N)] = N + 2 - \frac{P(N)}{2} \qquad (2)$$

since formation of each contact decreases $P(N)$ by two units. The perimeter P serves as a two-dimensional analog of the surface area exposed to solvent for a three-dimensional chain molecule in solution. When a lattice chain becomes more compact by forming more topological contacts, its perimeter necessarily decreases. Hence, the maximum number of topological contacts, $t_{\max}$, may be computed from Eq. 2 by setting P equal to the minimum, or *compact perimeter* P_c. The only shapes that have the minimum perimeter-to-area ratio on the square lattice are either squares or rectangles in which the lengths of the two sides differ by only one unit. Therefore for a chain of $N+1$ residues confined within such shapes, it can easily be deduced that

$$P_c(N) = 2(2m+1) \quad \text{for } m^2 < N+1 \le m(m+1)$$

$$P_c(N) = 4(m+1) \quad \text{for } m(m+1) < N+1 \le (m+1)^2 \qquad (3a)$$

hence

$$t_{\max} = N + 1 - 2m \quad \text{for } m^2 < N+1 \le m(m+1)$$

$$t_{\max} = N - 2m \quad \text{for } m(m+1) < N+1 \le (m+1)^2 \qquad (3b)$$

where m is a positive integer.

In this chapter, $\Omega_0(N)$, $\Omega^{(t)}(N)$, and related quantities are computed by exhaustive enumeration of every conformation that does not violate excluded volume contraints. Exhaustive simulation of short lattice chains has been a major theoretical tool in polymer science *(11)*. Orr *(1)* was the first to adopt the method in 1947 by considering $N \leq 8$ for square lattices and $N \leq 6$ for simple cubic lattices. With the invention of high-speed computers, a systematic effort to reach longer chain lengths was pioneered by Domb and his collaborators *(12–15)* in the beginning of the 1960s. That work is now recognized as the basis for most of the modern developments in polymer theoretical physics, including scaling law methods *(16)* and the modern path integral and renormalization theories *(17, 18)*. Inspiration from these analytic theories in turn channeled most of the more recent exhaustive simulation work into the determination of various scaling exponents and connective constants. A recent review of all the available data in two-dimensional lattices can be found in the work by Guttmann *(19)*. To our knowledge, the longest chain length on a square lattice for which $\Omega_0(N)$ has been successfully enumerated is $N = 25$, reported by Rapaport *(20)* in a study of the end-to-end distances of polymers.

Systematic exhaustive simulations have also been employed to address polymer problems involving intrachain contacts. The first was due to Orr *(1)*, who enumerated $\Omega^{(t)}(N)$ in a treatment of polymers in solution. Values for all possible t were provided by Orr. Fisher and Hiley *(21)* extended this work, and provided $\Omega^{(0)}(N)$ for $N \leq 14$ on square lattices, $N \leq 12$ on triangular lattices, and $N \leq 10$ on simple cubic lattices. Recently, Ishinabe and Chikahisa *(22)* have been able to enumerate conformations for longer chains, and obtained values for $\Omega^{(0)}(N)$ for $N \leq 22$ on square lattices and $N \leq 20$ on tetrahedral lattices. In both of the recent works cited above, much attention was devoted to chains with absolutely no self-contacts, the $t = 0$ case of polymers in "super-solvents." By comparison, much less attention was devoted to the compact configurations. These states have been of little interest for polymer solution theories: according to Fisher and Hiley *(21)*, calculations for the conformational freedom of a *single* chain with a high probability of self-contact formation are "not very significant since polymer molecules in solution then tend to attract one another so that coagulation and eventually phase separation set in." However, for proteins, these conformations are of fundamental importance.

We have determined $\Omega_0(N)$ on square lattices for $N \leq 17$, and $\Omega^{(t)}(N)$, $0 \leq t \leq t_{max}$, for $N \leq 15$ [see *(4)*]. Values for $\Omega^{(t)}(N)$ are listed in Table 1. In our enumeration, the two ends of a chain are considered to be distinct, and only conformations that are not related by translation, rigid rotation, or reflection are counted. With appropriate transformations to take care of minor differences in definitions *(23)*, our results for $\Omega^{(t)}(N)$ agree with those of Orr *(1)*. Values for $\Omega^{(0)}(N)$ in Table 1 also agree with those of Fisher and Hiley *(21)* except for a slight discrepancy *(24)* for $N = 14$. Also, our calculations for $\Omega^{(0)}(N)$ are in total agreement with those of Ishinabe and Chikahisa *(22)*.

The probability that a chain will form t contacts is dependent upon its chain length N. The maximum possible number of contacts, t_{max}, also depends on N, see Eqs. 2–3. We therefore define the *compactness*, ρ, as the ratio of the number of topological contacts of a conformation *relative* to the maximum number of contacts attainable for a given chain length.

$$\rho \equiv \frac{t}{t_{max}}, \quad 0 \leq \rho \leq 1 \tag{4}$$

Figure 2 is a histogram showing the distribution of the number of accessible conformations as a function of ρ for $N = 15$. From the data of Table 1, we can compute the average compactness

$$\langle \rho(N) \rangle \equiv \sum_{t=0}^{t_{max}} \frac{t}{t_{max}} \frac{\Omega^{(t)}(N)}{\Omega_0(N)} \tag{5}$$

which characterizes the tendency of a chain of length N to form contacts. Values for $\langle \rho(N) \rangle$ computed for $3 \leq N \leq 15$ show a very gradual decreasing trend with increasing N, with some slight oscillations, ranging from 0.24 to 0.30; see the inset of Fig. 2. A related quantity

Table 1. Number of conformations $\Omega^{(t)}(N)$ on the square lattice as a function of the number of contacts, t, and chain length, N.

		$\Omega^{(t)}(N)$									
N	$\Omega_{(0)}(N)$ (total)	$t = 0$	$t = 1$	$t = 2$	$t = 3$	$t = 4$	$t = 5$	$t = 6$	$t = 7$	$t = 8$	$t = 9$
3	5	4	1								
4	13	9	4								
5	36	21	11	4							
6	98	50	32	16							
7	272	118	92	43	19						
8	740	281	254	134	66	5					
9	2034	666	672	425	173	98					
10	5513	1584	1778	1229	576	298	48				
11	15037	3743	4622	3450	1944	803	444	31			
12	40617	8877	11938	9625	5718	2830	1262	367			
13	110188	20934	30442	26467	16736	9538	3722	1989	360		
14	296806	49522	77396	71570	48452	28297	13650	5655	2122	142	
15	802075	116579	194896	191814	138446	84607	45564	18733	8662	2705	69

$$q \equiv \sum_t (t/N)\, \Omega^{(t)}(N)/\Omega_0(N)$$

has been computed by Ishinabe and Chikahisa (22) for $N \leq 22$, and the limit $q \rightarrow 0.16$ was extrapolated for $N \rightarrow \infty$. Since $\langle \rho (N) \rangle = q(N/t_{max}) > q$ for finite N and $\langle \rho (N) \rangle \rightarrow q$ as $N \rightarrow \infty$, we estimate that $\langle \rho (N) \rangle$ is bounded between 0.16 and 0.24 for intermediate to very long chain lengths.

As shown in Table 1 and Fig. 2, the number of conformations is maximal at small ρ and decreases rapidly as ρ approaches unity. For $\rho = 1$, the number of maximally compact conformations

$$\Omega_c(N) \equiv \Omega^{(t_{max})}(N) \tag{6}$$

constitutes only a very small fraction of the full conformational space. For instance, $\Omega_c/\Omega_0 = 69/802075 = 8.6 \times 10^{-5}$ for $N = 15$. Throughout the rest of this chapter, "com-

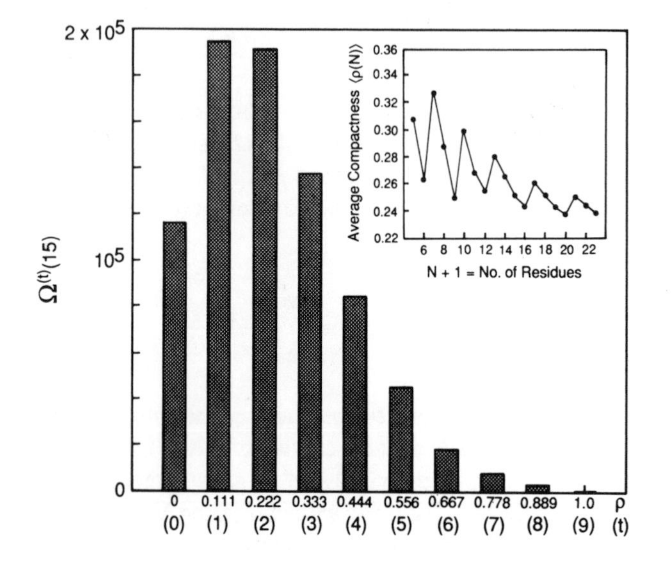

Fig. 2. Number of conformations as a function of compactness ρ, for $N = 15$, $\Omega_0(15) = 802075$. The number of topological contacts t are shown in parentheses below ρ. The inset shows the variation of average compactness $\langle \rho(N) \rangle$ as a function of chain length. Data for $N + 1 \geq 17$ is from Ishinabe and Chikahisa (22).

pact" will refer to conformations for which $\rho = 1$. Most real proteins are well represented as maximally compact, or nearly so *(25)*.

Due to the reduced conformational freedom in the compact states, longer chain lengths are more computationally accessible than for the open conformations. We have enumerated compact square-lattice chains with lengths $N + 1 \leq 30$ and $N + 1 = 36$. The numbers of compact conformations, $\Omega_c(N)$, are given in the second column of Table 2.

It is generally believed that Ω_c has an approximate exponential dependence on N; i.e., $\Omega_c(N) \sim \kappa^N$, where $\kappa \geq 1$ is known as the connective constant *(19)*. Since the dominant functional dependence for open chains is also approximately exponential, $\Omega_0(N) \sim N^{\gamma}\mu^N$ [see *(14, 26, 27)*], with a different connective constant $\mu > \kappa$, then $\ln(\kappa/\mu)$ represents the entropy loss *per segment* (residue) due to compactness, relative to the freedom of open configurations.

Considerable effort has been devoted to the determination of κ. According to the Flory approximation *(28)*, $\kappa \simeq (z-1)/e$, where z is the coordination number and $e = 2.7182...$ is the base of the natural logarithm. Substituting $z = 4$, this gives $\kappa = 1.1036$ for the square lattice. By taking into account the conditionality resulting from the vacancy of the adjoining site reserved for occupation by the preceding segment, the more refined Huggin approximation *(29, 30)* leads to $\kappa \simeq (z-1)/\alpha$, where $\alpha = (1-2/z)^{(1-z/2)}$; hence $\kappa = 1.5$ for the square lattice. Utilizing earlier results of Chang *(31)* and Miller *(32)* as bounds, together with exact enumeration results, Orr *(1)* estimated in 1947 that $\kappa \simeq 1.4$ for the square lattice and $\kappa \simeq 1.9$ for the simple cubic lattice.

Since these early efforts, Kasteleyn *(33)* was able to solve exactly by analytic methods the enumeration problem on two-dimensional oriented lattices. In contrast to the regular square lattice on which the bond angles $\pm 90°$ and 180° are allowed at each step except for excluded volume, he employed the so-called Manhattan lattice rules which restrict bond angles in a manner similar to one-way traffic intersections. The connective constant $\kappa = 1.3385...$ was computed exactly for such a lattice. Gordon, Kapadia, and Malakis *(34)* pointed out that this exact value is a lower bound for the value of κ which would be obtained for more realistic models of chain conformations on the square lattice. On the other hand, Domb *(35)* pointed out that an upper bound for κ can be obtained from the exact "square ice" value 1.539 due to Lieb *(36)*. Gujrati and Goldstein *(37)* combined these two

Table 2. Number of compact ($\rho = 1$) conformations $\Omega_c(N)$ on the square lattice.[a]

N ($N+1$)	$\Omega_c(N)$	S	N ($N+1$)	$\Omega_c(N)$	S
3 (4)	1	1	17 (18)	1673	30
4 (5)	4	8	18 (19)	544	8
5 (6)	4	2	19 (20)	503	2
6 (7)	16	22	20 (21)	11226	187
7 (8)	19	6	21 (22)	11584	68
8 (9)	5	1	22 (23)	4577	22
9 (10)	98	30	23 (24)	3997	6
10 (11)	48	8	24 (25)	1081	1
11 (12)	31	2	25 (26)	100750	238
12 (13)	367	68	26 (27)	52594	88
13 (14)	360	22	27 (28)	45238	30
14 (15)	142	6	28 (29)	16294	8
15 (16)	69	1	29 (30)	13498	2
16 (17)	1890	88	35 (36)	57337	1

[a] N is the number of bonds, $N+1$ is the number of residues, and S is the number of compact shapes.

observations and stated $1.338 \leq \kappa \leq 1.539$ as rigorous bounds. More recently, Schmalz, Hite, and Klein *(38)* estimated κ by counting compact closed loops (Hamiltonian circuits) using transfer matrix methods. They gave an estimate $\kappa \simeq 1.472$, and a lower bound $\kappa \geq 1.3904$, the latter of which is an improvement over the Manhattan lattice lower bound of 1.3385.

We now turn to our own simulations. From a casual inspection, it is clear that the values of $\Omega_c(N)$ in Table 2 do not follow any approximate exponential trend. Instead, as N increases, $\Omega_c(N)$ oscillates with ever-increasing amplitudes. This apparent anomaly can be resolved, however, by realizing that in each of the earlier calculations that predicted $\Omega_c(N) \sim \kappa^N$, it is assumed, either explicitly or implicitly, that only *one* overall shape is allowed for the compact chain conformations. However, isolated compact chains can adopt many different shapes. For certain chain lengths, *many* possible overall shapes are consistent with the minimum perimeter condition (Eq. 3*a*). Therefore, in the earlier efforts cited above, other authors have considered, in essence, the number of compact conformations *per shape*, or the shape average, $\Omega_c(N)/S$, where S is the total number of compact shapes for chain length N. As will be shown below, $\Omega_c(N)/S$ more closely approximates an exponential function of N than does $\Omega_c(N)$.

When the number of residues is a perfect square, $N + 1 = (m + 1)^2$, then the only maximally compact shape is a square; see Fig. 3a. When $N + 1$ is a "near-perfect square," $N + 1 = m(m + 1)$, the compact conformations can take two possible shapes, as shown in Fig. 3b, which are related by a rigid rotation of 90°. These two categories of numbers of residues $(N+1) = 2, 4, 6, 9, 12, 16, 20, 25, 30, 36, 42, 49, \ldots$ are special because essentially only one overall shape on the square lattice is accessible to their compact conformations. For convenience, these numbers of residues will be referred to below as "magic numbers." Obviously these magic numbers apply only to the square lattice and are lattice-dependent. Different sets of magic numbers will be found for other lattices. In general, for chain lengths not equal to the magic numbers, there will be a much larger number of geometric shapes consistent with the requirement of maximum compactness. For example, 238 shapes are possible for $N + 1 = 26$, a few of which are depicted in Fig. 3c. Values for S are listed in the third column of Table 2. The conventions *(39)* here for S are as follows: (i) All a priori shapes consistent with Eqs. 3 are counted by S, no matter whether a chain can be suitably configure to fill it (see Fig. 3d); (ii) Nonidentical rotations and reflections of shapes are counted as distinct. As an illustration, $S = 1$ for $N + 1 = (m + 1)^2$, while $S = 2$ for $N + 1 = m(m + 1)$; see Fig. 3a–b.

The logarithm of the *shape-averaged* number of compact conformations $\Omega_c(N)/S$ is plotted against $N + 1 \leq 30$ in Fig. 4. The data shows an overall exponential increase, with oscillations bounded approximately by two dotted lines of the same slope $\kappa \simeq 1.41$, such that

$$c_1\kappa^N \leq \Omega_c(N)/S \leq c_2\kappa^N \qquad (7)$$

where $c_1 \simeq 0.0787$ and $c_2 \simeq 0.318$. The global increasing trend (κ^N dependence) is attributable to the intrinsic conformational freedom associated with each chain segment, which leads to a general exponential increase of $\Omega_c(N)/S$ with chain length N. The oscillations, on the other hand, can be accounted for as corrections due to shape. Thus the number of conformations, and therefore the entropy, are comprised of two factors: (i) a conformational entropy per chain segment, and (ii) a shape entropy for the compact object. It is clear that all values on the upper dotted line in Fig. 4 ($c_2 \simeq 0.318$) correspond to the magic numbers; the values on the lower dotted line ($c_1 \simeq 0.0787$) represent chains for which the number of residues equals 1 *plus* a magic number. Intermediate cases fall between the two dotted lines. The magic number result $\Omega_c(35)/S = \Omega_c(35) = 57\,337$ for $N + 1 = 36$ is not plotted in Fig. 4 because data for $31 \leq N + 1 \leq 35$ are not yet available; nevertheless, it is easy to verify that $c_2\kappa^N$ gives a good estimation of this value to within 7%. Shapes for magic-numbered chains are square and near-square rectangles, but shapes for chains containing a number of residues between magic

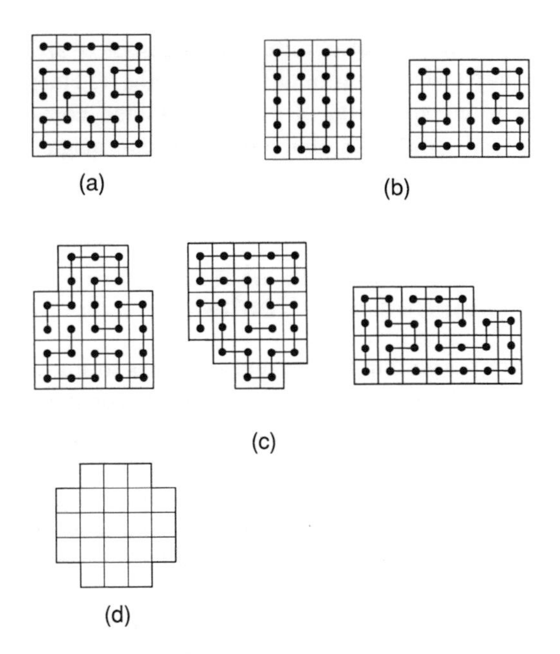

Fig. 3. Compact shapes on the square lattice with minimum perimeter determined by the number of sites $N+1$. (**a**) Only one shape is possible for $N + 1 = (m+1)^2$. (**b**) Exactly two shapes are possible for $N + 1 = m\,(m+1)$. (**c**) Three among a total of 238 possible shapes for $N + 1 = 26$. An example compact chain conformation is shown for each shape in (**a**), (**b**) and (**c**). (**d**) No continuous chain with $N + 1 = 21$ residues can be fit into this 21-site compact shape.

numbers are in general more articulated, as shown by examples in Fig. 3c. The most articulated shapes occur when the number of residues equals 1 plus a magic number, since they have the maximum freedom to take different shapes. The "articulateness" of a compact shape may be quantitated as the number of corners on the perimeter. Since squares and rectangles with 4 corners are the only shapes possible for magic-numbered chains, according to this standard, they have the minimum articulateness of 4; by comparison, the average number of corners for $N + 1 = 26$ compact chains is 7.67.

The more articulated is the shape of the box into which the chain is fit, the fewer are the configurations which can fill it precisely. In some cases of extreme articulation, there is no possible way that a chain can configure itself into that shape; for example see Fig. 3d, for $N + 1 = 21$. This observation explains the drop in $\Omega_c(N)/S$ when one residue is added to a magic-numbered chain: despite the intrinsic gain in conformational freedom due to an increased chain length, the shape effects due to increased articulateness dominate, resulting in a decrease in $\Omega_c(N)/S$. The local maxima of $\Omega_c(N)$ in Table 2 are those with $N + 1$ equal 1 plus a magic number, coinciding with the local minima of $\Omega_c(N)/S$. In other words, the local maxima in $\Omega_c(N)$ are caused by shape multiplicity, *in spite of* the fact that the corresponding shape-averaged number of conformations $\Omega_c(N)/S$ are at their local minima. Equation 7 best summarizes our findings here: the general exponential trend is verified with a connective constant $\kappa \simeq 1.41$, which is very close to that of Orr *(1)* and is consistent with the recent lower bound proposed by Schmalz, Hite, and Klein *(38)*. The principal correction is due to boundary effects: more articulated shapes are shown to be more restrictive to conformational freedom.

In some earlier works, such as that of Kasteleyn *(33)*, boundary effects were avoided by adopting periodic boundary conditions, which effectively configures the chain on the surface of a torus. More recently, the importance of boundary effects of free edges on $\Omega_c(N)$ has been pointed out by Gordon, Kapadia, and Malakis *(34)*. Numerical results were provided by Malakis *(40)* for Manhattan lattices, but only for boundaries of rectangular shapes.

In the more general case of interest here,

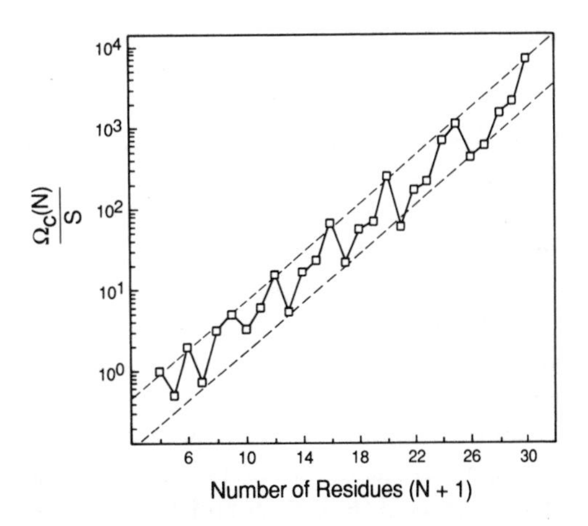

Fig. 4. Shape-average number of compact conformations $\Omega_c(N)\,/\,S$ versus number of residues $N+1$. The general trend is exponential increase. Oscillations are due to articulateness of the shapes; see text for details.

of the many possible compact shapes that can realize the P_c condition Eq. 3*a*, it would be of value to have a quantity to correct for articulateness. To characterize the variation within the class of $\rho = 1$ compact chains, we define the parameter

$$\sigma \equiv \frac{N+1}{[P_c/4]^2} \tag{8}$$

which is the ratio of the actual area occupied by chain residues to the area of a *hypothetical* square with the *actual* perimeter P_c. It is straightforward to show that

$$\left[\frac{2m}{2m+1}\right]^2 < \sigma \leq 1$$

$$\text{for } m^2 < N+1 \leq (m+1)^2 \tag{9}$$

and $\sigma = 1$ only if $N+1 = (m+1)^2$. As such, the deviation of σ from unity may be used to measure the deviation of a compact shape from a perfect square. This measure is consistent with the consideration of articulateness above: the σ values for the $N+1 = m(m+1)$ magic numbers are $4N/(4N+1)$, which is very close to 1 for large N with an $O(1/N)$ deviation. The most articulated compact chains with $N+1$ equal to one plus a magic number have a corresponding deviation of $O(1/\sqrt{N})$. We also found empirically that the exponential relation

$$\Omega_c(N)/S \simeq c\kappa^{\sigma(N+1)} \tag{10}$$

where $c \simeq 0.226$ and $\kappa \simeq 1.40$ seems to hold approximately. This is shown graphically in Fig. 5, in which the oscillations of Fig. 4 are dramatically suppressed by the use of this quantity σ.

σ may be considered as a refined measurement of compactness *within* the class of $\rho = 1$ compact chains. The closer the values of σ are to unity, the more "well-packed" are the compact chains, since they have a smaller perimeter to area ratio.

In this chapter, we study all accessible $\rho = 1$ compact chains. We include data from all chain lengths $N+1 \leq 30$ and $N+1 = 36$. This not only provides us with the chain-length dependence of various physical properties, but also allows us to probe the differential packing effects within the class of $\rho = 1$ compact chains, since different values of σ are sampled by chains of different lengths N. Thus a wider spectrum of data is obtained compared to studies that only consider magic-numbered chains.

Compact Chains with One Presumed Self-Contact

Our purpose in the present section is to study cyclization, or loop formation, in compact polymers. Cyclization is an important property which has previously been studied extensively in open chains. Using random-flight theory, Jacobson and Stockmayer first showed that the probability of spatial adjacency between monomers (residues) i and j is a diminishing function of their separation along the chain *(41)*. That theory has been augmented to take into account effects of local chain stiffness *(42)* and excluded volume *(4, 26, 43)*. In the present section, we explore the same question, but applied to only the compact conformations. The principal question addressed here is the following: how much is the compact conformational space restricted by the constraint that residues i and j are adjacent? We refer to this constraint *(i, j)* as a *presumed* contact, implying that it is specified a priori. This terminology is to distinguish that contact from the many other contacts in any

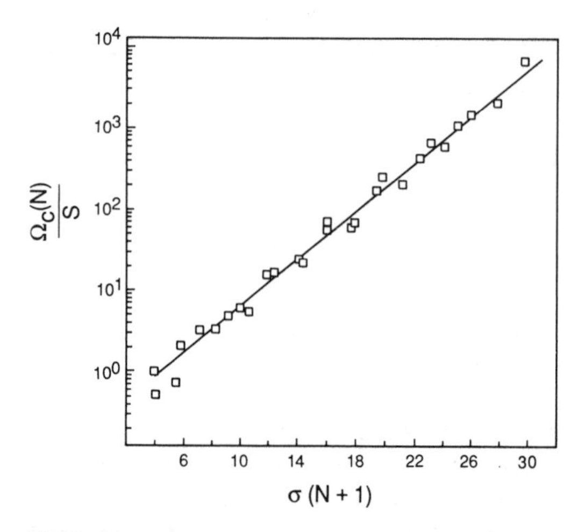

Fig. 5. Shape-average number of compact conformations $\Omega_c(N)/S$ versus the parameter $\sigma(N+1)$. The straight line is the best fit, $\Omega_c/S = c\kappa^{\sigma(N+1)}$, with $c \simeq 0.226$ and $k \simeq 1.40$.

particular compact conformation which arise simply as degrees of freedom of the system, not specified in advance. The effect of presuming a single contact pair (i, j) is measured by the reduction factor $R(N; i, j)$, defined to be the following ratio [see *(4)*]:

$$R(N; i, j) \equiv \frac{\Omega(N; i, j)}{\Omega_0(N)} \quad (11)$$

where $\Omega(N; i, j)$ is the number of conformations that have the contact pair (i, j) and $\Omega_0(N)$ is the total number of accessible conformations.

We are interested in the cyclization probability as a function of compactness. We define the following quantities,

$$\omega^{(t)}(N; i, j) \equiv \sum_{t'=t}^{t_{max}} \Omega^{(t')}(N; i, j) \quad (12a)$$

$$\omega_0^{(t)}(N) \equiv \sum_{t'=t}^{t_{max}} \Omega^{(t')}(N) \quad (12b)$$

$$R^{(t)}(N; i, j) \equiv \frac{\omega^{(t)}(N; i, j)}{\omega_0^{(t)}(N)} \quad (12c)$$

as a function of the compactness parameter, t. In Eq. 12*a*, $\Omega^{(t')}(N; i, j)$ is the number of conformations with a total of t' contacts that also satisfy the condition that the pair of residues (i, j) are in contact, and $\Omega^{(t')}(N)$ in Eq. 12*b* is defined in Eq. 1. In these equations, the superscript (t) in $\omega^{(t)}$, $\omega_0^{(t)}$, and $R^{(t)}$ denotes the *minimum* number of contacts in the chains under consideration. Hence $R^{(t)}(N; i, j)$ measures the effect of presuming the contact pair (i, j) in the collection of chains that have *at least* t contacts. In these collections, the average chain compactness $\langle\rho\rangle$, over all chains with number of contacts $t' \geq t (t' \leq t_{max})$, increases with t, while the size of the conformational space contracts with increasing t, since the conformational space of $R^{(t+1)}$ is a *subspace* of the conformational space of $R^{(t)}$. For $t = 0$, $R^{(0)}$ equals the open-chain reduction factor R in Eq. 11; on the other hand, when $t = t_{max}$, $R^{(t_{max})}$ becomes the *compact* reduction factor R_c that only takes into account the compact ($\rho = 1$) conformations, defined as

$$R_c(N; i, j) \equiv \frac{\Omega_c(N; i, j)}{\Omega_c(N)} = R^{(t_{max})}(N; i, j) \quad (13)$$

where $\Omega_c(N; i, j) = \Omega^{(t_{max})}(N; i, j)$ is the number of compact conformations that have the (i, j) contacts. $R^{(t)}$ therefore interpolates between the open and compact chains. Its behavior as a function of t is useful for studying how the formation of specific contacts are affected by the overall packing density of the chain.

The general restricting effects of contacts on conformational freedom may be characterized by the *average reduction factor* $\langle R^{(t)}(N)\rangle$ *per contact*, defined as

$$\langle R^{(t)}(N)\rangle \equiv \frac{\sum_{i<j} R^{(t)}(N; i, j)\omega^{(t)}(N; i, j)}{\sum_{i<j} \omega^{(t)}(N; i, j)} \quad (14)$$

In the above equation, the numerator is a weighted sum of the reduction factor $R^{(t)}$ over all possible contact pairs (i, j). The weight $\omega^{(t)}(N; i, j)$ is the number of configurations which have the (i, j) contact and compactness greater than or equal to t/t_{max}. The denominator

$$\sum_{i<j} \omega^{(t)}(N; i, j) = \sum_{t'=t}^{t_{max}} t'\Omega^{(t')}(N) \quad (15)$$

is simply the total number of *contacts* present in the same conformational space.

The average reduction factor $\langle R^{(t)}(N)\rangle$ is computed for chains with $N + 1 = 14$ residues and the data are shown as squares in Fig. 6. $\langle R^{(t)}(N)\rangle$ gradually increases with chain compactness towards $\rho = 1$. Physically, this implies that the a priori constraint (i, j) becomes less restrictive relative to all other constraints, as the number of total constraints increases. Although the constraint (i, j) is less restrictive of compact chains than of open chains, nevertheless its restriction on conformational freedom is still significant. For the short chain example $N + 1 = 14$, the average reduction factor for $\rho = 1$ compact chains when $t = t_{max}$ is $\langle R_c(13)\rangle = 0.224$, compared with $\langle R^{(0)}(13)\rangle = \langle R(13)\rangle = 0.098$ for open chains. The average reduction factor is only increased by a factor of 2.30 by packing. As a function of chain length, our calculations show that both $\langle R(N)\rangle$ and $\langle R_c(N)\rangle$ tend to decrease gradually as N increases. For example, $\langle R_c(N)\rangle$ for $N + 1 = 20$, 25, 30, and 36 are 0.177, 0.155, 0.138, and 0.127, respectively, indicating that an average contact is more restrictive in longer than in shorter chains.

The average reduction factor $\langle R^{(t)}(N; k)\rangle$ of contacts (loops) of order (size) k, lying along the kth diagonal on the contact map, can be computed by restricting the summations over i and j in Eq. 14 to $j - i = k$. $\langle R^{(t)}(N; k)\rangle$ for $N + 1 = 14$ are plotted in Fig. 6. The two thicker curves that envelop the $k = 3-13$ thinner curves represent the minimum and maximum reduction factor found in each conformational space labeled by t. As the chains increase compactness, $\langle R^{(t)}(N; k)\rangle$ exhibits more dramatic changes than the smooth behavior of $\langle R^{(t)}(N)\rangle$. The differences in average reduction factors for different contact orders becomes smaller, accompanied by a narrowing of range between the minimum and maximum reduction factor. At maximum compactness $\rho = 1$, $\langle R^{(t_{max})}(N; k)\rangle$ for all k are comparable, resulting mainly from the rapid increase for large k upon packing. The highest order $k = 13$ actually overtakes the lowest order $k = 3$ as the most favored contact order at $\rho = 1$.

This phenomenon is easy to understand physically. When the chains are more open, most of the contacts are concentrated at the lower orders close to the main diagonal on the contact map *(4)*, because they are less restrictive of conformational freedom. However, when chains become more compact by forming increasing numbers of contacts, the lower-order contacts become saturated and higher-order contacts must be formed. Consequently higher-order contacts are enhanced by packing.

A more convenient global representation of the contact pattern in compact chains is the topological contact free energy surface introduced in studies of open chains *(4)*. Here the *compact* reduction factor $R_c(N; i, j)$ of Eq. 13 is computed for every (i, j). The *contours* for the quantity $[-\ln R_c(N; i, j) + \text{constant}]$ are then plotted on the contact map. This quantity is a dimensionless free energy, in units of kT, due only to chain conformational entropy, for relative formation probabilities of the different contacts.

Figure 7 shows such a topological free energy plot for compact chains with $N + 1 = 30$ residues. As is also observed for open chains *(4)*, contours curve near the vertical and horizontal boundaries of the contact map, showing that there are end effects. Contact formation is more favored at the chain ends. The least-favored contacts (lightest-colored) are located in the middle of the contact map.

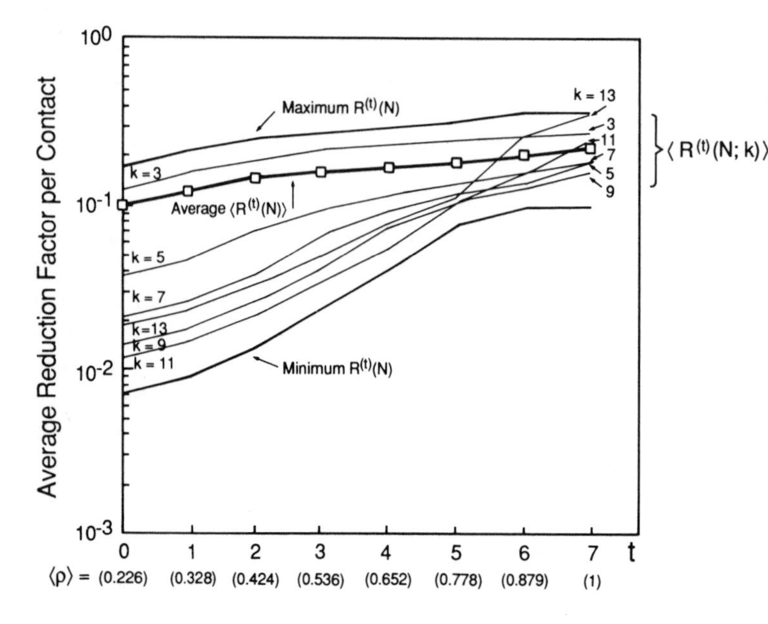

Fig. 6. Average reduction factor per contact for chains with $N + 1 = 14$ residues, as a function of increasing average compactness $\langle\rho\rangle$. On the horizontal axis, t refers to the *minimum* number of contacts in the chains.

In this example with $N + 1 = 30$, the least-favored contacts are those formed between residues that are approximately seven segments from the two chain ends.

For open chains, loops at either chain end are favored relative to loops internal in the chain, for two reasons. First, there is less volume excluded at the ends to interfere with their configurational freedom. Second, end segments can have one more topological neighboring residue than segments at midchain, because end segments have only one connected neighbor, whereas midchain segments have two connected neighbors. The least favored contacts in open chains are those between segments near the two chain ends, since contacts between those segments would cause the greatest restriction of conformational space. For compact chains also, the end effects arise from these same factors. However, since higher-order contacts are more favored in compact chains than in open chains, there are two differences. First, the least favored contacts in compact chains are for pairs of chain segments somewhat closer together in the sequence than for open chains (i.e., they are found on a lower-order diagonal). Second, the overall free energy landscape is much flatter for compact chains than for open chains (see also Fig. 6).

Figure 8 shows in somewhat more detail the end effects on the probability of cyclization of small loops in compact polymers. Exact compact reduction factors $R_c(N; i, j)$ are computed for contact orders $k = |j - i| = 3, 5$, and 7 at all possible locations along the chain, for various chain lengths $N < 30$. $R_c(N; i, j)$ is expected to depend on the two tail lengths $l_0 = i - 1$ and $l_0' = N - j - 1$, shown in Fig. 8a. In Fig. 8b, the *average* of $R_c(N; l_0 + 1, l_0 + k + 1)$ over different chain lengths, N, are plotted as a function of l_0. It is appropriate that only chain lengths that are substantially longer than the contact orders are included in the averaging. Data in Fig. 8b are obtained from averaging over $14 < N < 30$ for $k = 3$, $15 < N < 30$ for $k = 5$, and $16 < N < 30$ for $k = 7$. The variation in R_c for different N is recorded by the 1 standard deviation error bars in the plot. These variations are in general small; most of them are well below 10% of R_c. We believe that they are caused by the packing constraint, $\rho = 1$. By contrast, the corresponding variations are much smaller for

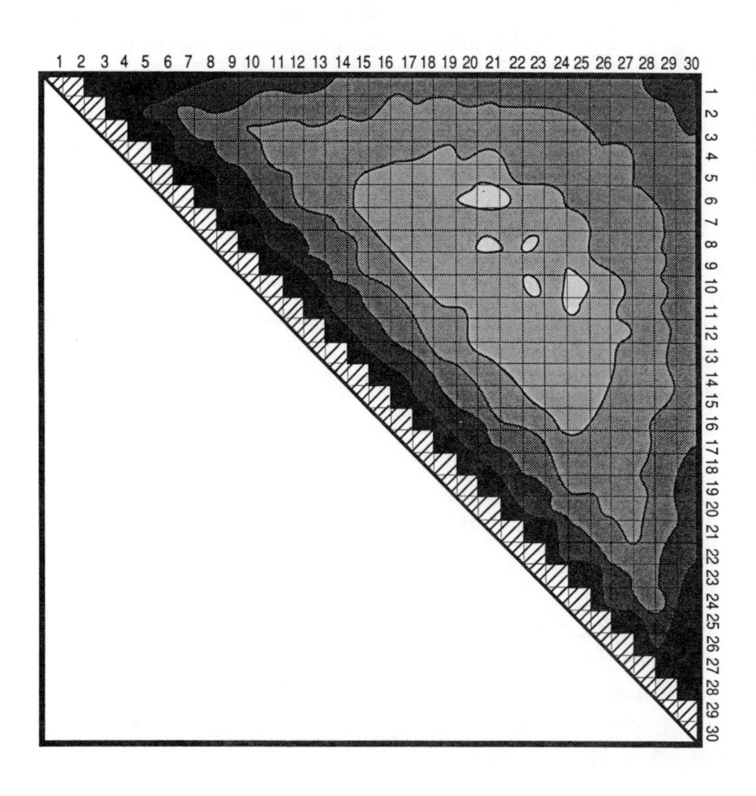

Fig. 7. Free energy contour plot for the entropy of contact formation in compact square lattice chains with 30 residues. Contours are given in $0.4kT$ steps.

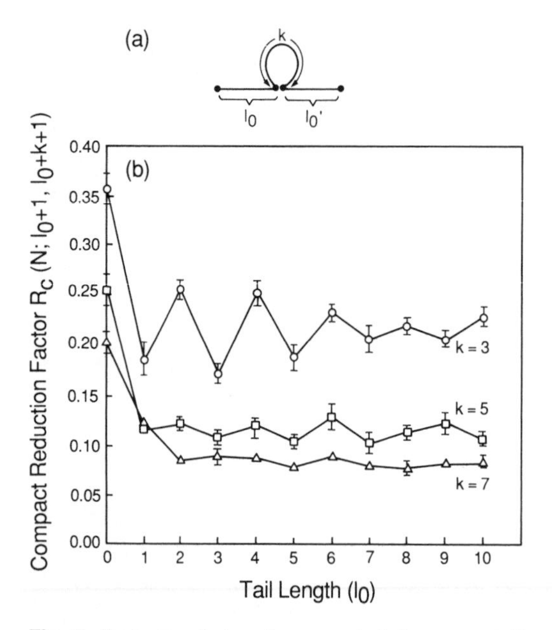

Fig. 8. Reduction factors in compact chains versus tail length l_0 defined in diagram **(a)**. Contacts are favored at chain ends.

open chains, in which case *(4)* limiting N-independent values of R are rapidly reached for $N \geq 16$.

Figure 8b confirms that there are end effects in compact chains, as in open chains *(4)*. For each of the three contact orders, $k = 3, 5$ and 7, R_c is highest at $l_0 = 0$, implying that contact (loop) formation is more preferred at the chain ends than in the middle. The large-amplitude even–odd oscillations of the $k = 3$ and $k = 5$ curves near $l_0 = 0$ is a feature peculiar to compact chains, not observed previously in open chains *(4)*. This phenomenon is a packing effect, since a loop with a tail chain comprised of an even number of residues is easier to pack into a compact conformation than one with an odd-numbered tail chain. On the topological free energy plot of Fig. 7, these oscillations are manifested by the wavy features of some contours near the main diagonal.

The compact reduction factors R_c differ from their open chain counterparts R in one important respect: R_c decreases much slower than R as the contact order k increases. In fact, at $l_0 = 1$, R_c values for $k = 5$ and $k = 7$ are approximately equal. At middle chain positions, the ratio between R_c of $k = 3$ and $k = 5$ is approximately $0.22/0.12 = 1.83$, and between $k = 5$ and $k = 7$ is approximately $0.12/0.08 = 1.50$. Both numbers are much smaller than the corresponding ratios of 4.42 and 2.27 for open chains *(4)*, hence the flattened landscapes in compact polymers as noted above.

Next we consider the power law dependence of the compact reduction factor $R_c(k; 1, k + 1)$, the cyclization probability that the chain forms a ring with no tails by making contact between the two ends, $k = N$. The data are plotted in Fig. 9, which shows that for sufficiently large k, this probability has an approximate $k^{-\nu}$ dependence. Our best estimate from the data points with $k \geq 11$ gives $\nu \simeq 0.82$, which is about half of the corresponding exponent $\nu \simeq 1.63$ for open chains *(4)*. Hence, for chains of the same length, cyclization probabilities are relatively higher in compact chains than in open chains. Indeed, the probability is still reasonably high even for compact chains with intermediate lengths, for example, $R_c(k; 1, k + 1)$ equals $2310/13498 = 0.171$ for $N + 1 = k + 1 = 30$ and $9648/57337 = 0.168$ for $N + 1 = k + 1 = 36$. In summary, the two ends are quite likely to be near each other if the whole chain is confined to a small space. This may account for the fact that the two ends of protein molecules are often observed to be close to each other in the native state *(44)*.

Our value of the exponent ν may be com-

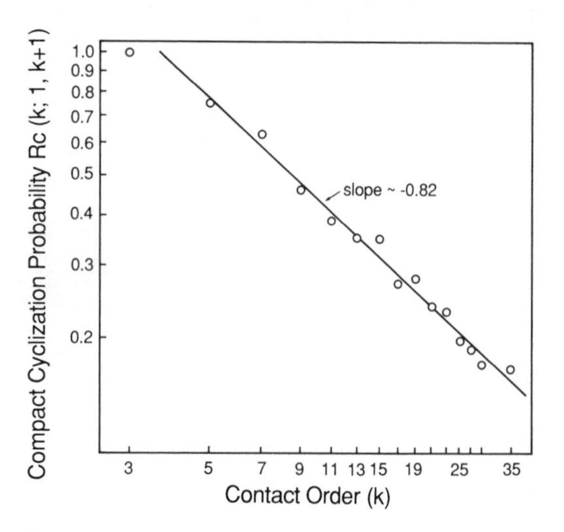

Fig. 9. Power law dependence of the compact cyclization probability.

pared with a recent study of the mean-square end-to-end distance R_N^2 for compact chains by Ishinabe and Chikahisa *(22)*. Since the ring formation probability in two dimensions is approximately proportional to R_N^{-2}, their exponent $\nu_c \simeq 0.43$ ($2\nu_c \simeq 0.86$) in the scaling relation $R_N^2 \sim N^{2\nu_c}$ is equivalent to an exponent $\nu \simeq 0.86$ for the ring-formation probability. In view of the fact that not all data points lie exactly on the best fit line, their result is consistent with our value of $\nu \simeq 0.82$.

Correlation among Contacts in Compact Chains

In the preceding section, we considered the occurrence of a single specified loop, or intrachain contact, within the ensemble of compact polymer conformations. In the present section, we consider pairs of loops, two contacts (i_1, j_1) and (i_2, j_2) within the same chain. This is of interest because a pair of loops, represented by a set of two dots on the contact map, is the most elementary building block for secondary structures: helices, parallel and antiparallel sheets, and turns. We have recently shown, through development of a theory for "topological" pair correlation functions for interactions among the loops, that the most probable conformations of open chains containing two intrachain contacts are helices and antiparallel sheets *(4)*. Our aim in the present section is to extend the application of this topological pair correlation function approach to the subset of compact conformations of polymers.

If two loops in a chain are near each other in the sequence, then their cyclization probabilities may not be independent. The interdependence can be defined as follows *(4)*. Suppose contact (i_1, j_1) is presumed to be given, indicated by the hollow circles on the contact maps in Fig. 10 and Fig. 11. Then if $R_c(N;\, i_1, j_1;\, i_2, j_2)$ is the reduction factor for simultaneously presuming two contacts (i_1, j_1) and (i_2, j_2) in compact chains of length N, the *conditional* probability of forming the second contact (i_2, j_2) subject to the presumed first contact is simply

$$\frac{R_c(N;\, i_1, j_1;\, i_2, j_2)}{R_c(N;\, i_1, j_1)} \tag{16}$$

The logarithm of this quantity is plotted as contours for all possible positions (i_2, j_2) on the contact map. The contours then represent the relative probabilities of forming any second contact for a given first contact. In this representation, the two-contact "interaction" is represented by this potential surface, analogous to the field description of particle–particle interactions in classical physics.

Figures 10 and 11 show two such potential surfaces for chains with $N + 1 = 30$ residues. In Fig. 10, the presumed contact is of order 3 at (14, 17). In Fig. 11, the presumed contact is of order 5 at (13, 18). The 30-residue compact chains are taken here as a typical example representative of intermediate-length chains. We have verified by exhaustive enumerations of various chain lengths that potential surfaces of comparable lengths have features similar to those shown in Figs. 10 and 11. The subtlety and complexity, which arise simply from excluded volume in the compact state, is quite remarkable.

The most important features of these free energy surfaces are the locations of the deepest minima. The deepest minima identify the most probable second loop, given a specific presumed first loop. It is clear from Fig. 10 that given the smallest possible presumed loop at (i_1, j_1) = (14, 17), the two most probable configurations in the compact conformational space are the antiparallel sheet (13, 18) and the helix (12, 15) and (16, 19). Those configurations are favored by $3.2kT$ and $2.8kT$, respectively, relative to the least favored contacts. Likewise, the same conclusion is drawn from Fig. 11. For the presumed contact of order 5 at (13, 18), the most probable second contact is that which "fills in" the turn for the antiparallel sheet (14, 17), $4.0kT$ more favorable than the least favored contact. The next most favorable contacts are those that may be associated with helices: (11, 14) or (17, 20), or extension of the antiparallel sheet (12, 19), all of which are equally favored by $3.2kT$ relative to the worst contacts. It is also clear that the end effects are as strongly favored as the second most prob-

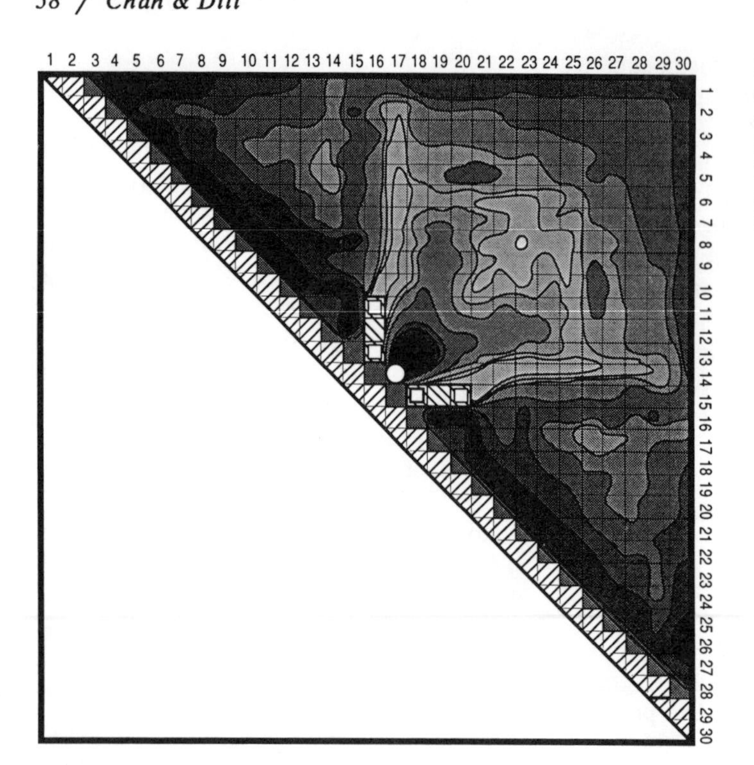

Fig. 10. Contour plot (0.4 kT steps) of inference potential surface for compact chains with 30 residues and a single presumed contact at (14, 17). Most favored second contacts occur at helix and antiparallel sheet positions.

able conformation in each case. These results are obtained through exhaustive search of every conformation accessible to the compact chain.

Similar results have been observed for open chains *(4)*. In that case also, given one presumed contact of order 3 or 5, the most probable second contact results from "zipping up" the helix or antiparallel sheet from that point. Hence there is a driving force for secondary structure formation when only two intrachain contacts have formed, $t = 2$. As shown above, the same driving force exists at $t = t_{max}$, i.e., in the compact state. In the next section, it is shown that this driving force persists for the formation of every additional contact for all densities increasing from $t = 2$ to $t = t_{max}$. Thus, steric forces act to drive the formation of secondary structures.

Certain other features of these free energy surfaces are also similar for compact and open chains. The hollow squares on some sites of the contact maps in Figs. 10 and 11 represent "implied blocks," contacts which could not be formed for *any* chain configuration, open or compact, for the given presumed contact. Clearly, therefore, these lines of blocks will appear identically for open or compact chains. Similarly, the extensions of these lines of implied blocks are disfavored conformations, for either open or compact conformations (see columns 16–18 or rows 12–14 in Figs. 10 and 11). For open conformations, the origins of these disfavored conformations in competing effects of excluded volume have been described elsewhere *(4)*.

The interdependences of cyclization of two loops within a chain can be described in terms of a correlation function $g_{c;k1,k2}$ for two contacts of orders $k_1 = |j_1 - i_1|$ and $k_2 = |j_2 - i_2|$

$$g_{c;k_1,k_2}(L) = \frac{R_c(N; i_1, j_1; i_2, j_2)}{R_c(N; i_1, j_1)\, R_c(N; i_2, j_2)} \qquad (17)$$

defined here for compact chains. This definition for correlation function differs from that given previously *(4)* only in that the subscript c indicates that it applies to the compact conformations. In Eq. 17, R_c refers to compact reduction factors defined above, and L is the separation between the two contacts under consideration, measured as the number of

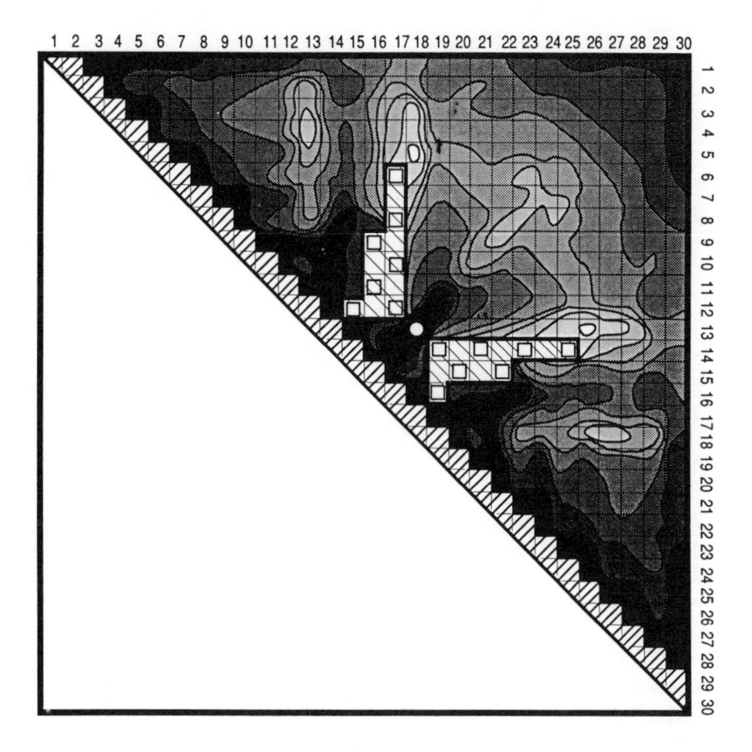

Fig. 11. Contour plot (0.4 *kT* steps) of inference potential surface for compact chains with 30 residues and a single presumed contact at (13, 18). For this order 5 presumed first contact, the most favored second contacts occur at antiparallel sheet position (14, 17).

bonds from the starting point on the larger loop to the corresponding point on the smaller loop *(4)*. g_c is the ratio of the actual number of compact conformations which satisfy the two presumed contacts to the number of compact conformations if the two contacts were independent. Hence it is a measure of the degree to which one loop hinders or enhances the formation of the other: enhancement is indicated when $g_c > 1$, hindrance when $g_c < 1$, and $g_c = 1$ implies independence.

Shown in Fig. 12 is the compact correlation function $g_{c;\,k_1,\,k_2}$ calculated for $k_1 = k_2 = 3$ and $k_1 = 3, k_2 = 5$ and $N + 1 = 30$ residues. Depending on the exact location of the two contacts along the chain, there is considerable variation even when the separation L between the two contacts remains constant. This is an effect of the compact packing. In open chains with the two contacts well embedded in the middle of long chains, such variations are negligible *(4)*. The *averages* of $g_{c;k_1,k_2}$ over all possible positions along the chain are shown as square boxes, while the variations are represented by 1 standard deviation error bars. The thinner continuous lines correspond to the open chain (all ρ) results obtained earlier *(4)*, reproduced here for comparison.

Secondary structure enhancement is apparent from Fig. 12, variations notwithstanding. Indeed, the average correlation g_c attains its maximum at $L = 2$ in Fig. 12a, which is the helix position, and at $L = 1$ in Fig. 12b, which is the antiparallel sheet postion. The open chain g also has a maximum at these positions *(4)*.

In addition to the secondary structure peaks, these correlation function plots also show much similarity for compact and open chains, indicating that the local steric restrictions are similar. The following are observed in both cases: (i) contacts are essentially independent (g, $g_c \simeq 1$) when the separation L is large compared to the contact orders; and (ii) there is always a local minimum when L equals the size of the larger of the two contacts *(4)*, which corresponds to $L = 3$ in Fig. 12a and $L = 5$ in Fig. 12b.

Two differences between the open and compact cases should be noted however. First, the even–odd oscillation in Fig. 12a is probably due to the same packing effect discussed above in conjunction with Fig. 8. Second, notice in Fig. 12b that the correlation function, g_c, is smaller at the peak value for compact than open chains. Although the absolute prob-

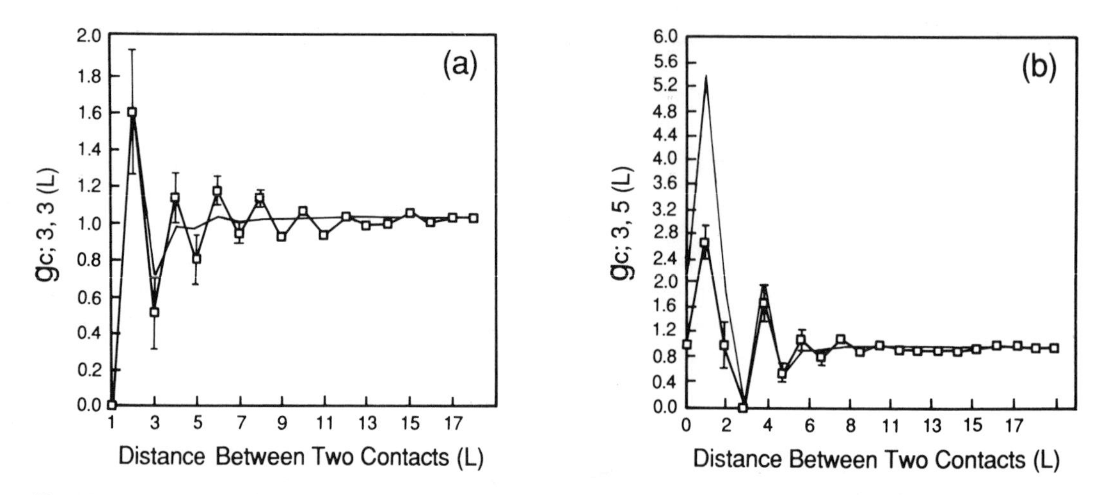

Fig. 12. Two-contact correlation $g_{c;k_1,k_2}(L)$ for compact chains. **(a)** Correlation between two order 3 contacts, $k_1 = k_2 = 3$. **(b)** Correlation between an order 3 and an order 5 contact, $k_1 = 3$ and $k_2 = 5$. The variations along the chain are represented by 1 standard deviation error bars. Error bars that lie within the hollow squares are not shown. In both cases, the peak positions indicate that certain secondary structures are favored.

ability of forming the antiparallel sheet is much higher in the compact chains (0.074) than in the open chains (0.014), recall that g_c is only a ratio, relative to the independent loops (see Eq. 17). In this case, it is simply that the single loops are also significantly enhanced by compactness.

Secondary Structures Driven by Packing

In the previous section, we found that steric constraints favor the association of two loops into helical or antiparallel sheet configurations in compact chains, in agreement with similar results for open chains *(4)*. In the present section, we ask the broader question of the joint distribution of *all* the possible loops within the molecule, including now also parallel sheets and turns. In addition, we study the distribution of secondary structural elements as a function of the compactness and length of the molecule, and of the position within the chain.

Using the definition introduced earlier (see "The Model"), the number of residues participating in various secondary structures are computed for each of the 802075 accessible conformations with $N + 1 = 16$ residues. Figure 13 shows the fractional participation, the total number of residues participating in secondary structures divided by $N + 1$, as a function of compactness, $\rho = t/t_{max}$. The principal conclusion from Fig. 13 is that the fraction of residues in secondary structures increases with compactness and is remarkably high in compact molecules. Clearly, since any secondary structure requires for its definition a minimum of $t = 2$ intrachain contacts, then no secondary structure can occur for $t = 0, 1$. Hence in the limit as $\rho \rightarrow 0$, for polymers in "super-solvents" or which are highly charged, the amount of secondary structure should be negligible. For maximally compact ($\rho = 1$) molecules, however, the fractional participation in secondary structures reaches 76.1% for the case shown. It is noteworthy that if the global conformational space of open chains (all possible ρ) is considered as a whole, the combined participation rate is about 15.3% for $N + 1 = 16$. Due to the near linearity of the combined participation curve, this rate is given approximately by the curve at $\rho = \langle\rho\rangle = 0.243$, indicated by an arrow in Fig. 13. It is interesting that proteins that are unfolded by "weak" denaturing agents, such as temperature or pH/salt in some cases, and have high-density unfolded states are then also predicted to have much secondary structure. This is in accord with experimental evidence *(45–46)*.

All four kinds of secondary structure are enhanced by compactness. In the present model, antiparallel sheets overtake helices at intermediate compactness as the most prominent type of structure. This phenomenon is straightforward to understand in terms of packing enhancement of higher-order contacts. Helices are confined to the order 3 diagonal, but sheets are not subject to this limitation. At low compactness, the order 3 diagonal is most favored. However, beyond a certain compactness, the order 3 diagonal will be saturated and higher-order contacts will have to be formed. Whereas Fig. 13 shows the behavior of the fractional number of *residues*, we have verified separately that the fractional number of *contacts* participating in sheets and helices has the same crossover at intermediate compactness.

Figure 14 shows the distribution of the number of conformations that each have a particular number of secondary structure participating residues for all compact 30-residue chains. Figure 14 shows the striking result that there is *no* compact conformation with less than 53% secondary structure for this chain length. The inset in Fig. 14 shows a randomly chosen example conformation with the *minimum* participation rate. The same general result holds for other chain lengths, though compact chains with non-magic numbers of residues tend to have broader distributions than those with magic numbers, and a few conformations may be able to configure without forming any secondary structures; see Fig. 15. This is not surprising because non-magic-numbered compact chains are not as well-packed as magic-numbered chains, and therefore are subject to less packing constraint. An overwheming majority of conformations accessible to compact chains are those with a high fraction of secondary structures. Compactness is simply inconsistent with most other arrangements of the chain. In Fig. 14, quite a few (847 conformations, 6.3%) compact chains in fact have 100% participation rates. Figure 16 depicts a few examples of how this is achieved.

The high proportion of secondary structure in compact $\rho = 1$ chains is not an artifact due to the shortness of the chains. Quite the contrary, longer compact chains have even larger fractions of residues participating in secondary structures. For instance, compact chains with 30 residues have a participation

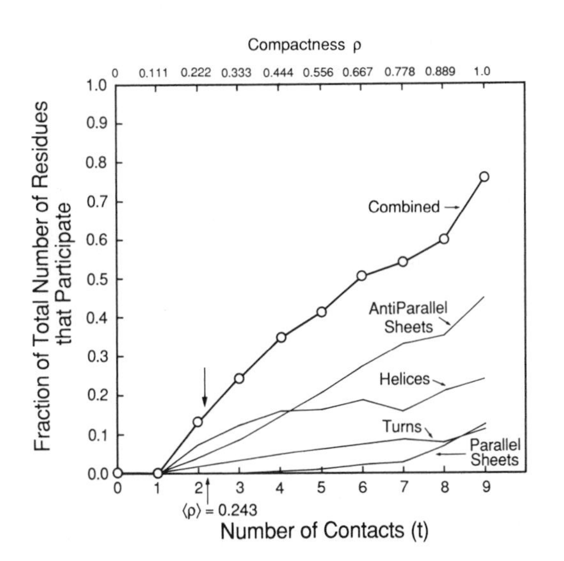

Fig. 13. Secondary structures in chains with 16 residues. All secondary structure types are enhanced by compactness. The arrow indicate the average compactness $\langle \rho \rangle$ of the global conformational space and the corresponding combined participation rate of 0.153.

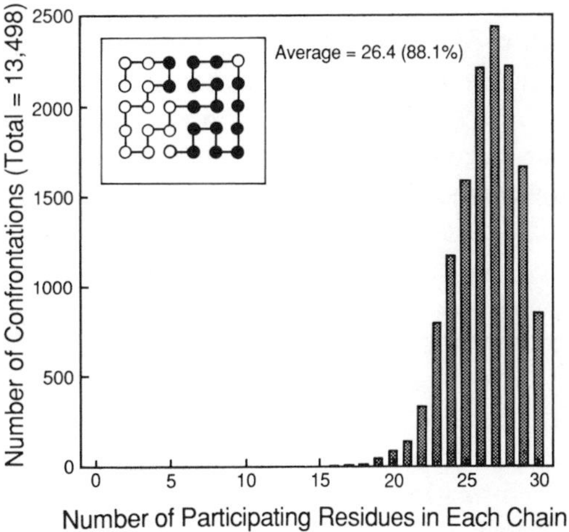

Fig. 14. Histogram showing how secondary structure is distributed throughout the compact conformational space, for chains with 30 residues. There is a lower cutoff at 16; no compact chains with 30 residues can configure without at least 16 residues participating in secondary structures. The inset shows an example of one conformation having this minimal participation. Residues participating in secondary structures are filled dots; others are represented as hollow circles.

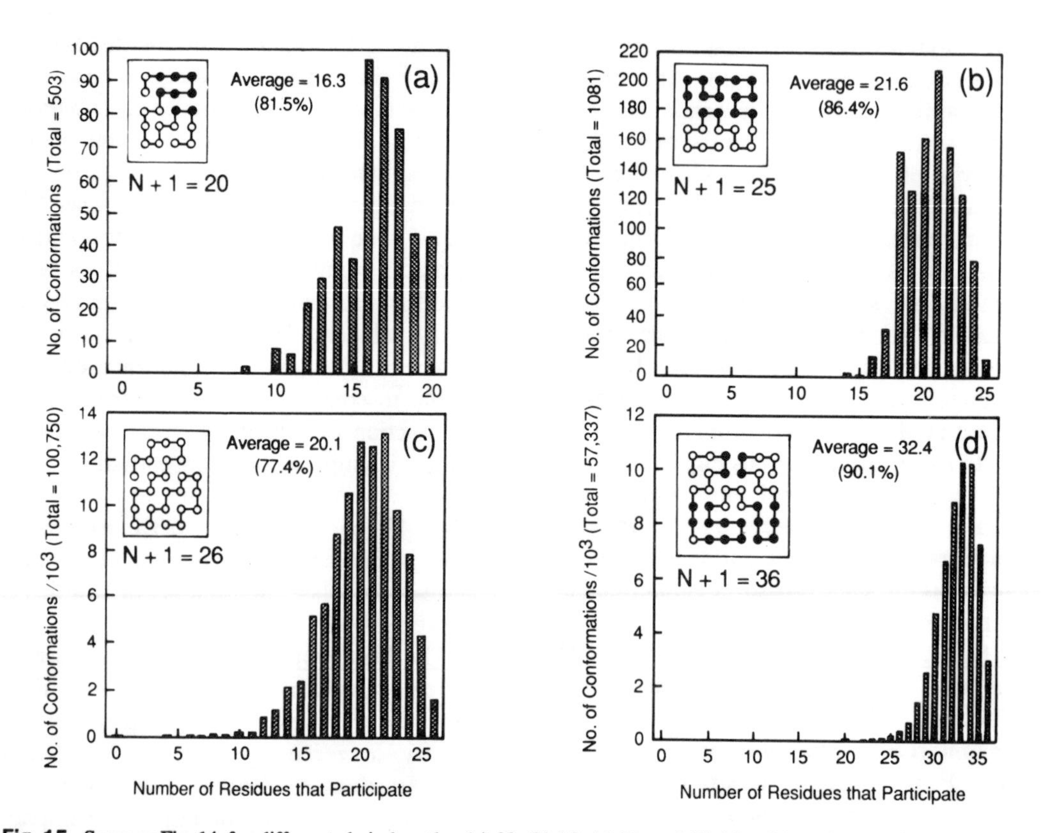

Fig. 15. Same as Fig. 14, for different chain lengths: **(a)** 20, **(b)** 25, **(c)** 26, and **(d)** 36 residues. Insets show examples of conformations with minimal participation. Residues participating in secondary structures are filled dots. Note that non-magic-numbered chains **(c)** have a broader distribution.

rate of 88.1%, whereas compact chains with 36 residues have an even higher participation rate of 90.1%. Figure 17 shows the result of our calculations for all compact chains with the number of residues $N + 1$ ranging from 14 to 30, demonstrating that the high participation rate is a universal phenomenon for two-dimensional compact chains. There are oscillatory variations with chain length. Secondary structure participation is largest for chain lengths with $N + 1$ equal to the magic numbers (defined in "Enumeration of the Chain Conformations") and is smallest when the number of residues is one unit longer than a magic number. Since magic-numbered chains are differentially more well-packed within the class of compact chains (see "Enumeration of the Chain Conformations"), then this measure of increased "packing" also correlates with increased secondary structure.

Figure 17 suggests that the secondary structure participation rate will be even higher for longer chains. In order to eliminate the magic-number-related oscillations and obtain a better extrapolation, we plot these participation rates of compact chains in Fig. 18 as a function of the perimeter/area ratio $P_c/(N + 1) = 4/[\sigma(N + 1)]^{1/2}$ (see Eq. 8), which tends to zero for chains of infinite length. From the general trend exhibited in Fig. 18, it is not unreasonable to conjecture that the combined participation rate in two dimensions may asymptotically approach unity as $N \rightarrow \infty$, i.e.,

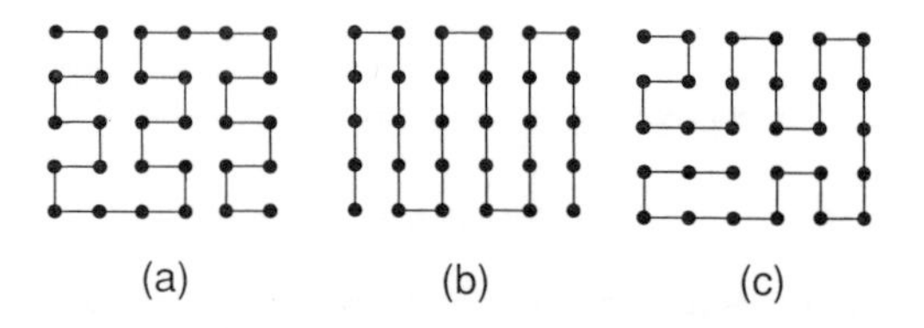

Fig. 16. Three examples of compact chains of 30 residues with 100% participation in secondary structure: **(a)** all helices; **(b)** all antiparallel sheets and turns; **(c)** an example that contains all four types of secondary structure.

the domination of secondary structure motifs may be complete at infinite chain length. Clearly, conformations with lower-than-unity participation rates are always possible for finite N, but the extrapolation indicates that those conformations form a diminishing fraction of all compact chains as the chain length N tends to infinity.

The distribution of various secondary structure types along the chain are shown in Fig. 19 for compact chains with 30 residues. Aside from end effects, the distributions are quite uniform. Without the incorporation of any specific interactions, secondary structure is predicted to be distributed uniformly along the chain. The end effects are also easy to understand: (i) The decrease in helix probability towards chain ends is a consequence of the fact that helices can form in only one direction at chain ends but two directions are available in the middle of the chain. (ii) There are no turns at chain ends simply because at least two residues have to be attached to each of the two residues comprising the turn. (iii) Parallel sheets are favored at chain ends because of the following: For two chain segments running in parallel, a chain segment of length considerably longer than the parallel segments is required to join the two parallel segments into a continuous chain. Clearly such connecting chain segments are more readily

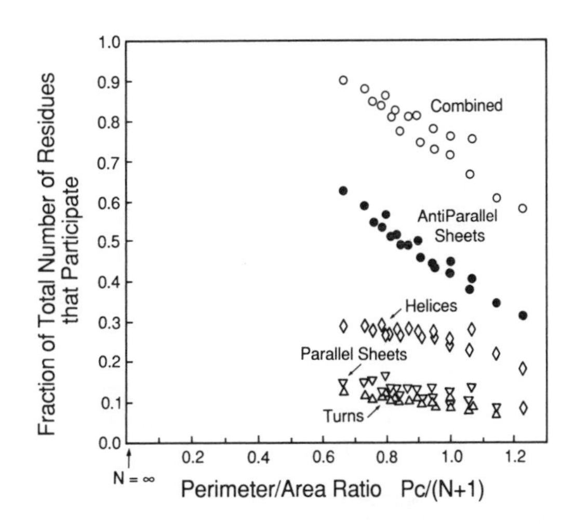

Fig. 18. Participation in secondary structure as a function of the perimeter/area ratio in two-dimensional compact square lattice chains. Data from chain lengths $N + 1 = 13-30$, and $N + 1 = 36$ are included. It is conjectured that the combined participation rate tends to unity as $P_c/(N + 1) \rightarrow 0$ or $N \rightarrow \infty$.

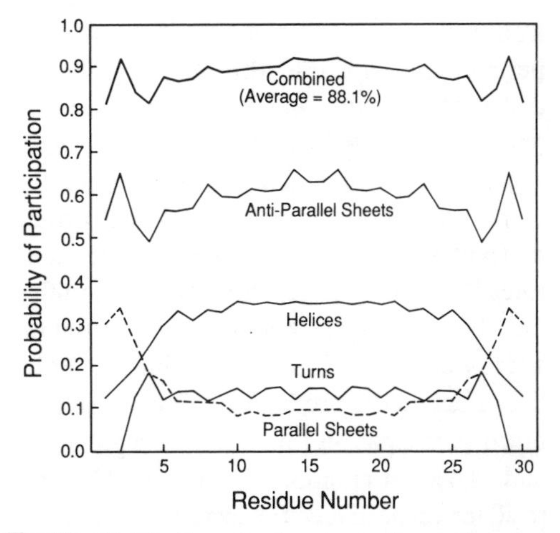

Fig. 19. Distribution of various types of secondary structure along the sequence for compact chains with 30 residues.

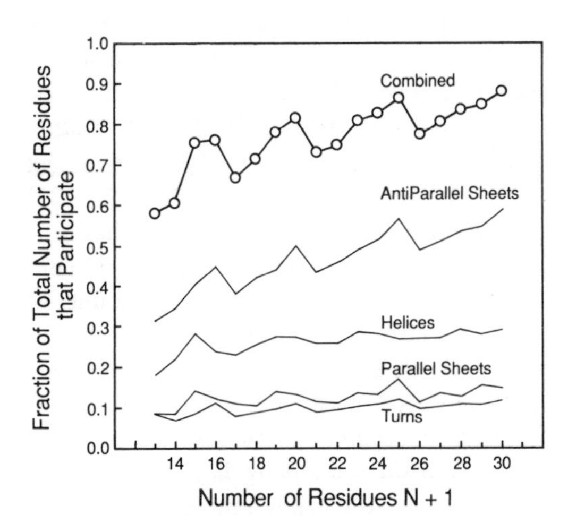

Fig. 17. Secondary structure participation in compact chains. Participation rates increase with chain length: magic-numbered chains have more secondary structure than others.

available if one of the parallel segments is located near a chain end.

To explore how sensitively our conclusions of secondary structure enhancement depend on the lattice definitions of secondary structure adopted in the Model, we have repeated all the above calculations with a modified and more restrictive definition. In this alternate approach, the definition for helices and turns (Fig. 1a, d) remained unchanged, but the minimal units of sheets (Fig.

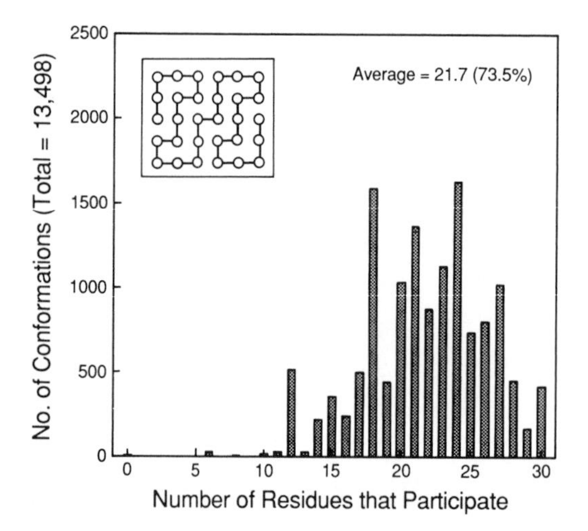

Fig. 20. Histogram of number of residues participating in secondary structure according to the more restrictive definition. Only one of the 13 498 compact chains with 30 residues configures without forming any secondary structure. This conformation is shown in the inset.

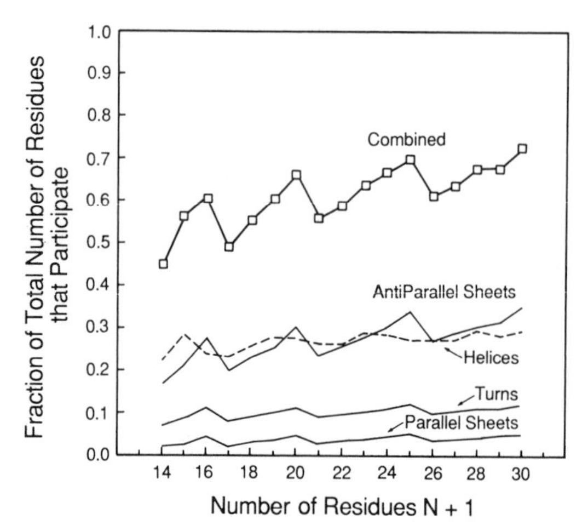

Fig. 21. Secondary structure participation in compact chains, according to the more restrictive definition.

1b, c) are modified from 2 to 3 contacts, or from 4 to 6 residues. One motivation for experimenting with this alternate definition is that it eliminates some of the possibilities of one residue participating in both a helix and a sheet *(47)*, which is allowed by the definition of the Model, but impossible in real protein molecules. Obviously, reduced participation rates for sheets will result from this modification. It will be shown below, however, that this altered definition has only a minor quantitative effect on the predictions above. The combined participation rates remain high: 72.5% and 74.4% for compact chains with 30 and 36 residues respectively, for example. These calculations using modified definitions of secondary structures serve the function of "control experiments." They show that reasonable variations in the criteria for cataloging secondary structures have little effect on the conclusions drawn from this model.

Figures 20–23 are equivalent to Figs. 14, 17–19, calculated using the alternate secondary structure definition instead of that given earlier (see "The Model"). It is clear that many qualitative features remained unchanged. Under the more restrictive secondary structure definition, a broader distribution of the number of participating residues is shown in the histogram of Fig. 20. In contrast to Fig. 14, there is *exactly one* conformation that is able to configure without forming any alternately defined secondary structure. This conformation is shown in the inset of Fig. 20. It is clear that even that particular structure has some order: an "antiparallel sheet" along the diagonal of the square lattice (see also Fig. 15c). Nevertheless, this is excluded from our present definitions of secondary structure. Indeed, this example illustrates that some form of ordered contact pattern is almost unavoidable in two-dimensional compact chains. Figures 21 and 22 show that the participation rate remains high (over 60%) for compact chains of intermediate lengths ($N + 1 \geq 22$) and continues to increase for longer chains.

Similar calculations have been carried out in three dimensions *(5)*. The amount of secondary structure driven by packing forces is only slightly smaller on a three-dimensional cubic lattice. There is greater freedom of definition of secondary structures on the cubic lattice than on the square lattice *(48)*, however, but even so there is a minimum of about 35–40% secondary structure observed in compact chains in three dimensions.

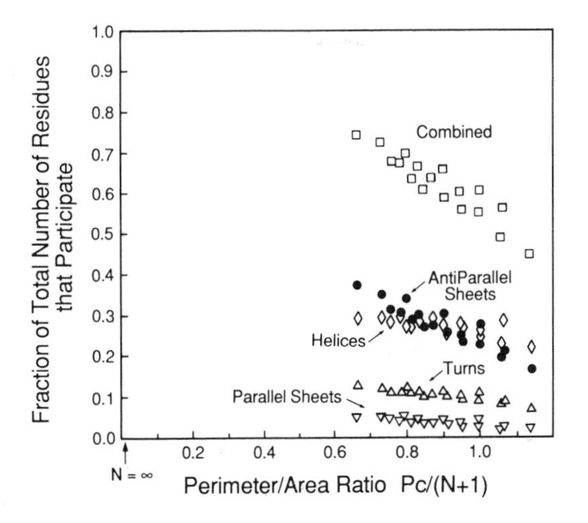

Fig. 22. Participation in secondary structure as a function of the perimeter/area ratio under the more restrictive definition. Data from chain lengths $N + 1 = 14-30, N + 1 = 36$ are included.

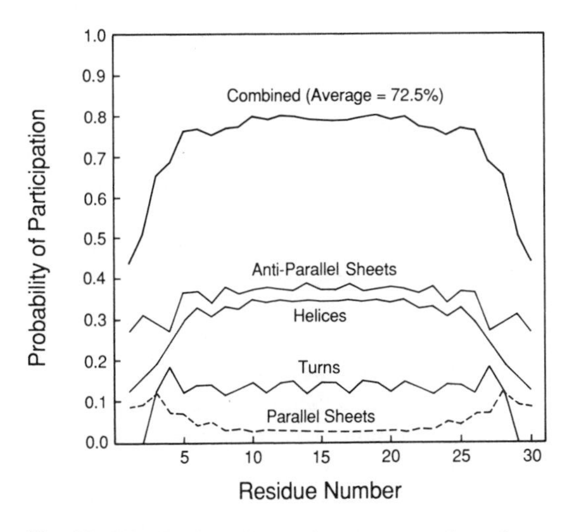

Fig. 23. Distribution of secondary structure along the sequence of compact chains with 30 residues, according to the more restrictive definition.

Hence our principal conclusion of packing enhancement of secondary structure is robust. It is valid under two different definitions of secondary structures in two dimensions and is valid for chains configured in both two and three dimensions. It appears that the basic principle that emerges here is that locally compact and nonarticulated substructures such as the secondary structures are favored in compact chains simply because they least obstruct the rest of the chain to configure into a compact conformation.

Acknowledgments

We thank K. F. Lau, J. Naghizadeh, and D. Yee for helpful discussions. We thank the NIH, the URI Program of DARPA, and the Pew Scholars Program in the Biomedical Sciences for support of this work.

References and Notes

1. Orr, W. J. C., *Trans. Faraday Soc.* **43,** 12 (1947).
2. Dill, K. A., *Biochemistry* **24,** 1501 (1985).
3. Lau, K. F., and K. A. Dill, *Macromolecules* **22,** 3986 (1989).
4. Chan, H. S., and K. A. Dill, *J. Chem. Phys.* **90,** 492 (1989).
5. Chan, H. S., and K. A. Dill, *J. Chem. Phys.* **92,** 3118 (1990); Chan, H. S., and K. A. Dill, *Proc. Natl. Acad. Sci. U.S.A.* **87,** 6388 (1990).
6. It can be shown by explicit construction that in a two-dimensional square lattice, a contact (dot on the contact map) *cannot* take part in more than one type of secondary structures.
7. Chou, P. Y., and G. D. Fasman, *Biochemistry* **13,** 211, 222 (1974).
8. Kabsch, W., and C. Sander, *Biopolymers* **22,** 2577 (1983).
9. Kabsch, W., and C. Sander, *Proc. Natl. Acad. Sci. U.S.A.* **81,** 1075 (1984).
10. Pullman, B., and A. Pullman, *Adv. Protein Chem.* **28,** 347 (1974).
11. Barber, M. N., and B. W. Ninham, *Random and Restricted Walks; Theory and Applications* (Gordon & Breach, New York, 1970) and references therein.
12. Domb, C., *Advan. Phys.* **9,** 149 (1960).
13. Sykes, M. F., *J. Math. Phys.* **2,** 52 (1961).
14. Domb, C., and M. F. Sykes, *J. Math. Phys.* **2,** 63 (1961).
15. Domb, C., *J. Chem. Phys.* **38,** 2957 (1963).
16. deGennes, P. G., *Scaling Concepts in Polymer Physics* (Cornell University Press, Ithaca, New York, 1979).
17. Edwards, S. F., *Proc. Phys. Soc. London* **85,** 613 (1965).
18. Freed, K. F., *Renormalization Group Theory of Macromolecules* (Wiley, New York, 1987) and references therein.
19. Guttmann, A. J., *J. Phys.* **A17,** 455 (1984).
20. Rapaport, D. C., *J. Phys.* **A18,** L39; L201 (1985).
21. Fisher, M. E., and B. J. Hiley, *J. Chem. Phys.* **34,** 1253 (1961). In this reference, Fisher and Hiley gave $t_{max} = N(z-2)/2$, which is an approxima-

tion; the full effect of excluded volume has not been taken into account. The deviation of their approximation to the exact value given by Eq. 3*b* of the text is of $O(\sqrt{N})$.

22. Ishinabe, T., and Y. Chikahisa, *J. Chem. Phys.* **85,** 1009 (1986).
23. The number of sites n in Orr's notation is equivalent to $N + 1$ in our notation. He counted rigid rotations and reflections of conformations as distinct but considered two ends of the chain as identical. Therefore, his $G(0)$ equals our $2[2\Omega^{(0)} - 1]$, and the coefficient of η^t in his $G(\eta)$ equals our $4\Omega^{(t)}$ for $t > 0$. The n of Fisher and Hiley and of Ishinabe and Chikahisa is equivalent to our N. In their convention, rigid rotations and reflections of conformations are counted as distinct, and two ends of the chain are also distinguished. Hence, their $c_n(0)$ equals our $4[2\Omega^{(0)}(N) - 1]$ for $n = N$.
24. Fisher and Hiley *(21)* gave $c_{14}(0) = 396\,204$, but our result for $\Omega^{(0)}(14)$ is 49 522, according to the translation rules given in the last footnote, our result implies that $c_{14}(0)$ should be 396 172.
25. Richards, F. M., *Ann. Rev. Biophys. Bioeng.* **6,** 151 (1977).
26. Hiley, B. J., and M. F. Sykes, *J. Chem. Phys.* **34,** 1531 (1961).
27. Guttmann, A. J., B. W. Ninham, C. J. Thompson, *Phys. Rev.* **172,** 554 (1968).
28. Flory, P. J., *J. Chem. Phys.* **10,** 51 (1942).
29. Huggins, M. L., *J. Phys. Chem.* **46,** 151 (1942).
30. Huggins, M. L., *Ann. N.Y. Acad. Sci.* **4,** 1 (1942).
31. Chang, T. S., *Proc. Roy. Soc.,* **A169,** 512 (1939).
32. Miller, A. R., *Proc. Cambridge Phil. Soc.* **38,** 109 (1942); **39,** 54 (1943).
33. Kasteleyn, P. W., *Physica* **29,** 1329 (1963).
34. Gordon, M., P. Kapadia, A. Malakis, *J. Phys.* **A9,** 751 (1976).
35. Domb, C., *Polymer* **15,** 259 (1974).
36. Lieb, E. H., *Phys. Rev. Lett.* **18,** 692, 1046 (1967).
37. Gujrati, P. D., and M. Goldstein, *J. Chem. Phys.* **74,** 2596 (1981).
38. Schmalz, T. G., G. E. Hite, D. J. Klein, *J. Phys.* **A17,** 445 (1984).
39. Our compact shapes constitute a subset of *compact lattice animals* commonly employed in lattice statistical mechanics. See, for example, Privman, V., and G. Forgacs, *J. Phys.* **A20,** L543 (1987).
40. Malakis, A., *Physica* **84A,** 256 (1976).
41. Jacobson, H., and W. H. Stockmayer, *J. Chem. Phys.* **18,** 1600 (1950).
42. Flory, P. J., and J. A. Semlyen, *J. Amer. Chem. Soc.* **88,** 3209 (1966).
43. Martin, J. L., M. F. Sykes, F. T. Hioe, *J. Chem. Phys.* **46,** 3478 (1967).
44. Thornton, J., *Nature* **335,** 10 (1988).
45. Goto, Y., and A. L. Fink, *Biochemistry* **28,** 945 (1989).
46. Dill, K. A., D. O. V. Alonso, K. Hutchinson, *Biochemistry* **28,** 5439 (1989).
47. We are grateful to S. Gutin and A. Badretdinov for bringing to our attention an example in which this overlap is still possible under the alternate definition of secondary structures.
48. The greater freedom of definition of lattice secondary structures on the cubic lattice is due to higher spatial dimension and the increased coordination number ($z = 6$ instead of $z = 4$ for the square lattice). Although further increase in coordination number should lead to increased accuracy in representation of secondary structures, and could therefore affect somewhat the predicted amounts of secondary structure, nevertheless the amount of secondary structure should not become diminishingly smaller with increased coordination number since there is a minimum resolution, or fraction of conformational space, required for its definition. Inasmuch as z for proteins has been estimated to be approximately 3.8 [see *(2)*], then the present model should not grossly misrepresent that fraction.

Part II

Relation of Amino Acid Sequence to Structure and Folding

Lila M. Gierasch

Several approaches have been fruitful in providing insights into the relationship between amino acid sequence and protein structure and folding. In this section, five different chapters describe methods as diverse as conformational analysis of highly constrained disulfide-stabilized peptides; statistical analysis of molecular evolution; two-dimensional (2-D) and 3-D nuclear magnetic resonance; and x-ray structural analysis of a large, multienzyme complex. Each can potentially shed light on the sequence/structure question.

Hardies and Garvin (chapter 5) examine sequences present in homologous genes to exploit molecular evolution in order to learn about the protein folding code. Their thesis is that sites that diverge most rapidly provide clues about "temporal clustering" of residue changes and possible "topological clustering" arising from structural interactions among residues. Several bacterial dihydrofolate reductase genes have been sequenced and compared to assess the importance of topological clustering. The authors conclude that clusters of four to six residues are supported by the comparative analysis, and now they are attempting to relate cluster data to structural results. Similar cluster analyses strategies can be used in saturation mutagenesis studies like those of Lim and Sauer *(1)*. Thus, the authors' analysis serves as a useful conceptual bridge between directed mutagenesis and molecular evolution.

The work by Wemmer and co-workers (chapter 6) compares conformations of several related peptides designed to include features of the bee venom peptide apamin and of the S-peptide from ribonuclease A. Nuclear magnetic resonance (NMR) shows these contributing domains to fold very similarly to native apamin, which emerges as a conformationally potent unit to nucleate secondary structure in an adjacent sequence. A further use of these hybrid peptides will be to pose questions about structure/function relationships for sequences of biological interest.

Bovine pancreatic trypsin inhibitor (BPTI) has been extensively used in folding studies because it is a structurally well-defined, stable, small globular protein. Its three disulfides were exploited by Creighton to isolate partially oxidized folding intermediates in his seminal work in the early 1970s *(2)*. As described by Kuntz and co-workers in chapter 7 of this section, it is now possible to involve site-specific mutagenesis in combination with high resolution NMR to dissect the detailed aspects of residue contributions to the structure of BPTI and its folding intermediates. They report the 3-D structure of a mutant lacking one disulfide (the 5–55 bridge), the Ala5, Ala55 mutant. They conclude, by combining their NMR data with a distance geometry approach, that most of the differences between the structure of the Ala5, 55 mutant and that of wild-type are *local.*

The next chapter in this section builds on the beautiful new structural data from x-ray diffraction of crystals of the $\alpha_2\beta_2$ multienzyme complex, tryptophan synthase. The x-ray structure of this molecule clarifies and confirms many previous biochemical and genetic studies. One of the most striking features is a "tunnel" from the active site of one subunit to that of another, which catalyzes the subsequent step. This structural feature provides a rationale for the higher-than-additive catalytic efficiency of the multisubunit complex, since the substrate is not allowed to diffuse away from the enzyme. Insights into its folding process are also provided by the structure. Miles elaborates these complementary results, while offering a model for how an apparently single-domain subunit with an alternating α, β TIM barrel fold can show clear evidence of two separately folding domains. Not only is the power of complete structural elucidation well-illustrated by the tryptophan synthase case, but also the need for a multipronged approach to understanding folding and function.

In the last chapter in this section, Gronenborn and Clore provide a lucid and timely overview of 2-D and 3-D NMR methods of protein structure determination. They describe examples of structure determinations of hirudin, a 65-amino acid anticoagulant protein produced in the leech, and of the 40 amino acid, cellulose binding domain of the large enzyme cellobiohydrolase. Clearly, the promise of NMR for protein structure determination is being realized.

References

1. Lim, W. A., and R. T. Sauer, *Nature* (London) **339**, 31 (1989).
2. Creighton, T. E., *Prog. Biophys. Molec. Biol.* **33**, 231 (1978).

5

Can Molecular Evolution Provide Clues to the Folding Code?

Stephen C. Hardies, Lorrie D. Garvin

The development of the polymerase chain reaction *(1)* has made it feasible to acquire sequences of homologous genes at a greatly accelerated rate *(2)*. Therefore, to the extent that suitable methods of analysis are available, it will be increasingly possible to address protein-structure questions by comparative methods. Most commonly, comparative analysis concentrates on those residues that are invariant or at least heavily conserved. Invariant residues yield information about those elements of the sequence that are so fundamental to the structure or function of the protein that they essentially can never change. It is necessarily true that the number of invariant residues can only diminish as additional homologous sequences are collected. Therefore, although analysis of invariant residues may help focus attention on key features of the sequence, it makes minimal use of the capacity to collect additional data.

We would like to concentrate on those residues that diverge most readily, because this approach makes maximum use of the capacity to collect additional sequences. Also, those residues that diverge most readily are found by comparison of closely related sequences, which are the easiest to obtain by hybridization techniques. Since the exact identities of rapidly evolving residues are obviously not crucial to the grand design of the protein, one can reasonably wonder if useful information can be obtained from them. Actually, the residues that change most readily have the greatest impact on molecular evolutionary theory because they determine the rate of divergence. A goal of this chapter is to find connections between aspects of evolutionary theory and issues in protein structure. We will argue that it is the detailed packing interactions among side chains that shape the patterns of replacements among the most rapidly diverging residues.

The considerations in this chapter were inspired by comparative observations on the sequences of bacterial genes for dihydrofolate reductase (DHFR). We have sequenced the genes for DHFR from *Citrobacter freundii* and *Enterobacter aerogenes*, two bacterial species that are closely related to *Escherichia coli*. Our analysis of these sequences *(3)*, in comparison with the known *E. coli* DHFR sequence and three-dimensional structure *(4)*, indicated that the most recent residue replacements in these proteins were both temporally and topologically clustered. Temporal clustering is related to variance in the rate of the molecular clock *(5–7)*, which we will discuss later in this chapter. Topological clustering of replacements has been reported for many different proteins. Sometimes clusters of replacements simply reveal regions of inherently greater structural flexibility such as in the C-peptide of insulin *(8)*. We are more interested in regions that are usually conserved but that show transient bursts of replacements, possibly reflecting revisions of the interactions among residues *(9)*. For example, in our examination of DHFR we saw that, starting from the same ancestral sequence, two descendants accumulated a cluster of replacements, each in mutually exclusive areas, whereas a third descendant remained the same as the ancestor, with the possible exception of a single replacement.

At any given time, a protein sequence is under constraints such that some replacements can be tolerated, but many cannot. For two descendants to accumulate replacements in different areas requires a change in those constraints. Of course, one can always suppose that a shift of constraints reflects some change in the external environment, necessitating an adjustment in the protein's structure or function. Changes driven by an external influence are an example of positive selection, which we will consider later in the chapter. First, we will discuss an alternate concept in which the constraints on individual residues change because the indentities of neighboring residues have changed. We will use the term "context effect" to mean that the acceptability of a specified mutation at one residue depends on the identity of surrounding residues. Then, changes of the surrounding residues may cause a shift in the context effect. Topological clustering is a natural result of theories of protein evolution that include a context effect.

Context effects routinely show up in the form of intragenic suppressors in genetic experiments *(10)*. A particularly thorough documentation of the context effect has been given by saturation mutagenesis of the λ repressor *(11)*. Intragenic suppression may work either through a local correction of a structural distortion caused by the first mutation or by some more global compensation for loss of stability or function of the protein *(10)*. The former is most easily envisioned in the form of the familiar packing interactions involved in maintaining structure. This chapter concentrates on context effects due to packing interactions because they should lead to topological clustering of the kind easiest to recognize. However, the reader should bear in mind that less localized interactions could similarly affect patterns of divergence.

With this background on our perspective, we will outline those aspects of molecular evolutionary theory that bear on the most rapidly diverging residues and on the interactions among them. It should be pointed out that the molecular evolution of proteins has been a contentious field and that, for each position, conflicting opinions can be found. It is beyond the scope of this chapter to expound on the various debates. Several reviews are available for the interested reader *(5–8, 12)*.

Protein Evolution and the Covarion Model

One of the first developments in the field of molecular evolution was the postulation of the molecular clock *(13)* in which the number of replacements accumulating in a protein is roughly proportional to the time that has passed. In other words, a given protein accumulates changes at an approximately constant rate. Examination of cytochrome c sequences taken from many different organisms was particularly supportive of the molecular clock *(14)*.

If the molecular clock is based on a constant accumulation of replacements, then the empirically observed clock should run down. This is because, after accepting some replacements, new replacements should start to fall on top of old replacements. Such redundant replacements do not increase the number of observable differences, so the clock should appear to slow down. One can do a mathematical correction for this effect up to a point *(6)*, but when there are no more new positions at which acceptable replacements may occur, the observable clock must come to a stop. Some proteins, for example, superoxide dismutase, do begin to saturate at long divergence times *(15)*. However, cytochrome c resists saturation over the entire eukaryotic kingdom.

The covarion model was proposed to account for the fact that some residues are more variable than others *(16)*. It says that only 12 codons in cytochrome c are free to diverge at any one time. These were called concomitantly variable codons, or covarions. When changes occur at some of these positions, they cause (i) some of the covariable codons to be locked in so that they cannot change any more and (ii) some new codons to be introduced into the covariable set. This has the effect of continually introducing fresh sites where differences can occur, and thus retarding the onset of saturation. The covarion model, therefore, invokes repeated context shifting to keep the molecular clock going.

Now that the DNA sequences of pseudogenes are available, one can appreciate how strongly context shifting influences the evolution of cytochrome c. Pseudogenes are vestigial genes that are no longer expressed, and they evolve at a predictable rate *(17)*. Their rate of evolution reveals how long it will take, on average, for a codon to change if it is selectively free to do so. A molecular clock based on mammalian pseudogenes saturates within the divergence time encompassed by the mammals. Therefore, without context shifting, nothing even remotely like the molecular clock would be observed. If we equate the context effect with side-chain packing, then the molecular clock is a protein structural clock.

Neutral versus Selective Theory

A large portion of the development of molecular evolutionary theory involved a conflict between neutral theory and selective theory for the evolution of proteins *(12)*. Neutral theory was developed by Kimura *(18)* who was trying to understand how mutations that had no significant effect on phenotype would drift in the population. Selective theory, on the other hand, follows from Darwin's theory of evolution by natural selection and states that replacements accrue in proteins because they improve the function of the protein and, hence, the fitness of the species. Neutral theory predicts that neutral changes to a protein's sequence will become fixed, meaning to spread to all members of the population, in a clock-like fashion. In contrast, changes favored by positive selection are expected to fix rapidly and at varying rates, depending on environmental conditions, the size of the population, and other factors. Naturally, the discovery of clock-like behavior in the evolution of proteins was cited as support for neutral theory. This led to the prediction that most changes in proteins are fixed by random drift, not by positive selection.

The idea that most residues in proteins were chosen by random drift was opposed by many scientists in favor of selective theory. Most residues in a protein seem to be optimized, or at least partially optimized, to support the structure and function of that protein. The idea that the identity of most residues are chosen at random seems wildly at odds with this perspective. Therefore, these scientists preferred the idea that most residues were chosen by natural selection because they improved the function of the protein.

Most molecular evolutionists now believe that proteins are subjected to a mixture of positive selection and neutral drift *(7)*, although there is still a considerable difference of opinion as to how much of each is involved. The context effect provides a way to integrate neutral drift and positive selection so that the best part of each is retained. The context effect means that each residue replacement that is a candidate to improve the protein will only be acceptable if the surrounding residues are compatible with it. If the context does not permit any improvements to occur, then random drift will ensue, causing the accumulation of changes that neither significantly improve nor detract from the function of the protein. Eventually, and by chance, the residues surrounding a candidate for positive selection will reach a combination that is compatible with that replacement. Then positive selection will rapidly fix the beneficial mutation, thus improving the function of the protein. In the process, the identity of the surrounding residues will be locked in. Since fixation by positive selection is much faster than by random drift, one can expect the protein to usually be in between selective events, waiting for random drift to produce a context compatible with the next improvement.

Repeated cycles of neutral drift coupled to positive selection presents a mode of evolution that is satisfactory from the standpoint of either neutral or selective theory. In an indirect sense, all of the replacements could be considered to have been selected for enhanced function of the protein. In this model, back-mutating the surrounding residues to their original identity after the selective event would damage the function of the protein, since they would then be incompatible with the positively selected residue. So the residues that had been fixed by random drift now appear as much optimized for function as does the residue that had been fixed by positive selection. Essentially, most of the positions in the protein can be envisioned as drifting among a set of acceptable residues until they get locked into combinations that are compatible with improved function.

On the other hand, most of the replacements will have been neutral at the time that they were fixed. Therefore, the rate of divergence will obey neutral theory and a molecular clock will be observed. Since those mutations that are truly fixed by positive selection have to wait on the neutrally fixed mutations, they too will be constrained to follow clock-like behavior. Therefore, a significant fraction of positively selected changes can be incorporated without disrupting the molecular clock.

Variance of the Molecular Clock

An important characteristic of the molecular clock is the degree of precision with which it runs. The clock is not expected to be perfectly uniform, since it is based on randomly occurring events. The minimum theoretical variance from uniformity in rate is given by the Poisson distribution and equals the number of differences counted between the sequences being compared *(5, 7)* with the standard deviation being the square root of the variance. Note that the application of this statistic implies that the replacements are independent events. We will return to this point below. The observed variance in the molecular clock actually runs about twice the theoretical minimum *(5, 7)*. This point has often been used to discredit neutral theory, arguing that the molecular clock must be affected by extrinsic selective pressures and, therefore, must be governed by selective processes rather than random drift.

A recent development in the theory of the molecular clock is its formulation as a doubly Poisson process *(19)* in which replacements are randomly chosen from a set of potential replacements which, itself, varies randomly in size. The doubly Poisson model fits the observed clock and has been taken to indicate selective processes at work *(20)*. However, it has also been reconciled with neutral theory *(21)*. The latter invokes a newer version of the covarion model, called "neutral space" *(22)* and includes the concept of a gradual degradation of the sequence by fixation of slightly deleterious replacements, interspersed with occasional improvements. All of these treatments incorporate the concept of interdependence of replacements.

If we acknowledge that the fixation of individual replacements are not independent of each other, this necessarily broadens the variance of the molecular clock. Therefore, any theory that incorporates context effects should be a better fit to the observed clock. Quantitatively, the variance of the molecular clocks indicates about half as many independent events as individual replacements, so we should think in terms of the average replacement being dependent on the prior occurrence of one other replacement.

The Behavior of Protein Sequences over Short Divergence Times

Our observations on bacterial DHFR suggest topological clusters of four to six residues. Therefore, we would like to think in terms of clusters of residues this size with interrelationships such that about half of the replacements are dependent on prior occurrence of the other half. This effort to break the evolution of the protein into quantized clusters is mainly a device to help us map the evolutionary process onto the three-dimensional structure of the protein. It is not intended to be a commentary on whether the process is episodic or nonepisodic *(7, 19–22)*. Even if the clusters meld one into the other in a continuous fashion, we still think that breaking the sum of the replacements into individual topological clusters will be a useful analytical procedure.

We speculate that topological clusters could develop whether or not selective pressures are at work. In either case, we would expect that some of the last replacements to take place in the cluster were dependent on the prior occurrence of some of the first replacements in the cluster. The later replacements might be positively selected, or they might be neutral replacements that, similarly, would be incompatible with back mutation of the first replacements in the series. A protein might then experience different proportions of each kind of cluster, depending on ambient external selective pressures. In this way, a protein evolving through periods with an extensive component of positive selection and other periods of pure neutral drift would not necessarily experience much change in its clock-like behavior. This leads to an important consequence of this model. The appearance of clusters is attributed to interactions in the internal environment of the protein, not to its interactions with the external environment. Therefore, we can hope to recognize clusters of replacements in diverging sequences and relate them to the structure of the protein without knowing how heavily positive selection is involved.

Ohta's model of molecular evolution *(21)* considers compensating interactions among residues to alleviate the accumulation of slightly deleterious but effectively neutral replacements. In the perspective that we have offered, the compensatory replacements occur first, so that the potential for deleterious effects of the later replacements are avoided. These two views are not contradictory because the former refers to the selective cost of the replacements, whereas the latter refers to the cost in terms of structural stability. We suspect that functional improvement usually comes at some cost to structural stability and vice versa. So the same series of replacements that might be slightly functionally deleterious could be building up a buffer of excess structural stability which could later finance a functional improvement.

Finally, we should consider the circumstances under which the context effect would be observable as a pattern of differences between two sequences. Firstly, the interaction among side chains must relate to some topological pattern. Packing is the obvious example, in that the interacting residues must be close together. However, other interactions such as conservation of surface charge or dipole might give rise to interpretable patterns. Secondly, there must be no more than one or two clusters under development at a time, or else the individual patterns will be obscured. Thirdly, the sequences must be separated by short divergence times to prevent sequential clusters from obscuring each other.

From Theory to Experiment

We have attempted to describe how molecular evolutionary theory may deal with the divergence patterns among closely related proteins. The theory described springs from the covarion model and various more recent considerations. What we have attempted to do is to convert abstract issues of rates and selective pressure into patterns of replacements mapped onto the three-dimensional structure of a protein. In our view, this fulfills a key requirement for making tangible application of evolutionary theory to issues of protein structure, namely, to associate specific residues with the hypothesis under consideration.

We do not know how consistently topological clustering will appear in different proteins. There could be consistency based on the commonality of packing forces dedicated to the common goal of producing stable structure. Alternatively, the individualized requirements of different proteins to carry out their respective functions may cause highly variable context effects. Also, different proteins vary widely in their overall rate of evolution, which might logically be reflected in the numbers of clusters undergoing development at a time.

It is our hope that patterns of replacements in closely related proteins will yield useful insights into protein structure and function. Admittedly, because of the random component in the data, it would be difficult to draw definitive conclusions based purely on sequence observations. However, sequence observations might prove valuable as a generator of specific hypotheses for subsequent testing by in vitro mutagenesis. For example, one might construct the double mutants illustrated in Fig. 1 either by saturation mutagenesis or by a directed site-specific mutagenesis of positions found within a topological cluster in the natural series.

Each rectangle in Fig. 1 represents two successive, interdependent amino acid replacements and the effect of each on the function of the protein. Going around the upper right corner of each rectangle, there is first a loss of function and, subsequently, a compensation by the second mutation. This is how a geneticist would isolate intragenic suppressors. The upper right-hand corner is forbidden to natural selection, so the required order during evolution is to go through the lower left-hand corner.

Comparison of the activities of the single mutants with the double mutants reveals information about the nature of the interaction between the two residues *(23)*. The symmetric rectangle A is the pattern of a global intragenic suppressor, while rectangle B is an idealized representation of an allele-specific suppressor. Whereas intragenic suppressors isolated by genetic methods may be of either type *(10)*, picking pairs of replacements from topologically clustered sets should tend to enrich for the more localized interactions characteristic of rectangle B. It is interesting to speculate whether evolution can traverse the lower left-hand corner of rectangle A. If Ab represents a truly superior protein, then natural selection may trap the gene in that configuration. Alternatively, if Ab represents an improvement in one aspect of the function at the cost of another, then both corners of rectangle A may be forbidden to evolution.

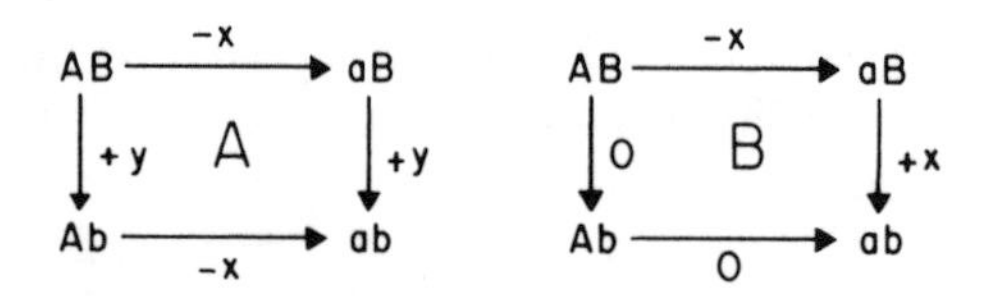

Fig. 1. Contrast in behavior between (A) a global and (B) an allele-specific intragenic suppressor.

The prospect that hidden functional aspects may have changed is a pitfall for any analysis of mutants. Whether sampling replacements from the historical series will help or hinder in this regard is a matter of speculation. If one considers formation of topological clusters to be driven by episodes of positive selection, then analysis of these replacements may indeed become mired in a web of subtle functional differences. Alternatively, if one considers natural selection to be more an exercise in staying even than in getting ahead, then mutations from the historical series may be better behaved than those isolated in the laboratory. In this perspective, evolution would have traversed rectangle B such that AB, Ab, and ab are held within acceptable limits for all aspects of the protein's function, including those as yet unknown. Then the mutation and intragenic suppressor, AB → aB → ab, might have the useful property that the suppressor restores all aspects of function to the original state. Thus, using replacements from the evolutionary series might protect the analysis from confusion with subtle or hidden functional deviations.

Any attempt to relate evolution to a mutagenesis experiment must be tempered by the realization that natural selection may be fussier than biochemists when it comes to judging mutations. The covarion model received one of its first experimental tests when chimeric cytochrome c molecules were constructed from peptides that originated from divergent parents *(24)*. In contrast to the prediction of the covarion model that such molecules would suffer from incompatibilities between the two parts, the chimeras were found to be functional. However, artificially generated mutants are seldom characterized at the same level of stringency that competitive growth in the wild must impose. So the question arises as to whether context effects will become less important among mutants where there is a less stringent criterion for tolerance of replacements.

Lim and Sauer *(11)* have commented on the decline in the importance of the context effect under less stringent selection. Their saturation mutagenesis of the hydrophobic core of λ repressor revealed those combinations of replacements that were consistent with continued function. Some replacements were found only in combination with others, in line with a context effect. However, when the requirement for function was reduced, the dependency on neighboring residues was also reduced. So this constitutes at least anecdotal evidence that the context effect becomes less important as the requirement for function is reduced. On the other hand, as protein engineers put more and more replacements into the same sequence, subtle side effects will become more important. We anticipate that genetic engineers will have to learn how to include compensatory replacements to offset the accumulation of minor functional losses in extensively engineered proteins.

Acknowledgment

Supported by grant AQ1107 from the Robert A. Welch Foundation and assisted by NIH grant HG00190.

References and Notes

1. Saike, R. K., *et al., Science* **230**, 1350 (1985).
2. Kocher, T. D., *et al., Proc. Natl. Acad. Sci. U.S.A.* **86**, 6196 (1989); for a general review see White, T. J., N. Arnheim, H. A. Erlich, *Trends in Genetics* **5**, 185 (1989).
3. Garvin, L. D., and S. C. Hardies, *Mol. Biol. and Evol.,* in press.
4. Smith, D. R., and J. M. Calvo, *Nucl. Acids Res.* **8**, 2255 (1980); Bolin, J. T., D. J. Filman, D. A. Matthews, R. C. Hamlin, J. Kraut, *J. Biol. Chem.* **257**, 13650 (1982).
5. Wu, T. T., W. M. Fitch, E. Margoliash, *Ann. Rev. Biochem.* **43**, 539 (1974).
6. Wilson, A. C., S. S. Carlson, T. J. White, *Ann. Rev. Biochem.* **46**, 573 (1977).
7. Zuckerkandl, E., *J. Mol. Evol.* **26**, 34 (1987).
8. Brown, H., F. Sanger, R. Kitai, *Biochem. J.* **60**, 556 (1955).
9. Topological clustering, potentially reflecting residue interactions, has been reported for ribonuclease: Wyckoff, in the discussion section of *Brookhaven Symp. Biol.* **21**, 252 (1968); cytochrome c: Margoliash, E., W. M. Fitch, E. Markowitz, R. E. Dickerson, in *Structure and Function of Oxidation-Reduction Enzymes*, A. Aakeson and A. Ehrenberg, Eds. (Pergamon Press, New York, 1972), pp. 5–17; myoglobin: Castillo, O., H. Lehmann, L. T. Jones, *Biochim. Biophys. Acta 491*, 23 (1977); and in depth for snake venom toxins: Breckenridge, R., and M. J. Dufton, *J. Mol. Evol.* **26**, 274 (1987).
10. Yanofsky, C., V. Horn, D. Thorpe, *Science* **146**, 1593 (1964); Hecht, M. H., and R. T. Sauer, *J. Mol. Biol.* **186**, 53 (1985); Hampsey, D., M., G. Das, F. Sherman, *J. Biol. Chem.* **261**, 3259 (1986); Klig, L. S., D. L. Oxender, C. Yanofsky, *Genetics* **120**, 651 (1988); Nagata, S., C. C. Hyde, E. W. Miles, *J. Biol. Chem.* **264**, 6288 (1989); Minor, P. D., *et al., J. Gen. Virol.* **70**, 1117 (1989).
11. Lim, W. A., and R. T. Sauer, *Nature (London)* **339**, 31 (1989).
12. King, J. L., and T. H. Jukes, *Science* **164**, 788 (1969); with rebuttal by Clark, B., *Science* **168**, 1009 (1970); Milkman, R., *Trends in Biochem. Sci.* **1**, N152 (1976); with rebuttal by Kimura, M., *ibid.*; Goodman, M., *Nature (London)* **295**, 630 (1982); with rebuttal by Li, W.-H., T. Gojobori, M. Nei, *ibid.*
13. Zuckerkandl, E., and L. Pauling, in *Evolving Genes and Proteins*, V. Bryson and H. J. Vogel, Eds. (Academic Press, New York, 1965), pp. 97–166.
14. Fitch, W. M. and E. Margoliash, *Science* **155**, 279 (1966).
15. Ayala, F. J., *J. Hered.* **77**, 226 (1986); and see comments by Kimura, M., *J. Mol. Evol.* **26**, 24 (1987).
16. Fitch, W. M., and E. Markowitz, *Biochem. Genetics* **4**, 579 (1970); Fitch, W. M., *J. Mol. Evol.* **1**, 84 (1971); Fitch, W. M., *J. Mol. Evol.* **8**, 13 (1976).
17. Li, W.-H., T. Gojobori, M. Nei, *Nature (London)* **292**, 237 (1981).
18. Kimura, M., *Nature* **217**, 624 (1968).
19. Gillespie, J. H., *Proc. Natl. Acad. Sci. U.S.A.* **81**, 8009 (1984).
20. Gillespie, J. H., *Mol. Biol. Evol.* **3**, 138 (1986).
21. Ohta, T., *J. Mol. Evol.* **26**, 1 (1987).

22. Takahata, N., *Genetics* **116**, 169 (1987).
23. Carter, P. J., G. Winter, A. J. Wilkinson, A. R. Fersht, *Cell* **38**, 835 (1984).
24. Wallace, C. J. A., G. Corradin, F. Marchiori, G. Borin, *Biopolymers* **25**, 2121 (1986).

6

Folding and Activity of Hybrid Sequence, Disulfide-Stabilized Peptides[1]

Joseph H. B. Pease, Richard W. Storrs, David E. Wemmer

Introduction

It is well known that in disulfide cross-linked proteins the relative positions of cysteines are strongly conserved. Among families of proteins containing similar disulfides, there tends to be a strong conservation of the three-dimensional structure, even when the overall sequence homology is rather low *(1)*. Thus, it is clear that a wide variety of amino acid sequences can be accomodated within the fold defined by the cystine bridges. We have taken advantage of this adaptability to form structured, hybrid sequence peptides. In these hybrids, the folding is largely controlled by formation of disulfides, involving cysteines with relative positions taken from a natural peptide. However, many of the noncysteine residues can be replaced with a sequence from another protein.

In the cases discussed here, the disulfides were derived from apamin *(2–4)*, and the noncysteine amino acids from the helical region of apamin were replaced with those from the S-peptide from ribonuclease-A (RNase-A) *(5)*. The S-peptide, consisting of the first 20 amino acids from RNase-A, has been extensively characterized and has been shown to have a helix-forming propensity *(6–8)* (existing as up to ca. 40% helix under optimal conditions). In addition, there was shown to be a helix stop signal present *(9, 10)* near the 13th or 14th residue. This signal persists even in trifluoroethanol (TFE) where the overall helical fraction is significantly increased *(11, 12)*. The S-peptide binds (with an association constant of $\sim 10^6$) to S-protein, the large fragment of RNase-A from which it was cleaved, thereby reactivating nuclease activity *(3)*. In the apamin/S-peptide hybrids, we show that the disulfides stabilize the helical segment at the beginning of the S-peptide completely, giving us the opportunity to investigate the propagation of the helix after essentially perfect nucleation.

Using two-dimensional nuclear magnetic resonance (2-D NMR) *(13)*, we show that the first 16 residues of the hybrid fold into a structure essentially identical to that of the parent apamin and that the helix in the hybrid containing the full length S-peptide begins to be disrupted at the same position as in the isolated S-peptide, although there is maintenance of some helical character for four or five residues. This indicates that the helix stop signal in this essentially isolated helix is not highly localized but rather is spread over several residues. We also show that the full hybrid peptide is immunogenic and elicits antibodies that cross react with intact RNase-A.

1 Reprinted from *Proc. Natl. Acad. Sci. U.S.A.* **87**, 5643 (1990).

Materials and Methods

The hybrid peptides were synthesized on an Applied Biosystems 430A synthesizer using t-boc chemistry. Hydrogen fluoride cleavage was performed by Applied Biosystems, yielding 1.25 g of crude material (not all peptide). Then 180 mg of the crude material was dissolved in 100 ml of 5 mM Tris, pH 8.0, and a molar excess of dithiothreitol. The peptide was allowed to air oxidize at room temperature for one week. The solution was reduced to 10 ml by lyophilization. High performance liquid chromatography (HPLC) was done on a Waters HPLC equipped with a Waters Delta Pak preparative column (C18-300 Å, 19 mm × 30 cm). A two buffer system was used: buffer A, 0.1% TFA/H_2O, and buffer B, 0.1% TFA/60% CH_3CN/40% H_2O. The run started at flow rate of 4 ml/min, with 100% A for five minutes, then a linear gradient to 30% B over 10 minutes, and finally a linear gradient to 70% B over 80 minutes. The hybrid-II peptide eluted at 58% B and was the major peak at 235 nm. Approximately 60 mg of purified peptide was obtained from the 180 mg of crude starting material. Hybrid I was synthesized and purified in the same way, but the overall yield was significantly lower. The correct identity of the peptides was confirmed by fast atom bombardment (FAB) mass spectroscopy as well as by NMR spectroscopy.

NMR measurements were carried out on a GN-500 spectrometer. Samples were dissolved in 100 mM NaCl, pH = 2.5, 2-D and spectra were run at 20°C. For the nuclear Overhauser effect spectrum (NOESY) shown, a mixing time of 450 ms was used. The hybrid-I sample was ca. 5 mM, while hybrid II was ca. 10 mM.

Circular dichroism (CD) measurements were carried out on a Jasco J500 spectropolarimeter. Samples were at approximately 10 μM concentration in the same buffer used for NMR, at ambient temperature, in 0.2 cm cells.

Studies of complementation of RNase-S by the peptides were carried out by mixing 2 μl of either authentic S-peptide (200 μM or 2 μM, Sigma), a hybrid (200 μM or 2 μM), or water (control) with 2 μl of S-protein (3.6 μM, Sigma). These were mixed and allowed to stand at room temperature for 10 min to allow association of the peptide. 10 μl of RNA (18S + 23S, Boehinger Mannheim) at 80 μg/ml in 0.1 M NaOAc, pH 5, 0°C was added and quickly mixed. After a reaction time of 15–60 sec, the reaction was quenched by the addition of 50 μl of 1:1 phenol/chloroform and mixing. The water layer was then separated, RNA was precipitated with cold ethanol, resuspended in 30% glycerol, run on 1.2% agarose gels, and visualized with ethidium bromide. Initial separate bands for the 18S and 23S RNA components disappeared and were replaced by broad bands of cleavage products running faster on the gel. Controls were also run with hybrid peptide alone, S-peptide alone, and S-protein alone. No cleavage was seen for the peptides; however, a slow cleavage (relative to RNase-A or the peptide complexes) was seen with S-protein alone (probably resulting from contamination with RNase-S).

Hybrid II (sequence given in Fig. 1) was also used to induce antibodies in rabbits. Following standard methods, three rabbits were immunized with the unconjugated peptide in Freund's adjuvant at one month intervals. To assay the activity of antibodies in the serum, microtiter wells were coated with RNase, then were treated with bovine serum albumin (BSA) to block remaining sites. Controls were just treated with BSA alone. Dilutions of serum were done with buffered saline, containing Triton X-100 and BSA. Although the background is high at low dilution, there is clear reactivity to RNase-A.

Results

The sequences of apamin, S-peptide and the two hybrids of apamin, and the S-peptide are shown in Fig. 1. The numbering of residues used is based on the apamin sequence. The first, hybrid-I, consists of only the residues of the S-peptide that are in a helical conformation in RNase and, correspondingly, are par-

```
             1   3   5   7   9  11  13  15  17  19  21  23  25
             |   |   |   |   |   |   |   |   |   |   |   |   |
S-peptide                K E T A A A K F E R Q H M D S S T S A A
apamin       C N C K A P E T A L C A R R C Q Q H
hybrid-I     C N C K A P E T A A C K F E C Q H M
hybrid-II    C N C K A P E T A A C K F E C Q H M D S S T S A A
```

Fig. 1. The sequences of S-peptide, apamin, and the two hybrids used in this work are given (using standard one letter codes), with the residue numbering indicated.

tially helical in the free peptide in solution as well. The second, hybrid II, has the full-length S-peptide incorporated. When the peptides were air oxidized in dilute solutions they fold spontaneously to the apamin-like structure (conditions under which synthetic apamin also folds into its active structure *(14–16)*). CD spectra of apamin and the two hybrid peptides are essentially identical in shape (data not shown), with a maximum negative ellipticity near 208 nm and a distinct shoulder arising from the helix at 220 nm, as has been seen previously for apamin *(17)*, providing evidence that the peptides fold similarly.

The solution structures of the peptides were characterized using 2-D NMR and distance-geometry methods *(18–20)*, as had been applied to apamin *(21, 22)* and many other proteins previously *(13)*. Relevant sections of the 2-D NOESY spectrum are shown in Fig. 2, and a summary of the sequential connectivities is shown in Fig. 3 for hybrid II. The pattern of sequential connectivities for residues 1–18 of both hybrids is identical to that observed for apamin, indicating that the overall folding is very similar. In moderate salt conditions such as those used, there is no evidence for aggregation of either S-peptide or apamin. Although extensive tests were not carried out on the hybrids, we have found no evidence for their aggregation either.

Peak intensities from NOESY spectra of hybrid II were used to derive interproton distance constraints. These were used with the distance geometry program DSPACE (Hare Research) to derive structures that are consistent with the observed data for residues 1–18. As discussed below, the residues beyond 18 are dynamically disordered, and it is not meaningful to try to derive a structure with this method. The results of these calculations are shown in Fig. 4, a superposition of five of the resulting structures which have a root-mean-square (rms) variation in backbone atom positions of ca. 0.5 Å for the first 16 residues. A detailed comparison of these structures with those calculated for apamin shows that there are only very minor differences. Superposition of the backbone atoms of apamin with those of the hybrid results in rms fits of ca. 1.0 Å. This similarity is not surprising, since there are relatively few amino acid differences, and they all occur in the helical region. The contacts that were previously identified between the turn and helix in apamin are also present in the hybrid. In spite of the small size of the peptide, the disulfides provide a high stability. Monitoring chemical shifts as a function of temperature revealed no unfolding transition for the helical residues up to 80°C, although there was a gradual shift with temperature for residues beyond C15, consistent with increased fraying of the helix at the C-terminal end.

Beyond residue 18 in hybrid II, there is a gradual change in the connectivity pattern. Beginning at residue 17, a decrease in the intensity of sequential NOEs associated with the α-helical conformation is seen, along with an increase in those NOEs associated with extended conformations, particularly from each amide proton to the preceding α proton. The latter increase gradually in intensity from being not observable for residue 16 to having an intensity almost as large as fully extended residues 1, 2, and 3 at residue 21. In addition, the long range (i, i+3 and i, i+4) helical NOEs *(23)* are also lost beyond residue 18. This pattern suggests strongly that the helix becomes dynamically disordered *(24, 25)*, beginning at residue 17, with the fraction of time spent in the helical conformation decreasing gradually from residue 17 through 21. This interpretation is further strengthened by

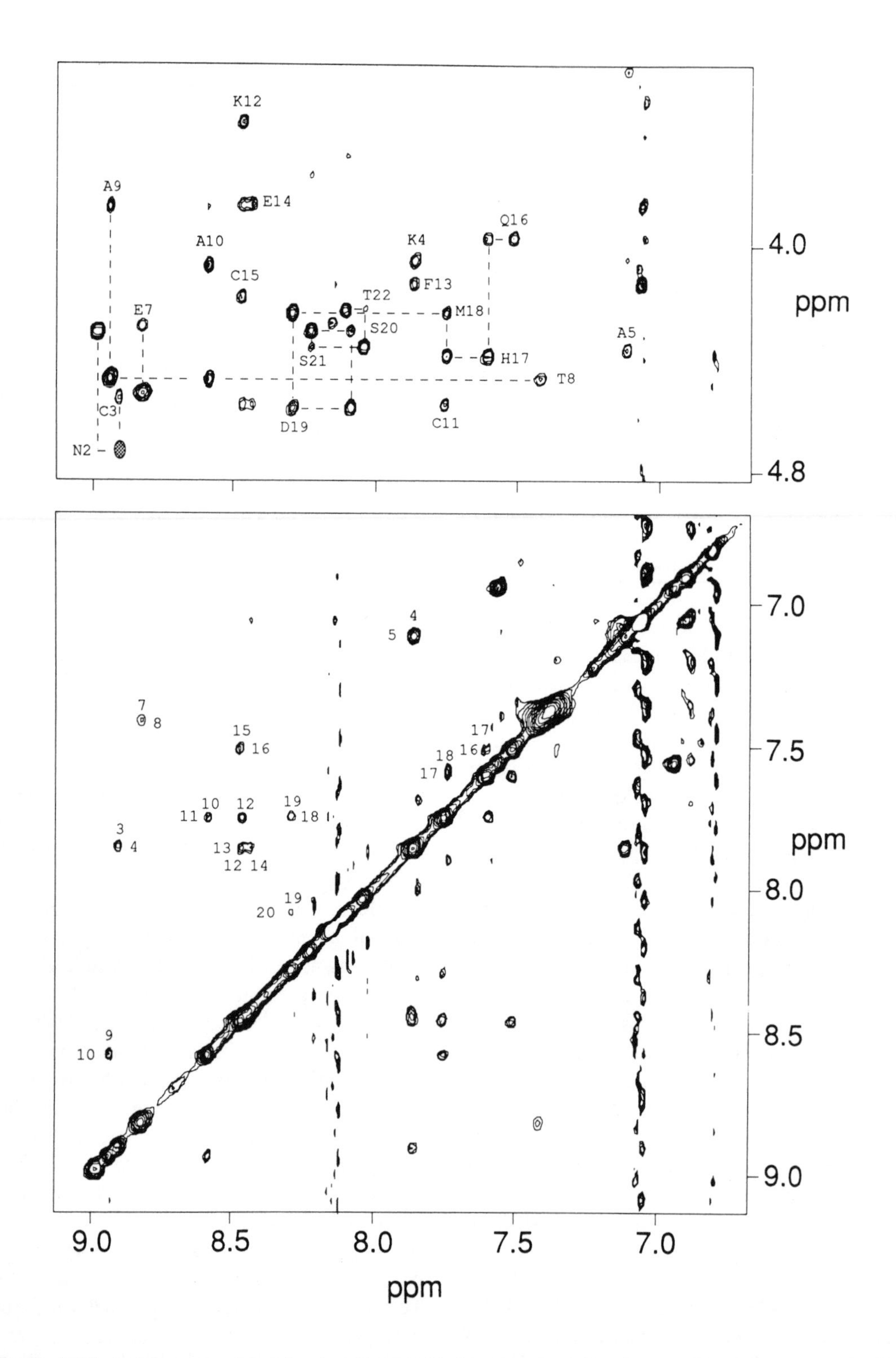

Fig. 2. Amide to alpha-proton (**upper**) and amide to amide (**lower**) cross peak regions of a NOESY spectrum of hybrid II in H_2O solution. In the upper region, peaks are labeled at the position of the intraresidue amide to α position. Dashed lines indicated amide to preceding α connectivities. In this spectrum the α-proton of residue 2 was under the solvent peak and was saturated. The position of the cross peak from C3 to N2 seen at lower temperature is indicated by a shaded oval. In the lower region, cross peaks are labeled with the two residues from which they arise.

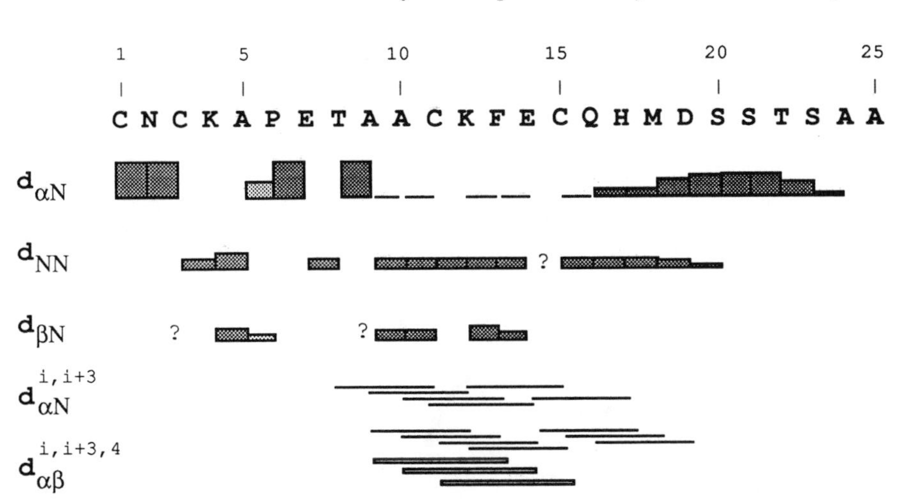

Fig. 3. Summary of sequential connectivities seen for hybrid II. The shaded bars for $d_{\alpha N}$, d_{NN} and $d_{\beta N}$ indicate observed connectivities, and the height of the bar indicates the intensity of the observed cross peak. Question marks indicate cross-peak positions that are either obscured by other peaks or too near the diagonal to be seen. For some peaks, the data in Fig. 2 were supplemented by a second NOESY run at 5°C, in which some peaks shifted sufficiently to remove degeneracies. The i, i+3 connectivities are indicated by lines that do not represent intensities. Additional *i, i*+4 cross peaks were seen between residues 9 and 13, 10 and 14, and 11 and 15.

the observation that all NOEs for residues 23–25 are greatly reduced in intensity, including those corresponding to fixed distances such as the αH to methyl on alanines 24 and 25. Since the other alanine residues have strong NOE cross peaks between these protons, the reduced intensity must come from a change in the effective correlation time for the motion of the internuclear vectors. The bulk of the peptide rotates sufficiently slowly that negative NOEs are seen in spite of the relatively low molecular weight. Just a moderately small decrease in correlation time could make these protons reach the zero crossing in NOE intensity. This observation is qualitatively similar to Rico *et al. (26)* who found a change in the sign of the NOE in this region of the isolated S-peptide and who also interpreted it as increased mobility of this segment. Our data are consistent with the C-terminal eight residues of the peptide being dynamically disordered, with a gradually decreasing correlation time for internal motion toward the C-terminus of the peptide.

These data show that the helix stop signal identified in this region is not abrupt, but rather represents a progressively lower fraction of time spent in helical conformations for successive residues over a stretch of five residues, about 1.5 helical turns. This pattern is qualitatively similar to what one would predict from the helix propagation values, *s*, derived from host-guest data, assuming that the helix is perfectly nucleated at residue 15. Using literature values *(1)*, the fraction helix that is predicted for each residue (assuming 100% helix at residue 15 and simply multiplying by the successive *s* values) is

15					20					25
C	Q	H	M	D	S	S	T	S	A	A
(1.00)	0.96	0.77	0.90	0.59	0.46	0.35	0.30	0.22	0.24	0.25

These values correlate fairly well with the observed NOE pattern, out to about residue 20. Beyond this, it is difficult to assess because of increased mobility, but the percentage helix probably drops more quickly than predicted. It is clear that the largest residue-to-residue difference in the NOEs is from M-18 to D-19, which is predicted by the helix propagation parameters. This behavior might be somewhat different for free S-peptide, since, in general, both nucleation and propagation will affect the fraction of helix at any particular residue, and the nucleation parameter, σ, is strongly dependent on amino acid.

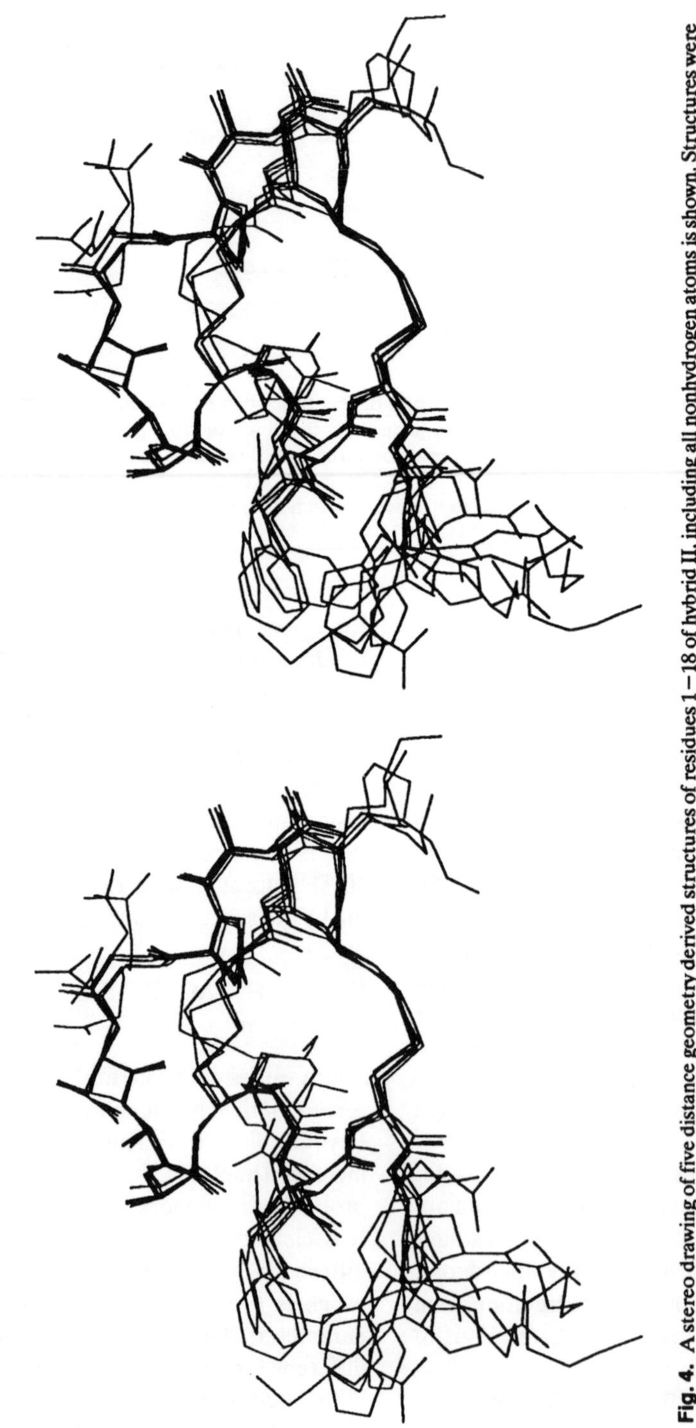

Fig. 4. A stereo drawing of five distance geometry derived structures of residues 1−18 of hybrid II, including all nonhydrogen atoms is shown. Structures were superimposed by minimizing rms deviations for backbone atom positions of residues 1−16. The calculations were carreid out as described in *(13)*. The helical segment, containing residues 9−16, is clearly visible at the left. The helix becomes disordered beyond this point, as discussed in the text.

Using the distance geometry structure of hybrid II, we carried out modeling of the complex of hybrid II with S-protein. The exposed face of the helix in this hybrid corresponds to the buried face of the S-peptide when it is bound in RNase. The extra residues, corresponding to the β turn and loop of apamin, protrude out from the surface of the enzyme. We carried out complementation studies of RNase-S with both hybrids by examining the cleavage of a mixture of 18S and 23S RNA, followed on agarose gels. We find that both peptides restored activity to a level essentially equivalent to that found with the S-peptide itself either with peptide in excess or with near stoichiometric mixtures (Fig. 5). Hybrid sequence peptides of RNase with angiogenin, which confer modified activity with RNAs, have also been made *(27)*. In this case, the intent was to transfer substrate specificity rather than to stabilize a particular secondary structure as in our case.

In two of the three rabbits immunized with hybrid II, antibodies were produced that cross react with native RNase-A using enzyme-linked immunosorbant assay (ELISA), (Fig. 6). Although these are polyclonal antibodies, the fairly good cross reaction of the antipeptide antibodies with native RNase is of interest. It suggests (though it does not prove conclusively) that the antibodies recognize the helical conformation of the hybrid peptide.

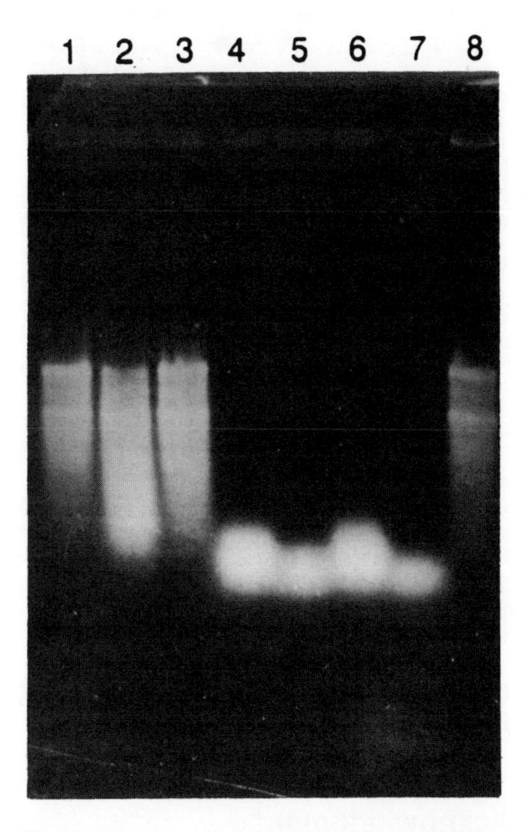

Fig. 5. Ethidium-stained agarose gel showing hydrolysis of a mixture of 18S and 23S RNA by reconstituted RNase, as described. Lane 1 (numbered left to right), RNA only; lane 2, S-protein only, 15 sec reaction; lane 3, S-protein only, 60 sec reaction; lane 4, S-protein + hybrid II, 15 sec reaction; lane 5, S-protein + hybrid II, 60 sec reaction; lane 6, S-protein + S-peptide, 15 sec reaction; lane 7, S-protein + S-peptide, 60 sec reaction; lane 8, hybrid II only, 60 sec reaction.

Discussion

These studies demonstrate clearly that the use of disulfides to stabilize a particular peptide fold does not interfere with activity, provided the contact face is appropriately exposed and that the stabilized secondary structure is the correct one. In the example discussed here, a helical conformation was required for activity. Hybrid peptides of this general construction should thus be useful for testing hypotheses about the active conformations of peptides. Similar approaches using nonpeptide covalent constraints have already been used in the design of various drugs, hormones, and other small, biologically active molecules. The use of disulfides to this end is convenient because of the ease of peptide synthesis. By studying natural disulfide-stabilized peptides, it should be possible to map a sequence of interest on a variety of different secondary structures. These hybrid peptides also provide excellent models for understanding certain aspects of protein folding such as propagation of nucleated helices, as discussed above. Since the helical segment of the peptides are very stable, these hybrids should provide a mechanism for targeting antibodies, both to a particular sequence and to a helical conformation, by

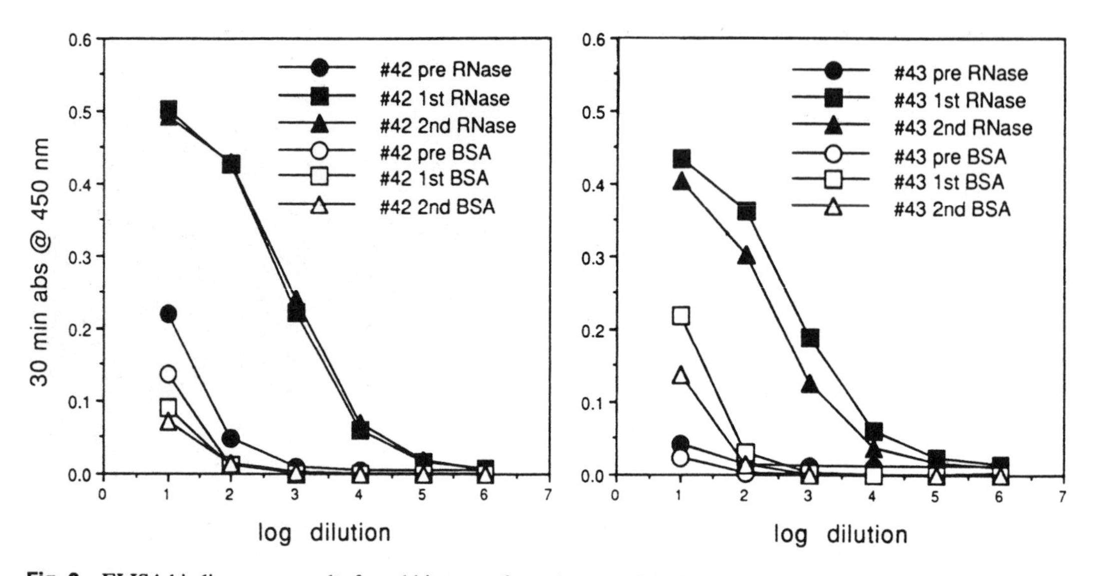

Fig. 6. ELISA binding assay results for rabbit serum from the two of the immunized rabbits. "Pre" indicates reaction from serum before the first immunization; "1st" indicates after the first boost; "2nd" after the second boost. The curves marked BSA arise from BSA treatment of the wells rather than RNase-A.

transferring a protein segment into this hybrid framework.

Acknowledgment

We would like to thank Mr. David Koh for help with peptide synthesis, Ms. Ann Caviani Pease for help with the RNase activity assays, Prof. I. Tinoco, Jr. for access to his CD instrument, Prof. Peter Kim for producing the antibodies and helpful discussions, and the National Science Foundation for financial support through grant DMB 8815998. We would also like to acknowledge equipment grants from the Department of Energy (DE F605 86ER75281) and NSF (DMB 8609035).

References

1. Creighton, T. E., *Proteins, Structures and Molecular Properties* (Freeman, New York, 1984).
2. Habermann, E., and K. G., Reiz, *Biochem. Z.* **343,** 192 (1965).
3. Callewaert, G.L., R. Shipolini, C. A. Vernon, *FEBS Lett.* **1,** 11 (1968).
4. Shipolini, R., A. F. Bradbury, G. L. Callewaert, C. A. Vernon, *Chem. Commun.* **679** (1967).
5. Richards, F. M., and P. J. Vithayathil, *J. Biol. Chem.* **234,** 1459 (1959).
6. Brown, J. E., and W. A. Klee, *Biochemistry* **10,** 470 (1971).
7. Bierzynski, A., P. S. Kim, R. L. Baldwin, *Proc. Natl. Acad. Sci. U.S.A.* **79,** 2470 (1982).
8. Kim, P.S., A. Bierzynski, R. L. Baldwin, *J. Molec. Biol.* **162,** 187 (1982).
9. Rico, M., *et al., FEBS Lett.* **162,** 314 (1983).
10. Kim, P. S., and R. L. Baldwin, *Nature* **307,** 329 (1984).
11. Nelson, J. W., and N. R. Kallenbach, *Proteins* **1,** 211, (1987).
12. Nelson, J. W., and N. R. Kallenbach, *Biochemistry* **28,** 5256 (1989).
13. Wuethrich, K., *NMR of Proteins and Nucleic Acids* (John Wiley, New York, 1986).
14. Granier, C., E. Pedroso Muller, J. van Reitschoten, *J. Eur. J. Biochem.* **82,** 293 (1978).
15. Cosand, W. L., and R. B. Merrifield, *Proc. Natl. Acad. Sci. U.S.A.* **74,** 2771 (1977).
16. van Reitschoten, J., C. Granier, H. Rochat, S. Lissitzky, F. Miranda, *Eur. J. Biochem.* **56,** 35 (1975).
17. Hider, R. C., and U. Ragnarsson, *Biochim. Biophys. Acta* **667,** 297 (1981).
18. Havel, T., I. D. Kuntz, G. M. Crippen, *Bull. Math. Biol.* **45,** 665 (1983).
19. Havel, T., and K. Wuethrich, *Bull. Math. Biol.* **46,** 673, (1984).
20. Wagner, G., *et al., J. Molec. Biol.* **196,** 611 (1987).
21. Wemmer, D., and N. R. Kallenbach, *Biochem.* **22,** 1901 (1983).

22. Pease, J. H. B., and D. E. Wemmer, *Biochem.* **27,** 8491 (1988).
23. Wuethrich, K., M. Billeter, W. Braun, *J. Molec. Biol.* **180,** 715, (1984).
24. Dyson, H.J., M. Rance, R. A. Houghten, R. A. Lerner, P.E. Wright, *J. Molec. Biol.* **201,** 161, (1988).
25. Dyson, H. J., M. Rance, R. A. Houghten, R. A. Lerner, P. E. Wright, *J. Molec. Biol.* **201,** 201 (1988).
26. Neito, J. L., *et al., FEBS Lett.* **239,** 83 (1988).
27. Harper, J. W., D. S. Auld, J. F. Riordan, B. L. Vallee, *Biochem.* **27,** 219 (1988).

7

^{1}H NMR Assignments and Three-Dimensional Structure of Ala14/Ala38 Bovine Pancreatic Trypsin Inhibitor Based on Two-Dimensional NMR and Distance Geometry

Hossein M. Naderi, John F. Thomason, Brandan A. Borgias
Stephen Anderson, Thomas L. James, Irwin D. Kuntz

Introduction

The first reported nuclear magnetic resonance (NMR) study of a protein was the publication in 1957 of a spectrum of bovine pancreatic ribonuclease A *(1)*, but the full power of this method for studies of the structure and function of proteins has only been realized during the last few years with the advent of two-dimensional (2-D) NMR. This approach has remarkable potential for determining the tertiary structures of proteins in noncrystalline environments, especially in aqueous solution *(2–7)* and has now led to the structures of a large number of small proteins and peptides [for example, see the references in *(8–9)*]. The determination of structure relies on a three-step process: (i) assignments of resonances of the hydrogen atoms to specific residues, (ii) measurements of internal distances and angles, and (iii) construction of the three-dimensional (3-D) structure using a variety of mathematical techniques. The specific assignments use experiments such as correlated spectroscopy (COSY), homonuclear Hartmann-Hahn (HOHAHA), and nuclear Overhauser effect spectroscopy (NOESY) *(10–13)*. Wüthrich and his colleagues were the first to use hydrogen-hydrogen through-space distances obtained from 2-D NMR NOESY experiments, combined with other constraints imposed by the covalent structure, along with distance geometry calculations *(7, 14–16)* to determine the 3-D structure of a peptide and a protein *(7, 17)*. These techniques are now widely used *(8, 9, 18)*.

With these successes, it becomes feasible to combine structural studies with site-directed mutagenesis to explore the roles played by specific amino acids and disulfide bonds as structural determinants in proteins. For such a project, bovine pancreatic trypsin inhibitor (BPTI) and its mutants are an ideal choice. BPTI has been extensively studied by NMR *(4, 19–23)*. It is also the first protein for which detailed folding kinetics became available *(24)*. BPTI has also been used for studies of molecular dynamics *(25–26)* and structural predictions *(27, 28)*.

The wild-type structure has three disulfide bonds: Cys14/Cys38, Cys30/Cys51, and Cys5/Cys55. It has been shown that when cysteines 14 and 38 are blocked chemically, the kinetics of refolding to the native structure in the presence of oxidized dithiothreitol is markedly affected *(29–31)*. To examine the role of the Cys14/Cys38 disulfide in folding, local structure, and global structure of BPTI, the mutant Ala14/Ala38 BPTI was expressed in *Escherichia coli*. The folding question has already been addressed *(32)*. At physiological temperature, the Ala14/Ala38 BPTI refolded quantitatively in vivo and in vitro to a native-like state.

In this chapter we report the ^{1}H NMR assignments of the Ala14/Ala38 mutant of BPTI and explore the chemical shift differences relative to native BPTI. We also present a 3-D structure of this mutant from the NMR NOESY data and distance geometry calculations. Finally, we describe the small conformational differences relative to the wild-type BPTI.

Materials and Methods

Materials. The BPTI mutant was prepared at Genentech, Inc. *(32)*. BPTI (trasylol) was obtained from Mobay Chemical Corp. We obtained chymotrypsin (Sigma Chemical Co.), CNBR-sepharose (Pharmacia), organic solvents (Fisher Scientific), deuterated solvents (Aldrich). All other reagents were obtained from Sigma Chemical Corporation.

The preparation and purification of Ala14/Ala38 BPTI. The mutant was expressed by *E. coli* by fusing the gene for the mutant to the coding sequence for the *E. coli* heat-stable enterotoxin II. It was expressed under the control of the alkaline phosphatase promoter, induced in the transformed *E. Coli* W3110 cells by growth under phosphate-limited conditions *(33–34)*. The protein was then purified from cell extracts, using a modification of the method of Kassel *(35)*. After removal of the TCA acid-insoluble proteins and neutralization *(34)*, extracts were passed over the chymotrypsin-Sepharose column at 4°C. The column was developed by first washing with tris-NaCl-$CaCl_2$ buffer and then with 0.01 M HCl, 0.5 M NaCl, and 10 mM $CaCl_2$. The collected protein peaks were lyophilized and then desalted on a Sephadex G-25 column with 0.5% ammonium acetate. The final purification was done on a Pharmacia fast protein, peptide, and polynucleotide liquid chromatography (FPLC) system with a Mono-S (cation exchange) column using a linear gradient of 0.05 to 1.0 M ammonium acetate, pH 4.0. The ammonium acetate was then removed by lyophilization. The purified mutant was stored at −20°C in lyophilized form. The concentration of the protein was determined using the extinction coefficient of the native protein (5400 l/cm-mol) at 280 nm *(35)*. The level of expression was approximately 20 mg/liter of cell suspension.

NMR samples were prepared at a final concentration of 6 mM protein in either 90% H_2O/ 10% D_2O or 100% D_2O. The chemical shift of the ε-protons of tyrosine 23, taken as 6.33 ppm *(36)*, was used as an internal reference. NMR data were acquired at 15, 25, 36, and 45°C at pH 4.6. No buffer was used. The D_2O samples were also adjusted to a reading of 4.6 using DCl or NaOD.

NMR Spectroscopy. All NMR experiments were carried out on a 500 MHz GN 500 spectrometer. The absolute value COSY spectra were acquired using the $(90\text{-}t_1\text{-}90\text{-}t_2)n$ pulse sequence *(10)*. The water proton signal was suppressed by presaturation during the relaxation decay *(13)*. A total of 512 by 2048 complex data points were recorded and processed with an unshifted sine-bell function in both domains before zero filling and Fourier transformation. The NOESY spectra used the phase-sensitive mode with quadrature detection *(37)*, with recycle times of 5–6 s. Acquisition of the HOHAHA spectra with MLEV-17 *(11)* required modification of the spectrometer by inclusion of a 6-W amplifier (Astron Corp, CA) as a transmitter with a γB_1 of 8 KHz. Mixing times of 30, 75, and 100 ms were used. A total of 512 by 2048 complex data points were expanded to a final real matrix of 1024 by 4096 through zero filling. For all 2-D NMR experiments, a spectral width of 7024 Hz was used in both dimensions. The carrier frequency was placed on the H_2O resonance, which was irradiated at all times except for the t_1 and t_2 periods. Data processing programs originated at the University of Groningen, Groningen, the Netherlands. Modifications and improvements were done by R. Scheek, S. Manogaran, and M. Day in our laboratory. The NOESY and HOHAHA spectra were base-line corrected in both dimensions *(38–39).*

The distance geometry program used was VEMBED *(8, 40),* a vectorized version of EMBED *(16);* the program was run at the San Diego Supercomputer Center. From the 1044 unique NOEs assigned, 1633 experimental distances were generated using an all-atom model for the molecular representation. When a single peak assignment may involve more than one pair of atoms, the appropriate correction factor is added as in the pseudo-atom method *(41).* This factor is 1.5 Å for methyl and nonstereospecifically assigned methylene protons. For similar protons on aromatic rings, the correction factor is 3.0 Å. The normal procedures of bound smoothing, structure embedding, and optimization were employed *(8, 42, 43).* On the average, a run of eight structures required 34 minutes of Cray XMP-48 time.

Calculated 2-D NOE intensities were obtained using the program CORMA *(44),* assuming a single effective isotropic correlation time of 3.2 ns. Dipolar relaxation rates are generated according to standard expressions *(45, 46).* The methyl group correlation times could be adjusted separately. The matrix of relaxation rates is then diagonalized and solved as described previously *(47).* This approach avoids the substantial errors inherent in the isolated spin pair approximation *(44).* The CORMA method is a relatively fast calculation, taking approximately 30 minutes on a Sun 3/160 (Sun Microsystems, Mountain View, CA) with floating point accelerator. An additional program, MARDIGRAS, was used to calculate distances after the first round of distance geometry *(48).*

The energy refinement was obtained using AMBER, a general purpose molecular mechanics program using empirical potential functions. There are provisions for positional constraints for groups of atoms and selective minimization of parts of the system *(49–51).*

Results

The assignment strategy. Assignments for Ala14/Ala38 BPTI were greatly facilitated by the similarity of many of the chemical shifts to those of wild-type BPTI *(36),* in agreement with partial assignments of chemically modified RCAM-BPTI in which the disulfide bond 14–38 is reduced and protected by carboxamidomethylation. Only a few significant points are presented here. This mutant shows the large upfield shift of the NH proton of Gly-37 *(52).* The NH, α-carbon proton cross-peaks are clearly visible in the COSY and HOHAHA spectra. The new resonances for Ala14 and Ala38 were readily identified by the normal combination of sequential and spin system assignments. The Thr-32 γ-methyl to α-carbon proton relay cross-peak was not observed in the HOHAHA spectra. The sequential assignments *(17)* are shown schematically in Fig. 1.

Chemical shift differences. The chemical shifts of Ala14/Ala38 BPTI at 36°C, pH 4.6 are given in Table 1, with identification of the shifts that deviate significantly from the wild-type protein. Hydrogen chemical shifts are thought to be very sensitive to the details of

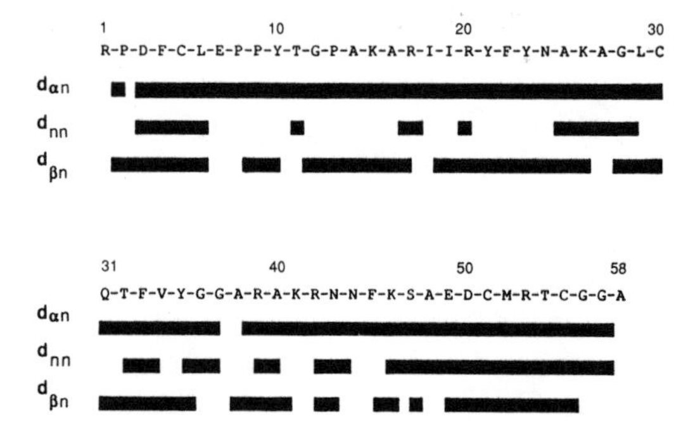

Fig. 1. Summary of the sequential NOE connectivities in the Ala14/Ala38 BPTI at pH 4.6, 36° and with a mixing time of 200 ms.

Table 1. Chemical shifts on the assigned H^1 NMR lines of Al 14/Ala38 BPTI, pH 4.6, 36° C, in parts per million.

Amino acid residue	δ of											
	NH	CαH		CβH		CγH		CδH		CεH		CζH
Arg-1		4.36		1.89	1.81	1.49	1.36	3.09	2.88	7.05		
Pro-2		4.34		2.05	0.92	1.87	1.61	3.73	3.59			
Asp-3	8.66	4.27		2.77								
Phe-4	7.82	4.59		3.34	2.96			7.00		7.37		7.31
Cys-5	7.42	4.34		2.86	2.75							
Leu-6	7.57	4.49		1.85	1.57	1.70		0.96	0.86			
Glu-7	7.49	4.60		2.27	2.21†	2.61	2.57					
Pro-8		4.63		2.42	1.83	2.11		3.98	3.72			
Pro-9		3.77		0.22	0.16†	1.22	0.12†	3.31	2.92			
Try-10	7.89†	4.86†		2.96				7.29		6.95‡		
Thr-11	8.75‡	4.51		4.11†		1.36						
Gly-12	7.09†	4.11‡	3.37‡									
Pro-13		4.35‡		2.25†	2.04†	1.99		3.64	3.59			
Ala-14	8.29‡	4.41‡		1.27‡								
Lys-15	8.09†	4.34†		1.92‡	1.67†	1.38	1.25	1.59		2.94		
Ala-16	8.09‡	4.33		1.19								
Arg-17	8.19	4.33		1.60		1.48‡	1.30	3.13	3.07			
Ile-18	8.25‡	4.26‡		1.86		1.35	0.98	0.72				
Ile-19	8.64	4.33		1.95		1.47	1.40	0.69				
						0.73						
Arg-20	8.39	4.74		1.84	0.86	1.37		3.48	3.04	7.43		
Tyr-21	9.18	5.69		2.71				6.70		6.78		
Phe-22	9.76	5.27		2.90	2.82			7.14		7.21	7.30	
Tyr-23	10.54	4.31		3.45	2.74			7.17		6.33		
Asn-24	7.78	4.60		2.86	2.18			7.89	7.09			
Ala-25	8.74	3.77		1.57								
Lys-26	7.91	4.07		1.90	1.54	1.74	1.46			3.04		
Ala-27	6.81	4.29		1.19								
Gly-28	8.13	3.93	3.63‡									
Leu-29	6.81	4.74		1.73	1.43	1.43		0.85	0.76			
Cys-30	8.40	5.60		3.67	2.67							
Gln-31	8.78	4.85		2.24†	1.73	2.15†	1.90			7.31	7.22†	
Thr-32	8.04	5.27		4.03		0.60						
Phe-33	9.37	4.88		3.11	2.97			7.13†		7.21†	7.07†	
Val-34	8.33	3.93		1.96		0.82	0.71					
Tyr-35	9.29†	4.82†	2.64	2.50				7.22‡	7.15‡	6.99	6.97	
Gly-36	8.42‡	4.35†	3.52†									
Gly-37	4.46	3.88‡	3.02‡									
Ala-38	7.91‡	4.70‡		1.31‡								
Arg-39	8.47‡	4.25‡		2.09‡		1.70	1.64†	3.30†		7.22†		
Ala-40	7.86‡	4.57‡		1.28†								
Lys-41	8.37†	4.45		2.32†	1.64	2.20†	1.14	1.58	1.34			
Arg-42	8.24†	3.66		1.05	0.41†	1.49	1.21	2.82	2.72	7.09		
Asn-43	7.19	5.04		3.32†	3.26			7.95	7.80			
Asn-44	6.77	4.89		2.75	2.56†			7.88	3.42			
Phe-45	9.93	5.14		3.41	2.78			7.36		7.87	7.64	

Table 1 *(continued)*

Amino acid residue	δ of										
	NH	CαH		CβH		CγH		CδH	CεH		CζH
Lys-46	9.89	4.39		2.10	1.98	1.62	1.48	2.88	3.05	7.04	
Ser-47	7.45	4.54		4.12	3.87						
Ala-48	8.13	3.15		1.04							
Glu-49	8.53	3.88		2.02	1.85	2.35	2.24				
Asp-50	7.83	4.29		2.86	2.73						
Cys-51	6.99	1.70		3.17	2.88†						
Met-52	8.56	4.15		2.06	1.97	2.69			2.16		
Arg-53	8.28	3.99		1.93	1.86	1.72	1.62	3.22	7.24†		
Thr-54	7.40	4.07		4.00		1.61					
Cys-55	8.21	4.62‡		2.24	2.02						
Gly-56	7.93	3.85	3.80								
Gly-57	8.18	3.97	3.84†								
Ala-58	7.92	4.02		1.31							

† These chemical shifts deviate by 0.05 – 0.10 ppm from those of the corresponding protons in wild-type BPTI *(36)*.
‡ These chemical shifts deviate by more than 0.10 ppm from those of the corresponding protons in a wild-type BPTI *(36)*.

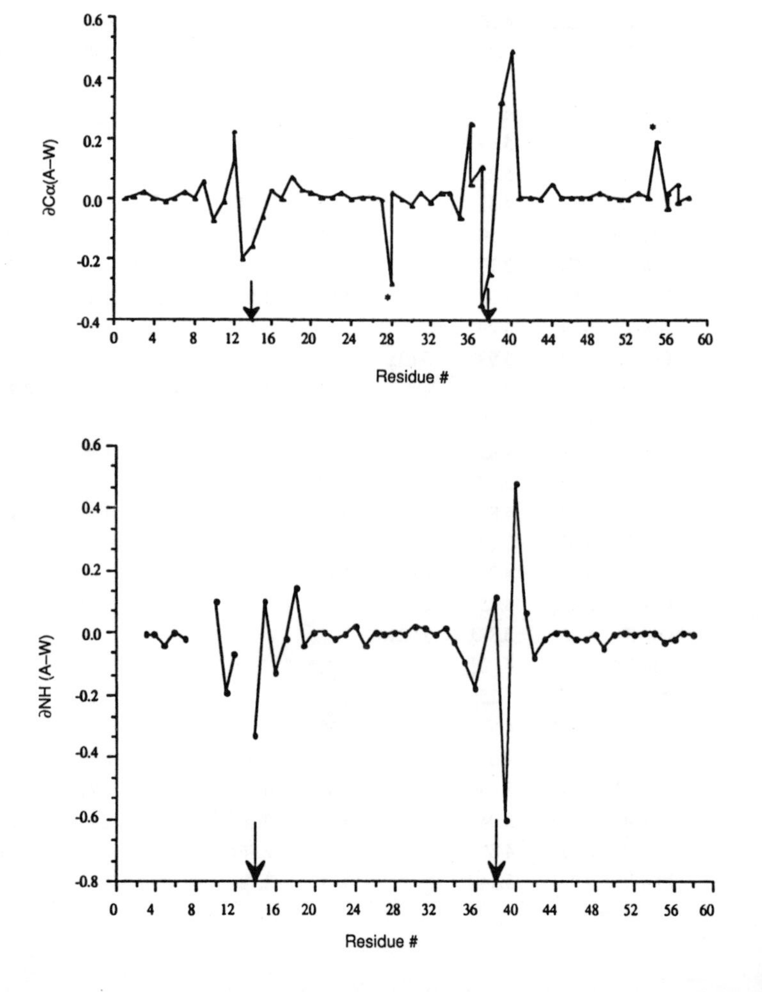

Fig. 2. Plot of the amide proton chemical shift differences between the Ala14/Ala38 BPTI and wild-type BPTI at pH 4.6 and 36° versus the residue numbers in the amino acid sequence. Positive values indicate that the resonances in the Ala14/Ala38 BPTI are at lower field.

Fig. 3. Plot of the *Cα*-proton chemical shift differences as in Fig. 2.

the 3-D structure *(9, 20)*. In Figs. 2 and 3 the chemical shift differences for α-carbon protons and amide protons are plotted versus residue number. There are relatively large changes in the vicinity of residues 9–19, 28, 35–42, and 55, in agreement with the assignments of RCAM-BPTI *(20, 53)*.

NOE measurements. We used an interactive computer-assisted approach to NOE assignments, based on the assignments of the wild-type protein and the coordinates of the neutron diffraction study of BPTI by Wlodawer and colleagues *(54)*. By ignoring assignments to protons greater than 5 Å apart in the crystal structure, we were able to make first pass, tentative assignments for about 1500 cross-peaks. These were each examined by direct inspection that included consideration of possible assignments of protons up to 10 Å apart in the neutron study. The final data set reported here includes 1044 assigned NOE cross-peaks.

Distance geometry constraints. All the NOEs were classified into two distance ranges, 2.0–3.0 Å and 2.0–4.0 Å, corresponding to strong and weak NOESY cross-peaks (see Appendix). We also identified 15 H bonds based on identification of regular secondary structures *(9, 36)* (Table 2). We used bounds of 1.8–2.0 Å for all H...O H-bond distances and 2.7–3.0 Å for the corresponding NH...O distances (Fig. 4; Table 3). The two disulfide bonds were fixed using a range of 2.0–2.1 Å for the S-S distance *(7)*. Using the NOE intensities and standard conformational analysis *(55)*, we have made preliminary assignments of 10 sets of β and three sets of γ-carbon protons (Table 4).

The three-dimensional structure of Ala14/Ala38 BPTI. The structure was computed in two steps. First, we used the VEMBED version of the distance geometry program *(8, 40)*. Second, the individual structures were minimized using AMBER (Table 5). We also computed the *average* coordinates for the set of distance geometry structures and minimized the energy of the average structure. As expected, all the distance geometry structures began at high potential energies but low violations of the distance constraints. Using

Table 2. H bonds involving the main chain that are included in the DG constraints.

Res1(NH)		Res2(O)
	α helix C-terminus	
Met-52	———	Ala-48
Arg-53	———	Glu-49
Thr-54	———	Asp-50
Cys-55	———	Cys-51
Gly-56	———	Met-52
	β sheet	
Ile-18	———	Tyr-35
Tyr-35	———	Ile-18
Arg-20	———	Phe-33
Phe-33	———	Arg-20
Phe-22	———	Gln-31
Gln-31	———	Phe-22
Asn-24	———	Leu-29
	3_{10}-helix N-terminus	
Cys-5	———	Pro-2
Leu-6	———	Asp-3
Glu-7	———	Phe-4

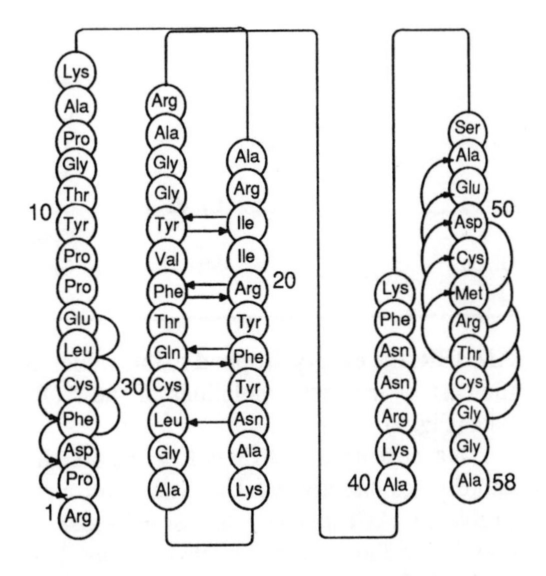

Fig. 4. Diagram of intermolecular hydrogen bonds in the Ala14/Ala38 BPTI as deduced from NMR data.

Table 3. NOE crosspeaks used to assign H-bond for the mutant Ala14/Ala38 BPTI.

H-bonds			NOE			
Res$_i$ (NH)		*Res$_j$* (O)	*Res*	*Atom*	*Res*	Atom
Cys-5	———	Pro-2	Cys-5	NH	Asp-3	HA
			Cys-5	HN	Pro-2	HA
Leu-6	———	Asp-3	Leu-6	HN	Phe-4	HA
			Leu-6	HN	Asp-3	HA
Glu-7	———	Phe-4	Glu-7	HN	Cys-5	HA
			Glu-7	HN	Phe-4	HA
Ile-18	———	Tyr-35	Ile-18	HN	Tyr-35	HN
			Ile-18	HN	Gly-36	HA
			Arg-17	HA	Gly-36	HA
Tyr-35	———	Ile-18	Tyr-35	HN	Ile-18	HN
			Tyr-35	HN	Ile-19	HA
			Val-34	HA	Ile-19	HA
Arg-20	———	Phe-33	Arg-20	HN	Phe-33	HN
			Arg-20	HN	Val-34	HA
			Ile-19	HA	Val-34	HA
Phe-33	———	Arg-20	Phe-33	HN	Arg-20	HN
			Phe-33	HN	Tyr-21	HA
Phe-22	———	Gln-31	Phe-22	HN	Gln-31	HN
			Phe-22	HN	Thr-32	HA
Gln-31	———	Phe-22	Gln-31	HN	Phe-22	HN
			Gln-31	HN	Tyr-23	HA
			Cys-30	HA	Tyr-23	HA
Asn-24	———	Leu-29	Asn-24	HN	Leu-29	HN
			Asn-24	HN	Cys-30	HA
			Tyr-23	HA	Cys-30	HA
Met-52	———	Ala-48	Met-52	HN	Glu-49	HA
			Met-52	HN	Ala-48	HA
			Ala-48	HA	Cys-51	HB
Arg-53	———	Glu-49	Arg-53	HN	Asp-50	HA
			Arg-53	HN	Glu-49	HA
			Glu-49	HA	Met-52	HB
Thr-54	———	Asp-50	Thr-54	HN	Cys-51	HA
			Asp-50	HA	Arg-53	HB
Cys-56	———	Cys-51	Cys-55	HN	Met-52	HA
			Cys-51	HA	Thr-54	HB
Gly-56	———	Met-52	Gly-56	HN	Arg-53	HA
			Gly-56	HN	Met-52	HA

AMBER, the energy dropped to low values at the cost of increased violations of the NOE data (Fig. 5).

We next used CORMA to compare the coordinates directly with the NOE intensities (Table 6). As expected, the use of separate correlation times for the methyl rotation had minimal effect *(47)*. It is convenient to calculate a function analogous to the R factor used in crystallography:

$$R = \frac{\Sigma \mid |I_o| - |I_c| \mid}{\Sigma |I_o|}$$

where $|I_o|$ is the absolute value of the scaled intensity of the NOE cross-peak and $|I_c|$ is the absolute value of the scaled intensity of the cross-peak calculated by CORMA. The normalization was based on summing the inten-

Table 4. Stereospecific assignment for β- and γ-carbon atoms that were included in the DG constraints as described in text.

β-carbons Res. no.	γ-carbons Res. no.
1- Phe-4	1- Gln-31
2- Phe-22	2- Val-34
3- Leu-29	3- Lys-41
4- Cys-30	
5- Arg-42	
6- Phe-45	
7- Glu-49	
8- Asp-50	
9- Cys-51	
10- Arg-53	

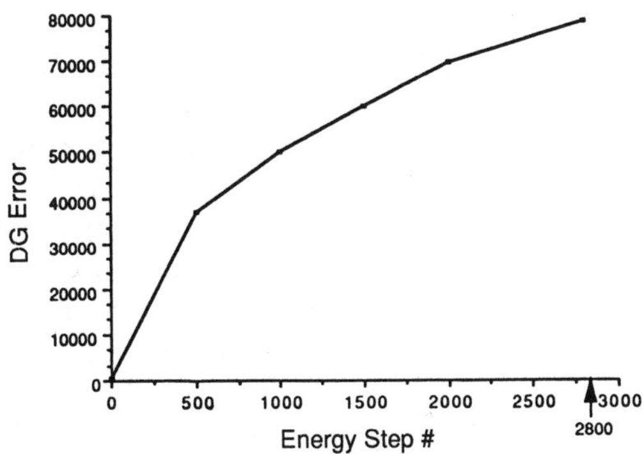

Fig. 5. Total root-mean-square displacement in Å of energy-minimized distance geometry structure plotted against the number of steps of minimization.

sity of equivalent geminal proton cross-peaks in each data set.

As noted above, the averaged coordinates from the cluster of DG structures were well behaved during minimization. This step allows us to provide a single structure to compare with crystallographic results. A comparison of the Ala14/Ala38 averaged structure with the x-ray results and NMR results *(56)* for wild-type BPTI are also given in Table 6. We calculated the phi and psi angles in the averaged, minimized structure of the mutant and compared them with those in the x-ray structure (Fig. 6). Most of the phi, psi angles of the NMR structure lie in acceptable regions of the Ramachandran map, but to emphasize differences, we also report the dot product of the angles of the C=O bonds in the backbone of the mutant and x-ray structures (Fig. 7). A direct root-mean-square (rms) comparison of the various backbone coordinate sets is given in Fig. 8. We also calculated the radius of gyration of the structures (Table 7). The NMR-derived structures are slightly more compact compared to the crystal structure.

The last approach we used to build a 3-D model of the mutant was to start with the crystallographic coordinates of 4PTI *(57, 58)*. These coordinates were simply optimized against the NMR constraints using the optimization section of VEMBED and then minimized using AMBER (Table 6; Fig. 8).

Table 5. Energy and DG error for eight DG structures after minimization, and the averaged, averaged-minimized, restrained-minimized, and BPTI-refined against NMR restraints minimized coordinate sets.

Structure no.	Energy (Kcal)	DG error	RMS vs. BPTI	RMS vs. DG average
1	–568	45282	1.45	1.00
2	–588	57893	1.75	0.89
3	–621	48238	1.21	0.84
4	–596	37548	1.55	0.91
5	–649	45474	1.61	0.98
6	–633	58974	1.52	1.05
7	–539	79994	1.50	1.17
8	–587	50470	1.57	0.95
Averaged	< 10,000	529	1.47	0.00
Averaged-minimized	–624	50058	1.26	0.71
Averaged-restrained and minimized	–83	9024	1.54	0.98
BPTI-refined minimized	–723	40194	0.83	1.10

Table 6. CORMA error result for Ala14/Ala38 BPTI mutant with identical correlation times for the methyl groups.

NOE type	R* methyl τ_c = 3.2 ns	R* methyl τ_c = 0.10 ns	R* BPTI† τ_c = 3.2 ns
Intraresidue	0.573	0.568	0.533
Interresidue	0.625	0.650	0.617
Total	0.595	0.603	0.569

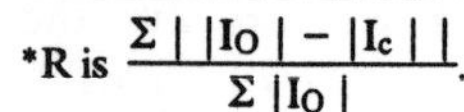

*R is $\frac{\Sigma \,|\, |Io| - |I_c| \,|}{\Sigma \,|Io|}$. †R for BPTI refined against NMR restraints vs. Ala14/Ala38 BPTI NOE intensities.

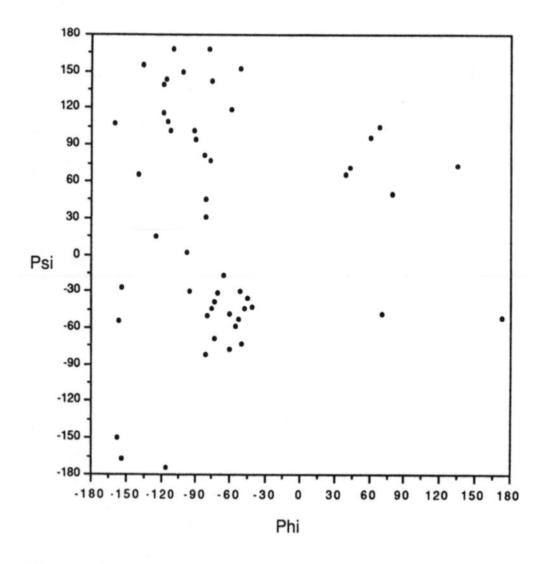

Fig. 6. Ramachandran plot of the ϕ and ψ angles for the restrained and minimized averaged DG structure of the Ala14/Ala38 BPTI.

Discussion

This investigation again affirms the utility of NMR in achieving good quality structures of small proteins in aqueous solution. Since attempts to crystallize Ala14/Ala38 BPTI have not been successful, NMR is the only technique presently capable of solving the 3-D structure of this interesting mutant. Generally, the structure is quite similar to other x-ray and NMR structures for BPTI. A detailed comparison is given below.

Chemical shifts. The chemical shift data (Figs. 2, 3; Table 8) indicate that there are noticeable changes in the vicinity of the mutations, but the unusually compact geometry involving Tyr-35, Gly-37, and Asn-44 is not greatly affected by removal of the disulfide bond. We base this latter conclusion on the small changes (ca. 0.15 ppm) in the large upfield shifts of the Gly-37 NH (4.3 ppm) and Asn-44 Nd_1 (3.5 ppm) protons *(52)*. This

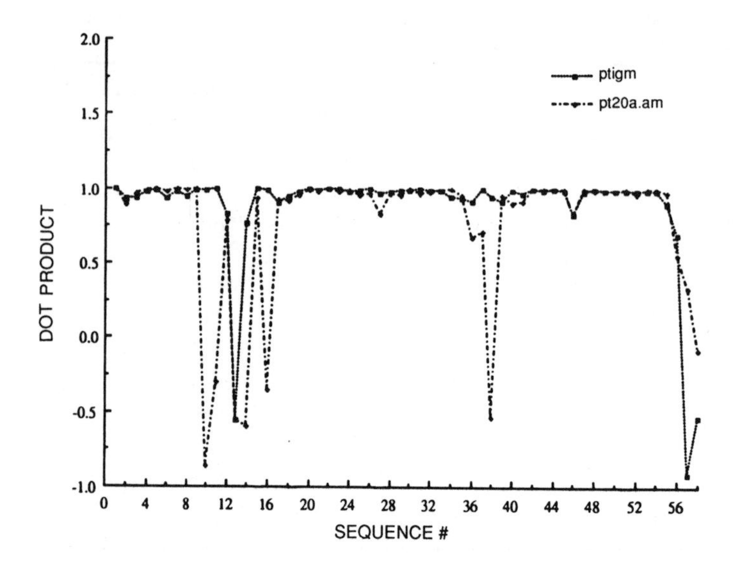

Fig. 7. The dot product of adjacent backbone carbonyl groups for the restrained and minimized averaged DG structure of Ala14/Ala38 BPTI compared to those for PBTI refined with NMR constraints (see text). Broken line, Ala14/Ala38 BPTI. Dotted line, refined BPTI.

Table 7. Radius of gyration for the DG-generated, averaged structure, its minimized, restrained-minimized Ala14/Ala38 BPTI mutant, and wild-type BPTI.

Structure name	Radius of gyration
Average	10.34
Average:minimized	10.76
Average:restrained and minimized	11.01
BPTI	11.18

"Average" is the DG-generated, averaged structure of the Ala14/Ala38 BPTI mutant. "Average:minimized" is the energy-minimized, DG-averaged structure. "Average:restrained and minimized" is the restrained and energy-minimized, DG-averaged structure.

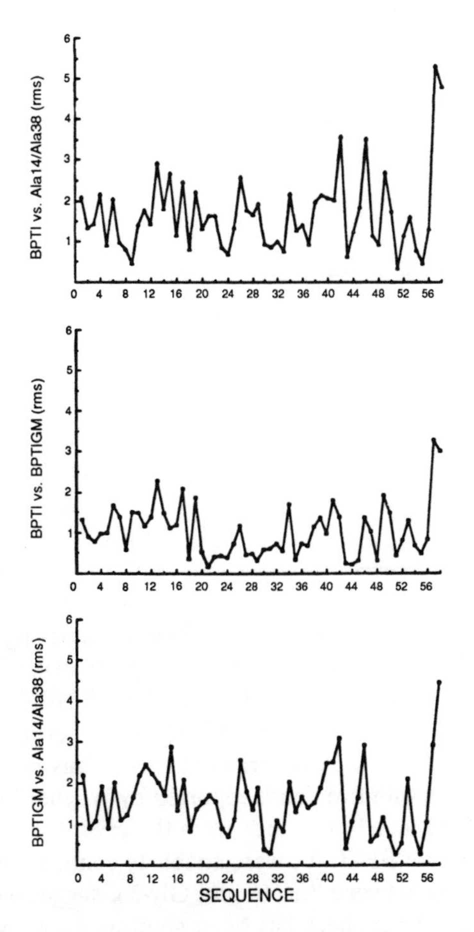

Fig. 8. Root-mean-square positional displacement for pairs of structure: (A) BPTI (4PTI, crystal structure) versus restrained, minimized averaged DG structure of Ala14/Ala38 BPTI. (B) BPTI versus BPTI, crystal structure refined against NMR restraints. (C) BPTI refined against NMR restraints versus Ala14/Ala38 BPTI. The displacement for all atoms are averaged for each residue.

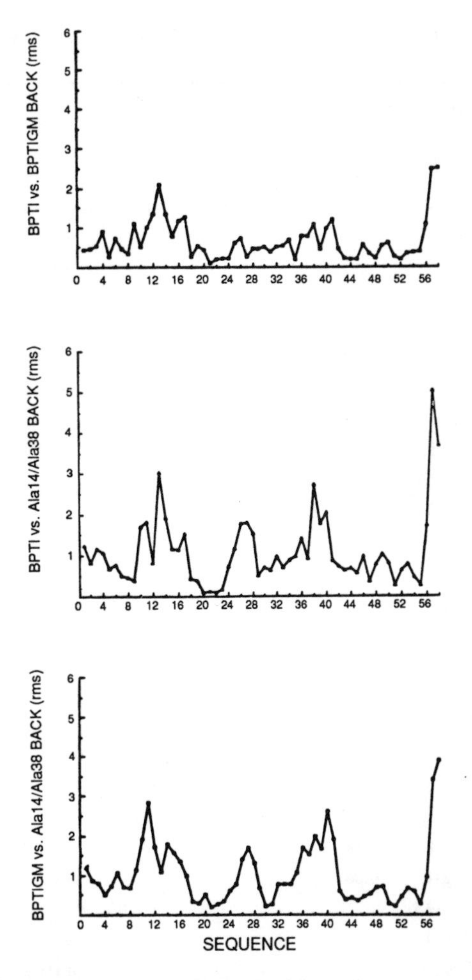

Fig. 9. Comparison of the rms positional displacement of backbone atoms as described in Fig. 8.

strongly suggests that only very small displacements of the three residues have occurred. The chemical shifts for Ala14/Ala38 are also very similar to those reported for RCAM-BPTI *(20)*. The two exceptions are the Gly-28 and Cys-55 α-protons. We see two separate peaks for the Gly-28 protons, while only one was reported for RCAM-BPTI. There is no simple explanation for the discrepancy in the Cys-55 α-proton resonance. Overall, we conclude that the chemical shift similarity over major parts of the structure supports the view that BPTI, RCAM-BPTI, and Ala14/Ala38 BPTI have very similar structures except at the immediate vicinity of residues 14 and 38. Small changes in ring geometry of Tyr-10, Tyr-21, Tyr-23, and Phe-45 could explain the

Table 8. Summary of chemical shift differences between Ala14/Ala38 BPTI and wild-type BPTI, pH 4.6, 36°C, in parts per million.

Amino acid residue	δ of											
	NH	CαH		CβH		CγH		CδH		CεH		CζH
Glu-7					+0.05							
Pro-9					+0.07		−0.05					
Tyr-10	+0.10	−0.07								−0.14		
Thr-11	−0.19			+0.06								
Gly-12	−0.07	+0.22	+0.12									
Pro-13		−0.20		+0.07	−0.06							
Ala-14	−0.33	−0.16		−1.51								
Lys-15	+0.10	−0.06		−0.17	+0.08							
Ala-16	−0.13											
Arg-17						+0.18						
Ile-18	+0.14	+0.07										
Gly-28			−0.28									
Gln-31				+0.08		−0.10					+0.06	
Phe-33								+0.06		+0.08		+0.08
Tyr-35	−0.09	−0.06						−0.55	+0.46	+0.11	+0.19	
Gly-36	−0.18	+0.05	+0.29									
Gly-37		−0.41	+0.11									
Ala-38	+0.11	−0.26		−1.72								
Arg-39	−0.60	+0.32		−0.19			+0.05	+0.05		−0.08		
Ala-40	+0.48	+0.49		+0.09								
Lys-41	+0.06			+0.08		+0.05						
Arg-42	−0.08				+0.05							
Asn-43				−0.06								
Asn-44					+0.07							
Cys-51					+0.07							
Arg-53										−0.05		
Cys-55		+0.19										
Gly-57			+0.05									

majority of the chemical shift differences that are seen. In Fig. 10 we show the chemical shift differences (with respect to wild-type BPTI) as a function of the distance from the mutation site.

Secondary structures

Helices. For the N-terminal 3_{10}-helix we report three i, i+3 hydrogen bonds in agreement with the x-ray structure *(54)* for BPTI. The C-terminal helix is not as regular as the 10 residue α-helix seen in the x-ray work because we were not able to establish a H bond between Ser-47 and Cys-51. The distances $d_{\alpha N}$ *(47, 51)* and d_{NN} *(47, 48)* are larger than 4.4 Å, also suggesting irregularity in the helix *(9)*.

β sheet. We were able to assign seven H bonds for the two-stranded sheet. This compares with five bonds reported by Wagner for BPTI and 10 H bonds from the crystal structure for BPTI. We were unable to confirm a H bond between Ala-16 and Gly-36, suggesting that the β sheet has been shortened by one residue on each side near the site of the mutation. A short distance is observed between the amide protons of Tyr-21 and Phe-45 supporting the addition of a third strand to the sheet *(54)*.

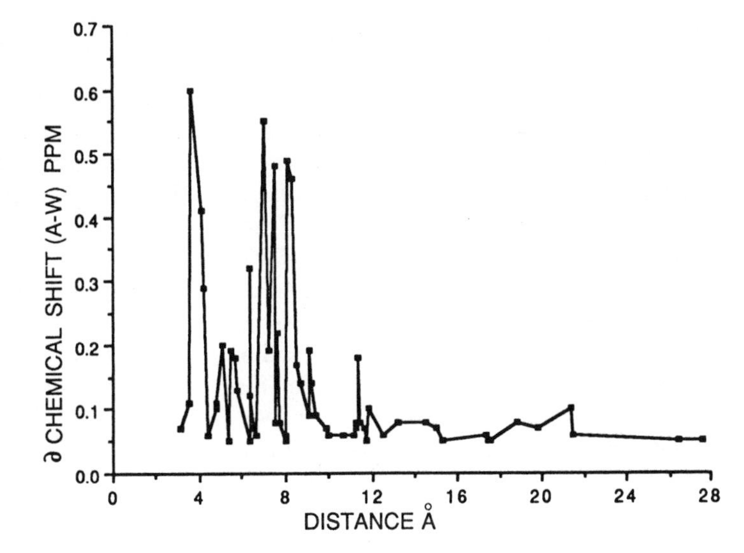

Fig. 10. Comparison of the chemical shift differences (∂) between Ala14/Ala38 BPTI and wild-type BPTI *(36)* plotted against the distance in angstroms from the hydrogen atoms to the mutation site. The mutation site is taken as the average of the Ala-14, Cα an Ala-38, Cα coordinates.

Three-dimensional structural comparisons. To a good approximation, the 3-D structure of Ala14/Ala38 BPTI is the same as the wild-type protein except for a slight opening at the very end of the β sheet in the region of residues 14 and 38. The refined distance geometry structures suggest some rearrangement of the backbone dihedral angles at these positions and, at the most, small displacements of Tyr-21, Tyr-35, and Phe-45. (Coordinates are available on request.)

An important issue for any structural technique is the assessment of its systematic and random errors. It is often difficult to do this assessment from first principles. One standard used in the NMR community is the rms deviations among randomly generated structures. For Ala14/Ala38 BPTI, we find a value of 0.85 Å for the well-determined part of the backbone and 1.55 Å for these backbone and side-chain atoms. Such values are typical for many NMR solution structures *(8)* and suggest coordinate errors of 1–2 Å. A second criterion for the quality of a structure is its internal energy. Using AMBER, we find low energy minima near each of the individual DG conformations and near the averaged DG structure as well. The slight compaction of the DG structures, before minimization, probably arises from the use of only short distance constraints, a characteristic of all NMR structure determinations. Energy minimization in vacuo cannot recover the full elongation of the structure. Minimization in the presence of molecular water would be more desirable.

It is known that NMR restraints are not well suited for precise determination of backbone dihedral angles for peptides *(8, 59, 60)*. These angles are largely fixed by the position of the carbonyl oxygen, which is relatively poorly placed in NMR structures in the absence of H-bonding restraints. We observe a number of phi, psi angles at considerable variance to the crystal structure results for wild-type BPTI. To ask if these divergencies are *required* by the NMR constraints, we used the simple expedient of starting a calculation using the 4PTI crystal structure *(58)* and refining the coordinates using NMR constraints directly in the VEMBED and AMBER programs. The resulting structure was slightly closer to the (starting) crystal structure. More importantly, only dihedrals at residues 12, 13, 14, and 46 were significantly altered. Thus, the dihedral angle changes at residues 10, 11, 16, 25, 32, 36–39, and 57–58 are consistent with, but *not required* by, the NMR constraints. We conclude from this test that while the overall accuracy of the NMR coordinates can approach 1 Å, the peptide dihedral angles remain relatively uncertain if only constrained by hydrogen-hydrogen distances generated by

NOE experiments. Better results could be obtained by using coupling constant data to restrict the phi, psi angles.

Finally, the use of an "R" factor to evaluate NMR structures is a direct way to compare the quality of fits of coordinates to experimental NOE data. The averaged and minimized structure for Ala14/Ala38 BPTI has an R factor of about 60%. There are several reasons for the large value of this parameter. First, the NOE intensities depend on the inverse sixth power of the interproton distances. This highly nonlinear relation overweights short distances enormously. Without complete stereospecific assignments, it is unlikely that the R factor can be significantly reduced. Second, there is a major scaling issue involving the very high intensity diagonal that dominates the calculation. This problem could be mitigated by calculating R values for only the off-diagonal elements, a strategy analogous to that used in the crystallographic community. Third, the values reported here have not been optimized. Such efforts are under way for other structures *(61)*. Finally, we note that a "random" structure has a much higher R factor in NMR (>100%) than in crystallography (ca. 60%).

We conclude that NMR structures based on this quality of data (i.e., three NOEs per resolved hydrogen and limited stereospecific assignments) have coordinate accuracies of 1.0–2.0 Å. This is consistent with the relatively poor determination of backbone dihedral angles in regions outside of the regular secondary structure. For our data, it proves appropriate to report a single "averaged" conformation that has been subjected to restrained energy minimization following the DG calculations.

Acknowledgments

Our thanks to Dr. Phyllis Anne Kosen for many useful suggestions in the critical phases of this work. The project was supported in part by Research Resources Grant RR-01695 and by NIH Grants GM-31497 (IDK), GM-19267 (IDK), and GM-39247 (TLJ). We gratefully acknowledge the use of the University of California, San Francisco, Computer Graphics Laboratory (Director, R. Langridge, RR-01081). Some of the calculations and code development were performed at the San Diego Supercomputer Center.

References

1. Saunders, M., A. Wishnia, T. G. Kirkwood, *J. Am. Chem. Soc.* **79,** 3289 (1957).
2. Arseniev, A. S., V. F. Bystrov, V. T. Ivanov, Y. A. Vchinnikov, *FEBS Lett.* **165,** 51 (1984).
3. Kline, A. D., W. K. Braun, K. Wüthrich, *J. Mol. Biol.* **189,** 377 (1986).
4. Kosen, P. A., *et al., Biochemistry* **25,** 2356 (1986).
5. Nagayama, K., and K. Wüthrich, *Eur. J. Biochem.* **114,** 365 (1981).
6. Wagner, G., A. Kumar, K. Wüthrich, *Eur. J. Biochem.* **114,** 375 (1981).
7. Williamson, M. P., T. F. Havel, K. Wüthrich, *J. Mol. Biol.* **182,** 295 (1985).
8. Kuntz, I. D., J. F. Thomason, C. M. Oshiro, in *Methods in Enzymology,* Oppenheimer, N. J., and T. L. James, Eds. (Academic Press, San Diego, 1989), pp. 159–203.
9. Wüthrich, K., in *NMR of Proteins and Nucleic Acids* (Wiley, New York, 1986).
10. Aue, W. P., E. Bartholdi, R. R. Ernst, *J. Chem. Phys.* **64,** 2229 (1976).
11. Bax, A., and D. G. Davis, *J. Magn. Reson.* **65,** 355 (1985).
12. Bodenhausen, G., R. Freeman, D. L. Turner, *J. Magn. Reson.* **27,** 511 (1977).
13. Kumar, A., G. Wagner, R. R. Ernst, K. Wüthrich, *Biochem. Biophys. Res. Commun.* **96,** 1156 (1980).
14. Braun, W., and N. Go, *J. Mol. Biol.* **186,** 611 (1985).
15. Havel, T. F., G. M. Crippen, I. D. Kuntz, *Biopoly.* **18,** 73 (1979).
16. Havel, T. F., I. D. Kuntz, G. M. Crippen, *Bull. Math. Biol.* **45,** 665 (1983).
17. Wüthrich, K., G. Wider, G. Wagner, W. Braun, *J. Mol. Biol.* **155,** 311 (1982).
18. Braun, W., *Q. Rev. Biophys.* **19,** 115 (1987).
19. Pardi, A., G. Wagner, K. Wüthrich, *Eur. J. Biochem.* **137,** 445 (1983).
20. Stassinopoulou, C., G. Wagner, K. Wuthrich, *Eur. J. Biochem.* **145,** 423 (1984).
21. Wagner, G., and K. Wüthrich, *J. Mol. Biol.* **160,** 343 (1982).

22. Wagner, G., and K. Wüthrich, *J. Mol. Biol.* **155,** 347 (1982).
23. Wüthrich, K., and G. Wagner, *FEBS Lett.* **50,** 265 (1975).
24. Creighton, T. E., *Prog. Biophys. Mol. Biol.* **33,** 231 (1978).
25. Brooks, B., and M. Karplus, *Proc. Natl. Acad. Sci. U.S.A.* **80,** 6571 (1983).
26. Levitt, M., *J. Mol. Biol.* **168,** 595 (1983).
27. Levitt, M., and A. Warshel, *Nature* **253,** 694 (1975).
28. Tanaka, S., and H. A. Scheraga, *Proc. Natl. Acad. Sci. U.S.A.* **72,** 3802 (1975).
29. Creighton, T. E., *J. Mol. Biol.* **95,** 167 (1975).
30. Creighton, T. E., *J. Mol. Biol.* **113,** 275 (1977).
31. Creighton, T. E., and D. P. Goldenberg, *J. Mol. Biol.* **179,** 497 (1984).
32. Marks, C. B., H. Naderi, P. A. Kosen, I. D. Kuntz, S. Anderson, *Science* **235,** 1370 (1987).
33. Inouye, H., S. Michaelis, A. Wright, J. Beckwith, *J. Bact.* **146,** 668 (1981).
34. Marks, C. B., M. Vasser, P. Ng, W. Anzel, S. Anderson, *J. Biol. Biochem.* **261,** 7115 (1986).
35. Kassell, B., *Methods Enzymol.* **11,** 844 (1970).
36. Wagner, G., *et al., J. Mol. Biol.* **196,** 611 (1987).
37. States, D. J., and R. A. Haberkorn, *J. Magn. Reson.* **48,** 286 (1982).
38. Basus, V. J., *J. Magn. Reson.* **60,** 138 (1984).
39. Pearson, G. A., *J. Magn. Reson.* **27,** 265 (1977).
40. Thomason, J., Ph.D. thesis, University of California, San Francisco (1991).
41. Wüthrich, K., M. Billeter, W. Braun, *J. Mol. Biol.* **169,** 949 (1983).
42. Crippen, G. M., *Distance Geometry and Conformational Calculations* (Wiley, Chichester, UK, 1981).
43. Crippen, G. M., and T. F. Havel, *Distance Geometry and Molecular Conformation,* (Research Studies Press, Taunton, UK, 1988).
44. Borgias, B. A., and T. L. James, *J. Magn. Reson.* **79,** 493 (1988).
45. Macura, S., and R. R. Ernst, *Mol. Phys.* **41,** 95 (1980).
46. Solomon, I., *Phys. Rev.* **99,** 559 (1955).
47. Keepers, J. W., and T. L. James, *J. Magn. Reson.* **57,** 404 (1984).
48. Borgias, B. A., and T. L. James, *J. Magn. Reson.* **87,** 475 (1990).
49. Singh, U. C., and P. A. Kollman, *J. Comp. Chem.* **5,** 129 (1984).
50. Weiner, S. J., *et al., J. Am. Chem. Soc.* **106,** 765 (1984).
51. Weiner, S. J., P. A. Kollman, D. T. Nguyen, D. A. Case, *J. Comp. Chem.* **7,** 230 (1986).
52. Tuchsen, E., and C. Woodward, *Biochemistry* **26,** 1918 (1987).
53. Yoshioki, S., H. Abe, T. Noguti, N. Go, K. Nagayama, *J. Mol. Biol.* **170,** 1031 (1983).
54. Wlodawer, A., J. Walter, R. Huber, L. Sjoelin, *J. Mol. Biol.* **180,** 301 (1984).
55. Basus, V. J., in *Methods in Enzymology,* Oppenheimer, N., and T. James, Eds. (Academic Press, San Diego, 1989) pp. 132–144.
56. Wagner, G., *et al., Eur. J. Biochem.* **157,** 275 (1986).
57. Marquart, M., J. Walter, J. Deisenhofer, W. Bode, R. Huber, *Acta Crystallogr., Sect. B* **39,** 480 (1983).
58. Wlodawer, A., J. Deisenhofer, R. Huber, *J. Mol. Biol.* **193,** 145 (1987).
59. Clore, G. M., D. K. Sukumaran, M. Nilges, A. M. Gronenborn, *Biochemistry* **26,** 1732 (1987).
60. Teeter, M. M., and M. Whitlow, *Proteins* **4,** 262 (1988).
61. Yip, P., and D. A. Case, *J. Magn. Reson.* **83,** 643 (1989).
62. Wüthrich, K., in *Methods in Enzymology,* Oppenheimer, N., and T. James, Eds. (Academic Press, San Diego, 1989) pp. 125–131.

Appendix. Distance constraints for Ala-14–38 BPTI (in Å) obtained from NOESY spectra.

	Sequential backbone					Medium-range backbone and long-range backbone					Constraints with sidechains				
				UP	LO				UP	LO				UP	LO
Arg-1															
											HB1	P2	HD1	4.0	2.0†
											HB1	P2	HD2	4.0	2.0†
											HB2	P2	HD1	3.5	2.0†
											HB2	P2	HD2	4.0	2.0†
											HD1	C55	HA	4.0	2.0†
											HD1	Y23	HE	6.0	2.0†
											HD2	A25	HA	4.5	2.0†
											HD2	D3	HA	4.5	2.0†
											HD2	L6	HD	5.5	2.0†
											HD2	Y23	HE	7.0	2.0†
											HE	A25	HB	5.5	2.0†
											HE	C5	HB2	4.0	2.0†
											HE	Y23	HE	7.0	2.0†
											HG1	C5	HB2	4.5	2.0†
											HG2	C5	HB2	4.5	2.0†
											HG2	L6	HD	6.0	2.0†
Pro-2															
	HA	D3	HN	3.6	2.1	HB1	D3	HB	6.0	2.0†	HB1	F4	HE	7.0	2.0
	HB1	D3	HN	4.5	1.8	HB1	F4	HB1	4.0	2.0†	HB2	F4	HE	6.0	2.0
	HB2	D3	HN	4.8	3.0	HB1	F4	HN	4.5	2.0	HB2	F4	HZ	4.0	2.0
						HB1	C5	HN	4.0	2.0	HG1	F4	HD	7.0	2.0†
						HB2	F4	HN	4.0	2.4	HG1	C5	HN	4.0	2.0†
						HD2	C55	HA	3.5	2.0	HG1	F4	HZ	3.8	2.0†
						HA	F4	HN	olap		HG2	F4	HE	7.0	2.0†
											HG2	C5	HN	4.0	2.0†
											HG	C55	HA	N	
Asp-3															
	HN	P2	HA	3.6	2.1						HA	L6	HG	5.5	2.0†
	HN	P2	HB1	4.5	1.8	HA	C5	HN	4.0	2.0	HB	F4	HD	7.0	2.0†
	HN	P2	HB2	4.8	3.0	HA	L6	HN	4.0	2.0†	HB	L6	HD2	6.0	2.0†
	HB1	F4	HN	4.5	2.0	HB	P2	HB1	6.0	2.0†	HB	L6	HN	5.5	2.0†
	HB2	F4	HN	4.5	2.0	HB	F4	HB2	5.5	2.0†					
	HN	F4	HN	3.0	2.0	HB	L6	HN	5.5	2.0†					
	HA	F4	HN	olap											
											HA	L6	HD	olap	
Phe-4															
	HN	D3	HB1	4.5	2.0†	HN	P2	HB1	4.0	2.0	HB1	R42	HD2	4.0	2.0†
	HN	D3	HB2	4.5	2.0†	HN	P2	HB2	4.0	2.4	HB2	E7	HG1	4.5	2.0†
	HN	D3	HN	3.0	2.0	HB2	D3	HB	5.5	2.0†	HD	C5	HA	6.0	2.0
	HA	C5	HN	3.6	2.2†	HA	L6	HB1	4.5	2.0†	HD	N43	HA	6.0	2.0

Appendix *(continued)*

	Sequential backbone				Medium-range backbone and long-range backbone					Constraints with sidechains				
			UP	LO				UP	LO				UP	LO
HN	C5	HN	3.0	2.0†	HA	L6	HN	4.0	2.0†	HD	N43	HB1	7.0	2.0†
					HB1	R42	HB1	4.0	2.0†	HD	R42	HB	7.0	2.0†
					HB1	R42	HB2	4.0	2.0†	HD	E7	HG1	7.0	2.0†
					HB1	N43	HN	4.0	2.0†	HD	F45	HZ	7.0	2.0†
					HB2	R42	HB2	4.0	2.0†	HE	C5	HA	7.0	2.0
										HE	N43	HA	6.0	2.0†
										HE	T54	HB	7.0	2.0†
										HE	R42	HD1	7.0	2.0†
										HE	F45	HE	7.0	2.0
										HE	T54	HG	6.0	2.0
										HE	F45	HZ	6.0	2.0†
										HZ	T54	HB	4.5	2.0†
										HE	P2	HB1	7.0	2.0
										HE	P2	HB2	6.0	2.0
										HZ	P2	HB2	4.0	2.0
										HD	P2	HG1	7.0	2.0
										HZ	P2	HG1	3.8	2.0
										HE	P2	HG2	7.0	2.0
										HA	E7	HG	olap	
Cys-5														
HN	F4	HA	3.6	2.2†	HA	E7	HN	4.0	2.0†	HB1	L6	HD1	5.5	2.0†
HN	F4	HN	3.0	2.0	HB2	Y23	HB2	4.0	2.0†	HB1	Y23	HD	7.0	2.0†
HA	L6	HN	3.7	2.2†	HN	P2	HB1	4.0	2.0	HB2	L6	HD2	5.5	2.0†
HB1	L6	HN	4.0	2.0†	HN	D3	HA	4.0	2.0	HB2	N43	HD1	3.0	2.0†
HN	L6	HN	3.0	2.0	HN	P2	HB1	4.0	2.0	HB2	N43	HD2	4.0	2.0†
										HB2	Y23	HE	7.0	2.0†
										HN	L6	HG	5.5	2.0†
Leu-6														
HN	C5	HN	3.0	2.0	HA	A25	HB	4.5	2.0†	HD1	Y23	HE	6.5	2.0
HN	C5	HA	3.7	2.2†	HB1	F4	HA	4.5	2.0†	HD1	A25	HN	5.5	2.0†
HN	C5	HB1	4.0	2.0†	HN	D3	HA	4.0	2.0†	HD2	A25	HB	5.0	2.0
HN	C5	HN	3.0	2.0†	HN	D3	HB	5.5	2.0†	HN	Y23	HD	7.5	2.0†
HA	E7	HN	3.6	2.2	HN	F4	HA	4.0	2.0†	HA	E7	HG	N	
HB2	E7	HN	4.0	2.0						HD	D3	HA	olap	
HN	E7	HN	3.0	2.0										
Glu-7														
HN	L6	HA	3.6	2.2	HB1	N43	HB1	4.0	2.0†	HA	P8	HG	6.0	4.3†
HN	L6	HB2	4.0	2.0	HB1	N43	HB2	4.0	2.0†	HB1	N43	HD1	4.0	2.0†
HN	L6	HN	3.0	2.0	HB1	N43	HN	4.0	2.0	HG1	F4	HD	7.0	2.0†
HA	P8	HD1	4.0	2.2	HN	C5	HA	4.0	2.0†	HG	L6	HA	N	
HA	P8	HD2	3.8	1.9						HN	F4	QD	N	
										HG	F4	HA	olap	
Pro-8														
HD1	E7	HA	4.0	2.2										
HD2	E7	HA	3.8	1.9										
HA	P9	HD1	4.0	2.1										
HA	P9	HD2	3.8	1.8										
Pro-9														
HD1	P8	HA	4.0	2.1	HA	N43	HB1	4.5	2.0†	HA	F22	HD	7.0	2.0
HD2	P8	HA	3.8	1.8	HA	N43	HB2	4.5	2.0†	HA	F33	HE	7.0	2.0†
HA	Y10	HA	4.5	2.9†	HB1	F22	HB1	4.0	2.0†	HB1	F22	HD	6.0	2.0
HA	Y10	HN	3.6	2.1	HB1	F22	HB2	4.0	2.0†	HB1	F33	HZ	4.0	2.0
HB1	Y10	HN	4.8	3.0	HB2	F22	HB2	4.0	2.0†	HB2	F22	HD	7.5	2.0

Appendix *(continued)*

Sequential backbone					Medium-range backbone and long-range backbone					Constraints with sidechains				
			UP	LO				UP	LO				UP	LO
										HB2	F33	HD	7.5	2.0
										HD1	F22	HD	6.0	2.0†
										HD1	F22	HE	6.5	2.0†
										HD2	F22	HD	7.5	2.0†
										HD2	N43	HD2	4.8	2.0†
										HD2	F22	HE	7.5	2.0†
										HG1	F22	HB1	4.0	2.0†
										HG1	F22	HD	7.0	2.0
										HG1	F22	HE	7.0	2.0
										HG1	F22	HZ	4.5	2.0
										HG2	F22	HB1	4.0	2.0†
										HG	F33	QR	olap	
Tyr-10														
HA	P9	HA	4.5	2.9†	HB	K41	HN	5.5	2.0†	HD	A40	HA	6.0	2.0†
HN	P9	HA	3.6	2.1	HB	N44	HB1	6.0	2.0†	HD	K41	HB1	7.0	2.0†
HN	P9	HB1	4.8	3.0†						HD	R39	HB1	7.0	2.0
HA	T11	HN	3.6	2.2						HD	K41	HD1	7.0	2.0†
HB	T11	HN	4.7	2.0						HD	Y35	HE1	7.0	2.0†
										HD	K41	HG1	6.0	2.0†
										HD	K41	HN	7.0	2.0†
										HE	A40	HA	6.0	2.0
										HE	R39	HA	7.5	2.0†
										HE	G12	HA2	6.0	2.0†
										HE	R39	HB1	6.0	2.0
										HE	P13	HD1	7.0	2.0†
										HE	P13	HD2	7.0	2.0†
										HE	K41	HG1	7.0	2.0†
										HE	R39	HG1	7.0	2.0†
										HE	K41	HG2	7.0	2.0†
										HN	F33	HE	7.0	2.0†
										HD	T11	HN	olap	
Thr-11														
HN	Y10	HA	3.6	2.2	HA	G36	HN	3.5	2.0†	HG	G36	HA2	5.5	2.0†
HN	Y10	HB	4.7	2.0						HG	V34	HB1	6.0	2.0†
HA	G12	HN	3.6	2.2†						HG	F33	HD	7.5	2.0†
HN	G12	HN	3.0	2.0						HG	Y35	HD1	5.5	2.0†
HB	G12	HN	olap		HA	Y35	HA	W		HG	V34	HG2	4.5	2.0†
										HG	G12	HN	5.0	2.0†
										HG	G36	HN	4.5	2.0
										HN	F33	HE	7.0	2.0†
										HA	F33	QR	olap	
										HA	Y35	QR	olap	
										HN	Y10	HD	olap	
Gly-12														
HN	T11	HA	3.6	2.2†	HA2	R39	HB	5.0	2.0†	HA2	Y10	HE	6.0	2.0†
HN	T11	HN	3.0	2.0						HA2	P13	HG	6.0	2.0†
HA1	P13	HD1	4.0	2.2†						HN	T11	HG	5.0	2.0†
HA1	P13	HD2	3.8	1.9†										
HA2	P13	HD1	4.0	2.2†										
HA2	P13	HD2	3.8	1.9†										
HN	P13	HD2	4.5	2.0†										
Pro-13														
HD1	G12	HA1	4.0	2.2†						HD1	Y10	HE	7.0	2.0†

Appendix *(continued)*

	Sequential backbone				Medium-range backbone and long-range backbone					Constraints with sidechains				
			UP	LO				UP	LO				UP	LO
HD2	G12	HA1	3.8	1.9†						HD2	Y10	HE	7.0	2.0†
HD1	G12	HA2	4.0	2.2†						HD2	Y35	HE1	4.0	2.0†
HD2	G12	HA2	3.8	1.9†						HG	G12	HA2	6.0	2.0†
HD2	G12	HN	4.5	2.0†						HG	A14	HB	5.5	2.0†
HB1	A14	HN	4.5	1.8†						HG	A38	HB	5.5	2.0†
HB2	A14	HN	4.8	3.0†						HG	R39	HG1	5.5	2.0†
										HG	A14	HN	5.5	2.2†
										HD1	A38	HB	olap	
										HD1	R39	HB	olap	
										HD2	R39	HB	olap	
Ala-14														
HN	P13	HB1	4.5	1.8†	HB	A16	HN	5.0	2.0†	HB	P13	HG	5.5	2.0†
HN	P13	HB2	4.8	3.0†	HB	G36	HA1	5.5	2.0†	HN	P13	HG	5.5	2.2†
HA	K15	HN	3.6	2.2	HB	G36	HA2	4.5	2.0†					
HB	K15	HN	4.8	2.0†	HB	G36	HN	5.5	2.0†					
					HB	G37	HA1	5.5	2.0†					
					HN	A38	HB	5.5	2.0†					
Lys-15														
HN	A14	HA	3.6	2.2	HN	G36	HA1	4.0	2.0†					
HN	A14	HB	4.8	2.0†										
HA	A16	HN	3.6	2.2†										
HB2	A16	HN	4.5	2.0†										
Ala-16														
HN	K15	HA	3.6	2.2†	HB	I18	HN	5.5	2.0†					
HN	K15	HB2	4.5	2.0†	HB	G37	HA1	5.5	2.0†					
HA	R17	HN	3.6	2.2	HB	G37	HA2	5.0	2.0†					
HB	R17	HN	4.5	2.0	HN	A14	HB	5.0	2.0†					
					HN	G36	HA1	3.0	2.0†	HB	I18	HD	olap	
Arg-17														
HN	A16	HA	3.6	2.2						HB	V34	HG1	5.0	2.0†
HN	A16	HB	4.5	2.0						HB	I19	HG2	6.0	2.0†
HB1	I18	HN	4.8	2.0†						HB	V34	HG2	6.3	2.0†
HB2	I18	HN	4.8	2.0†						HG1	V34	HG2	4.5	2.0†
HA	I18	HN	olap							HG2	V34	HG1	4.5	2.0†
										HN	I18	HD	6.0	2.0†
										HN	I18	HG2	5.5	2.0†
Ile-18														
HN	R17	HB1	4.8	2.0†	HB	G37	HA1	4.5	2.0†	HA	I19	HG3	6.3	2.0†
HN	R17	HB2	4.8	2.0†	HN	A16	HB	5.5	2.0†	HB	Y35	HD2	4.0	2.0†
HA	I19	HN	3.6	2.2	HN	V34	HA	4.0	2.0	HB	R20	HE	4.0	2.0†
HN	I19	HN	4.0	2.0†	HN	Y35	HN	4.0	2.0	HD	G37	HA1	4.5	2.0
										HD	G37	HA2	5.5	2.0
										HD	Y35	HB2	5.5	2.0†
										HD	Y35	HE1	4.5	2.0
										HG1	G37	HA1	5.5	2.0†
										HG1	G37	HA2	5.5	2.0†
										HG1	Y35	HB2	5.5	2.0†
										HG1	R20	HD2	5.5	2.0†
										HG1	Y35	HD2	5.5	2.0
										HG1	R20	HE	5.0	2.0†
										HG1	Y35	HE1	5.5	2.0
										HG1	I19	HN	4.5	2.0†
										HG2	R20	HB1	5.5	2.0†

Appendix *(continued)*

	Sequential backbone				Medium-range backbone and long-range backbone					Constraints with sidechains				
			UP	LO				UP	LO				UP	LO
										HG2	R17	HN	5.5	2.0†
										HN	V34	HG1	4.5	2.0
Ile-19														
HN	I18	HA	3.6	2.2	HA	F33	HN	4.0	2.0†	HA	R20	HG1	4.0	2.0†
HN	I18	HN	4.0	2.0†	HA	V34	HA	3.0	2.0	HG2	T32	HG	4.5	2.0†
HA	R20	HN	3.6	2.2	HA	Y35	HN	4.0	2.0	HG3	R17	HB	6.0	2.0†
										Hg3	I18	HA	6.3	2.0†
										HG3	R20	HA	6.0	2.0†
										HG3	Y21	HD	7.5	2.0†
										HG3	Y21	HE	7.5	2.0†
										HG3	T32	HG	4.5	2.0†
										HG3	F33	HN	5.3	2.0†
										HN	I18	HG1	4.5	2.0†
										HN	R20	HG1	4.5	2.0†
										HA	T32	HG	N	
										HG	R20	HN	N	
										HG	T32	HA	N	
										HG	T32	HB	N	
Arg-20														
HN	I19	HA	3.6	2.2	HB2	Y35	HB2	4.0	2.0†	HA	I19	HG3	6.0	2.0†
HB2	Y21	HN	3.0	2.0	HB2	Y35	HN	4.8	2.0†	HB1	I18	HG2	5.5	2.0†
HN	Y21	HN	4.0	2.0†	HB2	N44	HA	4.5	2.0†	HB1	F33	HD	7.0	2.0†
					HB2	F45	HN	4.0	2.0	HB1	Y35	HD1	4.5	2.0†
					HN	T32	HA	4.5	2.0†	HB2	Y21	HD	7.5	2.0†
					HN	F33	HN	3.0	2.0	HB2	Y35	HD1	4.8	2.0†
					HN	V34	HA	4.0	2.0†	HD1	Y21	HD	7.5	2.0†
					HN	Y35	HN	4.5	2.0†	HD2	I18	HG1	5.5	2.0†
HA	Y21	HN	W		HB1	F45	HN	N		HG1	I18	HB		
										HG1	I19	HA	4.0	2.0†
										HG1	Y35	HD2	4.0	2.0†
										HE	I18	HB	4.0	2.0†
										HE	I18	HG1	5.0	2.0†
										HE	K46	HA	4.0	2.0†
										HD	K46	HA	olap	
										HN	F33	QR	N	
										HN	I19	HG	N	
Tyr-21														
HN	R20	HB2	3.0	2.0	HA	Q31	HN	4.0	2.0†	HA	F22	HD	6.0	2.0†
HN	R20	HN	4.0	2.0†	HA	F33	HN	4.0	2.0	HA	T32	HG	5.5	2.0
HA	F22	HN	3.6	2.2	HB	F22	HA	5.5	2.0†	HD	I19	HG3	7.5	2.0†
HB	F22	HN	4.5	2.0	HB	C30	HA	5.5	2.0†	HD	R20	HB2	7.5	2.0†
					HB	T32	HA	5.5	2.0†	HD	R20	HD1	7.5	2.0†
					HB	F45	HN	5.5	2.0	HD	F22	HN	6.8	2.0†
					HB	K46	HN	5.5	2.0†	HD	C30	HB2	7.0	2.0
					HB	A48	HB	6.0	2.0†	HD	Q31	HA	7.5	2.0†
					HB	C51	HB1	4.5	2.0†	HD	Q31	HN	7.0	2.0†
					HN	F45	HN	3.0	2.0	HD	T32	HA	7.0	2.0
					HA	T32	HN	N		HD	T32	HG	6.0	2.0
										HD	K46	HA	7.0	2.0
										HD	S47	HN	6.5	2.0†
										HD	A48	HA	7.0	2.0
										HD	A48	HB	6.0	2.0
										HD	A48	HN	7.0	2.0†

Appendix *(continued)*

	Sequential backbone				Medium-range backbone and long-range backbone					Constraints with sidechains				
			UP	LO				UP	LO				UP	LO
										HE	I19	HG3	7.5	2.0†
										HE	C30	HB2	7.0	2.0†
										HE	Q31	HA	7.0	2.0†
										HE	T32	HA	7.0	2.0
										HE	T32	HG	6.0	2.0
										HE	T32	HN	7.0	2.0†
										HE	S47	HA	6.0	2.0
										HE	S47	HB2	7.0	2.0
										HE	A48	HA	7.0	2.0
										HE	A48	HB	6.0	2.0
										QR	K46	HB	N	
Phe-22														
HN	Y21	HA	3.6	2.2	HA	Y21	HB	5.5	2.0†	HB2	F33	HE	7.0	2.0
HN	Y21	HB	4.5	2.0	HA	Y23	HB1	4.5	2.0†	HB2	N43	HD1	3.0	2.0†
HA	Y23	HN	3.6	2.0	HA	N44	HA	3.0	2.0†	HD	P9	HA	7.5	2.0
HB1	Y23	HN	4.0	2.0	HA	F45	HN	4.0	2.0†	HD	P9	HB1	6.0	2.0
HB2	Y23	HN	4.0	2.0	HB1	P9	HB1	4.0	2.0†	HD	P9	HB2	7.5	2.0
					HB2	P9	HB1	4.0	2.0†	HD	P9	HD1	6.0	2.0†
					HB2	P9	HB2	4.0	2.0†	HD	P9	HD2	7.5	2.0†
					HN	C30	HA	4.8	2.0†	HD	P9	HG1	7.5	2.0
					HN	Q31	HN	3.0	2.0	HD	Y23	HA	7.0	2.0†
										HD	Y23	HB2	7.0	2.0†
										HD	Y23	HN	6.0	2.0†
										HD	N24	HA	7.0	2.0
										HD	N24	HN	7.0	2.0†
										HD	Q31	HN	7.0	2.0†
										HD	F33	HB2	6.0	2.0†
										HD	F33	HN	7.0	2.0†
										HD	N43	HD1	6.0	2.0†
										HD	N43	HD2	6.0	2.0†
										HE	P9	HD1	6.5	2.0†
										HE	P9	HD2	7.5	2.0†
										HE	P9	HG1	7.0	2.0†
										HE	N24	HB1	6.0	2.0
										HE	F33	HA	7.5	2.0†
										HE	F33	HB2	6.0	2.0
										HE	N43	HD1	7.0	2.0
										HZ	P9	HG1	4.5	2.0†
										HZ	N24	HB2	4.0	2.0
										HA	N43	HD1	4.5	2.0
										HB1	F45	HD	7.0	2.0
										HA	F45	QR	olap	
Tyr-23														
HN	F22	HA	3.6	2.2	HA	C30	HA	3.0	2.0	HA	F22	HD	7.0	2.0†
HN	F22	HB1	4.0	2.0	HA	Q31	HN	4.0	2.0†	HA	N43	HD1	3.0	2.0†
HN	F22	HB2	4.0	2.0	HB1	F22	HA	4.5	2.0†	HB2	F22	HD	7.0	2.0†
HA	N24	HN	3.6	2.0	HB1	C55	HB2	4.5	2.0†	HB2	N43	HD1	4.5	2.0†
HB1	N24	HN	4.0	2.0	HB2	C5	HB2	4.0	2.0†	HD	C5	HB1	7.0	2.0†
					HB2	C55	HB1	4.5	2.0†	HD	L6	HN	7.5	2.0†
					HB2	C55	HB2	4.0	2.0†	HD	A25	HA	7.0	2.0
										HD	A25	HB	7.0	2.0
										HD	C30	HA	7.0	2.0†
										HD	C30	HN	7.0	2.0†

Appendix *(continued)*

	Sequential backbone				Medium-range backbone and long-range backbone					Constraints with sidechains				
			UP	LO				UP	LO				UP	LO
										HD	M52	HE	6.5	2.0†
										HD	C55	HB1	7.0	2.0†
										HE	R1	HD1	6.0	2.0†
										HE	R1	HD2	7.0	2.0†
										HE	R1	HE	7.0	2.0†
										HE	C5	HB2	7.0	2.0†
										HE	L6	HD1	6.5	2.0
										HE	A25	HN	7.0	2.0†
										HE	G28	HA1	7.0	2.0†
										HE	L29	HN	7.0	2.0†
										HE	M52	HE	6.0	2.0†
										HN	F22	HD	6.0	2.0†
										HN	N43	HD1	3.0	2.0
										HN	N43	HD2	4.0	2.0
										QR	L29	HA	olap	
										QR	G56	HA	olap	
Asn-24														
HN	Y23	HA	3.6	2.2	HA	A27	HN	4.5	2.0†	HA	F22	HD	7.0	2.0
HN	Y23	HB1	4.0	2.0	HB1	A27	HB	5.5	2.0†	HB1	F22	HE	6.0	2.0
HA	A25	HN	3.7	2.3	HB2	A27	HB	5.0	2.0†	HB1	L29	HD1	5.5	2.0†
HB1	A25	HN	4.5	2.0†	HN	L29	HN	4.5	2.0†	HB1	Q31	HE1	3.0	2.0†
					HN	C30	HA	3.0	2.0	HB1	Q31	HE2	4.0	2.0†
					HN	Q31	HN	4.0	2.0†	HB2	F22	HZ	4.0	2.0
HN	A25	HN	olap							HB2	Q31	HE1	4.0	2.0†
										HB2	Q31	HE2	4.0	2.0†
										HD1	A27	HB	4.5	2.0†
										HD1	L29	HD2	5.5	2.0†
										HD1	Q31	HG2	4.0	2.0†
										HD2	A27	HB	4.5	2.0†
										HD2	A27	HN	4.0	2.0†
										HD2	Q31	HG2	3.0	2.0†
										HD2	Q31	HE1	3.0	2.0†
										HD2	Q31	HE2	3.5	2.0†
										HN	F22	HD	7.0	2.0†
										HN	L29	HD2	6.0	2.0†
										HD2	Q31	HN	N	
Ala-25														
HN	N24	HA	3.7	2.3	HA	A27	HN	4.5	2.0†	HA	R1	HD2	4.5	2.0†
HN	N24	HB1	4.5	2.0†	HA	G28	HA2	4.5	2.0†	HA	Y23	HD	7.0	2.0
HA	K26	HN	3.6	2.2	HA	G28	HN	4.0	2.0	HB	R1	HE	5.5	2.0†
HB	K26	HN	4.5	2.0	HA	L29	HN	4.5	2.0†	HB	Y23	HD	7.0	2.0
HN	K26	HN	3.0	2.0	HN	A27	HN	4.0	2.0†	HN	L6	HD1	5.5	2.0†
					HN	G28	HN	4.0	2.0†	HN	Y23	HE	7.0	2.0†
										HN	N43	HD1	4.8	2.0†
										HN	N43	HD2	4.0	2.0†
Lys-26														
HN	A25	HA	3.6	2.2	HA	G28	HN	4.0	2.0					
HN	A25	HB	4.5	2.0	HB2	A27	HB	5.5	2.0†					
HN	A25	HN	3.0	2.0										
HA	A27	HN	4.8	2.9†	HN	A28	HN	N						
HB2	A27	HN	3.6	2.2										
HN	A27	HN	3.0	2.0										
Ala-27														

Appendix *(continued)*

Sequential backbone					Medium-range backbone and long-range backbone					Constraints with sidechains				
			UP	LO				UP	LO				UP	LO
HN	K26	HA	4.8	2.9†	HB	N24	HB1	5.5	2.0†	HB	N24	HD1	4.5	2.0†
HN	K26	HB2	3.6	2.2	HB	N24	HB2	5.0	2.0†	HB	N24	HD2	4.5	2.0†
HN	K26	HN	3.0	2.0	HB	K26	HB1	5.5	2.0†	HB	L29	HG	4.5	2.0†
HA	G28	HN	3.6	2.2†	HN	N24	HA	4.5	2.0†	HB	Q31	HG1	5.5	2.0†
HB	G28	HN	4.5	2.0	HN	A25	HA	4.5	2.0†	HB	Q31	HE2	5.5	2.0†
HN	G28	HN	3.5	2.0	HN	A25	HN	4.0	2.0†	HN	N24	HD2	4.0	2.0†
										HB	L29	HD	olap	
Gly-28														
HN	A27	HA	3.6	2.2†	HA2	A25	HA	4.5	2.0†	HA1	Y23	HE	7.0	2.0†
HN	A27	HB	4.5	2.0	HN	A25	HA	4.0	2.0	HN	L29	HG	6.0	2.0†
HN	A27	HN	3.5	2.0	HN	A25	HN	4.0	2.0†					
HA1	L29	HN	3.6	2.2†	HN	K26	HA	4.0	2.0†					
HA2	L29	HN	3.6	2.2†										
HN	L29	HN	3.5	2.0										
Leu-29														
HN	G28	HA1	3.6	2.2†	HN	N24	HN	4.5	2.0†	HB1	M52	HE	6.0	2.0†
HN	G28	HA2	3.6	2.2†	HN	A25	HA	4.5	2.0†	HD1	N24	HB1	5.5	2.0†
HN	G28	HN	3.5	2.0						HD1	C30	HN	4.5	2.0†
HA	C30	HN	4.5	2.0						HD1	Q31	HA	5.5	2.0†
HB1	C30	HN	3.0	2.0						HD1	Q31	HG1	5.5	2.0†
HB2	C30	HN	4.0	2.0						HD1	Q31	HG2	4.5	2.0†
										HD2	N24	HD1	4.5	2.0†
										HD2	N24	HN	6.0	2.0†
										HD2	C30	HN	4.5	2.0†
										HD2	Q31	HG1	5.5	2.0†
										HD2	Q31	HG2	4.5	2.0†
										HD2	Q31	HN	5.5	2.0†
										HG	A27	HB	4.5	2.0†
										HG	G28	HN	6.0	2.0†
										HN	Y23	HE	7.0	2.0†
										HA	Y23	QR	olap	
										HD	A27	HB	olap	
Cys-30														
HN	L29	HA	4.5	2.0	HA	Y21	HB	5.5	2.0†	HA	Y23	HD	7.0	2.0†
HN	L29	HB1	3.0	2.0	HA	F22	HN	4.8	2.0†	HB2	Y21	HD	7.0	2.0
HN	L29	HB2	4.0	2.0	HB1	A48	HA	3.5	2.0†	HB2	Y21	HE	7.0	2.0
HA	Q31	HN	3.7	2.0	HB1	A48	HB	4.5	2.0†	HN	Y23	HD	7.0	2.0†
HB1	Q31	HN	4.0	2.0	HB2	A48	HA	4.0	2.0†	HN	L29	HD1	4.5	2.0†
HB2	Q31	HN	4.0	2.0	HB2	C51	HB2	4.0	2.0†	HN	L29	HD2	4.5	2.0†
					HA	Y23	HA	3.0	2.0	HN	Q31	HG1	4.3	2.0†
					HA	N24	HN	3.0	2.0	HN	Q31	HG2	5.0	2.0†
										HN	M52	HE	5.5	2.0†
										HN	M52	HG	5.0	2.0†
Gln-31														
HN	C30	HA	3.7	2.2	HN	Y21	HA	4.0	2.0†	HA	Y21	HD	7.5	2.0†
HN	C30	HB1	4.0	2.0	HN	F22	HN	3.0	2.0	HA	Y21	HE	7.0	2.0†
HN	C30	HB2	4.0	2.0	HN	Y23	HA	4.0	2.0†	HA	L29	HD1	5.5	2.0†
HA	T32	HN	3.7	2.2	HN	N24	HN	4.0	2.0†	HG1	A27	HB	5.5	2.0†
										HG1	L29	HD1	5.5	2.0†
										HG1	L29	HD2	5.5	2.0†
										HG1	C30	HN	4.3	2.0†
										HG1	T32	HN	4.8	2.0†
										HG2	N24	HD1	4.0	2.0†

Appendix *(continued)*

	Sequential backbone				Medium-range backbone and long-range backbone					Constraints with sidechains				
			UP	LO				UP	LO				UP	LO
										HG2	N24	HD2	3.0	2.0†
										HG2	L29	HD1	4.5	2.0†
										HG2	L29	HD2	4.5	2.0†
										HG2	C30	HN	5.0	2.0†
										HE1	N24	HB1	3.0	2.0†
										HE1	N24	HB2	4.0	2.0†
										HE1	N24	HD2	3.0	2.0†
										HE2	N24	HB1	4.0	2.0†
										HE2	N24	HB2	4.0	2.0†
										HE2	N24	HD2	3.5	2.0†
										HE2	A27	HB	5.5	2.0†
										HN	Y21	HD	7.0	2.0†
										HN	F22	HD	7.0	2.0†
										HN	L29	HD2	5.5	2.0†
										HN	N24	HD2	N	
Thr-32														
HN	Q31	HA	3.7	2.2	HA	R20	HN	4.5	2.0†	HA	Y21	HD	7.0	2.0
HA	F33	HN	3.6	2.2	HA	Y21	HB	5.5	2.0†	HA	Y21	HE	7.0	2.0
HB	F33	HN	3.0	2.0	HB	F33	HA	4.5	2.0†	HA	F33	HD	6.5	2.0†
HN	F33	HN	4.0	2.0†						HG	I19	HG2	4.5	2.0†
					HA	Y21	HA	N		HG	I19	HG3	4.5	2.0†
										HG	Y21	HA	5.5	2.0
										HG	Y21	HD	6.0	2.0
										HG	Y21	HE	6.0	2.0
										HG	F33	HN	5.5	2.0†
										HN	Y21	HE	7.0	2.0†
										HN	Q31	HG1	4.8	2.0†
										HA	I19	HG	N	
										HB	I19	HB	N	
										HG	I19	HA	N	
Phe-33														
HN	T32	HA	3.6	2.2	HN	I19	HA	4.0	2.0†	HA	F22	HE	7.5	2.0†
HN	T32	HB	3.0	2.0	HN	R20	HN	3.0	2.0	HB2	F22	HD	6.0	2.0
HN	T32	HN	4.0	2.0†	HN	Y21	HA	4.0	2.0	HB2	F22	HE	6.0	2.0
HA	V34	HN	3.6	2.2										
HB1	V34	HN	3.0	2.0										
HB2	V34	HN	3.0	2.0						HD	P9	HB2	7.5	2.0
										HD	T11	HG	7.5	2.0†
										HD	R20	HB1	7.0	2.0†
										HD	T32	HA	6.5	2.0†
										HD	V34	HN	6.7	2.0†
										HD	Y35	HA	7.0	2.0†
										HD	Y35	HB1	7.0	2.0†
										HE	P9	HA	7.0	2.0†
										HE	Y10	HN	7.0	2.0†
										HE	T11	HN	7.0	2.0†
										HE	F22	HA	7.0	2.0†
										HE	F22	HB2	7.0	2.0†
										HE	Y35	HA	6.5	2.0†
										HE	N44	HA	5.5	2.0†
										HE	N44	HB2	7.0	2.0
										HZ	P9	HB1	4.0	2.0†
										HZ	Y35	HA	4.0	2.0

Appendix *(continued)*

	Sequential backbone				Medium-range backbone and long-range backbone					Constraints with sidechains				
			UP	LO				UP	LO				UP	LO
										HZ	N44	HA	3.0	2.0†
										HN	I19	HG3	5.3	2.0†
										HN	F22	HD	7.0	2.0†
										HN	T32	HG	5.5	2.0†
										QR	P9	HG	olap	
										QR	T11	HA	olap	
										QR	R20	HN	N	
										QR	F45	HN	N	
										QR	F22	HN	olap	
										QR	F22	HB	7.0	2.0
Val-34														
HN	F33	HA	3.6	2.2	HA	I18	HN	4.0	2.0	HB	T11	HG	6.0	2.0†
HN	F33	HB1	3.0	2.0	HA	I19	HA	3.0	2.0	HG1	R17	HB	5.0	2.0†
HN	F33	HB2	3.0	2.0	HA	R20	HN	4.0	2.0†	HG1	R17	HG2	4.5	2.0†
HA	Y35	HN	3.6	2.2	HA	Y35	HB1	4.8	2.0†	HG1	I18	HN	4.5	2.0
HB	Y35	HN	4.0	2.0†	HA	Y35	HB2	4.8	2.0†	HG1	Y35	HN	4.5	2.0
										HG2	T11	HG	4.5	2.0†
										HG2	R17	HB	6.3	2.0†
										HG2	R17	HG1	4.5	2.0†
										HG2	Y35	HN	5.5	2.0
										HG2	G36	HA2	5.5	2.0†
Tyr-35														
HN	V34	HA	3.6	2.2	HB1	V34	HA	4.8	2.0†	HA	F33	HD	7.0	2.0
HN	V34	HB	4.0	2.0†	HB2	R20	HB2	4.0	2.0†	HA	F33	HE	6.5	2.0†
HA	G36	HN	3.6	2.2	HB2	V34	HA	4.8	2.0†	HA	F33	HZ	4.0	2.0†
					HN	I18	HN	4.0	2.0	HB1	F33	HD	7.0	2.0†
					HN	I19	HA	4.0	2.0	HB1	N44	HD1	4.0	2.0†
					HN	R20	HB2	4.8	2.0†	HB2	I18	HD	5.5	2.0†
					HN	R20	HN	4.5	2.0†	HB2	I18	HG1	5.5	2.0†
					HA	T11	HA	W						
										HB2	N44	HD1	4.0	2.0†
										HD1	T11	HG	5.5	2.0†
										HD1	R20	HB1	4.5	2.0†
										HD1	R20	HB2	4.8	2.0†
										HD1	N44	HD2	3.0	2.0†
										HD2	I18	HB	4.0	2.0†
										HD2	I18	HG1	5.5	2.0
										HD2	R20	HG1	4.0	2.0†
										HD2	G36	HN	3.5	2.0†
										HD2	G37	HN	4.0	2.0†
										HD2	A40	HB	6.0	2.0
										HD2	N44	HD1	4.5	2.0†
										HD2	N44	HD2	4.5	2.0†
										HE1	Y10	HD	7.0	2.0†
										HE1	P13	HD2	4.0	2.0†
										HE1	I18	HD	4.5	2.0
										HE1	I18	HG1	5.5	2.0
										HE1	G36	HN	4.5	2.0†
										HE1	G37	HA1	3.5	2.0
										HE1	G37	HA2	4.8	2.0
										HE1	G37	HN	4.5	2.0†
										HE1	R39	HN	4.5	2.0†
										HE1	A40	HA	3.0	2.0

Appendix *(continued)*

	Sequential backbone				Medium-range backbone and long-range backbone					Constraints with sidechains				
			UP	LO				UP	LO				UP	LO
										HE1	A40	HB	4.5	2.0
										HE1	A40	HN	3.0	2.0†
										HE1	N44	HD1	4.5	2.0†
										HE2	G36	HN	4.8	2.0†
										HE2	G37	HA2	4.5	2.0†
										HE2	N44	HD2	4.5	2.0†
										HN	V34	HG1	4.5	2.0
										HN	V34	HG2	5.5	2.0
										QR	T11	HA	olap	
Gly-36														
HN	Y35	HA	3.6	2.2	HA1	A14	HB	5.5	2.0†					
HN	G37	HN	4.0	2.0†	HA1	K15	HN	4.0	2.0†	HA2	V34	HG2	5.5	2.0†
					HA1	A16	HN	3.0	2.0†	HN	T11	HG	4.5	2.0
					HA2	A14	HB	4.5	2.0†	HN	Y35	HD2	3.5	2.0†
					HN	T11	HN	3.5	2.0†	HN	Y35	HE1	4.5	2.0†
					HN	A14	HB	5.5	2.0†	HN	Y35	HE2	4.8	2.0†
Gly-37														
HN	G36	HN	4.0	2.0†	HA1	A14	HB	5.5	2.0†	HA1	I18	HD	4.5	2.0
					HA1	A16	HB	5.5	2.0†	HA1	I18	HG1	5.5	2.0†
					HA1	I18	HB	4.5	2.0†	HA1	Y35	HE1	3.0	2.0
					HA2	A16	HB	5.0	2.0†	HA2	I18	HD	5.5	2.0
										HA2	I18	HG1	5.5	2.0†
										HA2	Y35	HE1	4.8	2.0
										HA2	Y35	HE2	4.5	2.0
										HN	Y35	HD2	4.0	2.0†
										HN	Y35	HE1	4.5	2.0†
Ala-38														
HB	R39	HN	4.5	2.0†	HB	A14	HN	5.5	2.0†	HA	R39	HG1	4.5	2.0†
					HB	R39	HB	6.0	2.0†					
HA	R39	HN	W							HB	P13	HG	5.5	2.0†
										HB	R39	HG1	4.5	2.0†
										HB	R39	HG2	5.5	2.0†
										HB	R39	HE	5.5	2.0†
										HB	P13	HD1	olap	
										HB	Y10	QR	olap	
Arg-39														
HN	A38	HB	4.5	2.0†	HA	A40	HB	5.5	2.0†	HA	Y10	HE	7.5	2.0
HA	A40	HN	3.7	2.2	HB	G12	HA2	5.0	2.0†	HB	Y10	HD	7.0	2.0†
HB	A40	HN	4.8	2.0†	HB	A38	HB	6.0	2.0†	HB	Y10	HE	6.0	2.0†
HN	A40	HN	3.0	2.0						HB	P13	HD1	olap	
										HB	P13	HD2	olap	
										HG1	Y10	HE	7.0	2.0†
										HG1	P13	HG	5.5	2.0†
										HG1	A38	HA	4.5	2.0†
										HG1	A38	HB	4.5	2.0†
										HG2	A38	HB	5.5	2.0†
										HE	A38	HB	5.5	2.0†
										HN	Y35	HE1	4.5	2.0†
Ala-40														
HN	R39	HA	3.7	2.2	HB	R39	HA	5.5	2.0†	HA	Y10	HD	6.0	2.0
HN	R39	HB	4.8	2.0†	HB	N44	HB1	5.5	2.0†	HA	Y10	HE	6.0	2.0
HN	R39	HN	3.0	2.0	HB	N44	HB2	5.5	2.0†	HA	Y35	HE1	3.0	2.0
HA	K41	HN	3.6	2.2						HB	Y35	HD2	6.0	2.0

Appendix *(continued)*

	Sequential backbone				Medium-range backbone and long-range backbone					Constraints with sidechains				
			UP	LO				UP	LO				UP	LO
HB	K41	HN	4.5	2.0						HB	Y35	HE1	4.5	2.0
										HB	N44	HD1	4.5	2.0†
										HB	N44	HD2	4.5	2.0†
										HN	Y35	HE1	3.0	2.0†
Lys-41														
HN	A40	HA	3.6	2.2	HN	Y10	HB	5.5	2.0†	HB1	Y10	HD	7.0	2.0†
HN	A40	HB	4.5	2.0						HD1	Y10	HB	5.5	2.0†
HA	R42	HN	3.6	2.2						HD1	Y10	HD	7.0	2.0†
HB	R42	HN	olap							HD2	N43	HN	4.0	2.0†
										HG1	Y10	HB	5.5	2.0†
										HG1	Y10	HD	6.0	2.0†
										HG1	Y10	HE	7.0	2.0†
										HG1	R42	HN	4.0	2.0†
										HG2	Y10	HB	5.5	2.0†
										HG2	Y10	HE	7.0	2.0†
										HN	Y10	HD	7.0	2.0†
										HN	N44	HD2	4.0	2.0†
Arg-42														
HN	K41	HA	3.6	2.2	HB1	F4	HB1	4.0	2.0†	HB1	F4	HD	7.0	2.0
HA	N43	HN	3.7	2.2†	HB2	F4	HB1	4.0	2.0†	HD1	F4	HD	7.0	2.0
HB2	N43	HN	4.0	2.0†	HB2	F4	HB2	4.0	2.0†	HD1	F4	HE	7.0	2.0†
HN	N43	HN	4.0	2.0						HD2	F4	HB1	4.0	2.0†
										HN	K41	HG1	4.0	2.0†
										HN	F4	QR	N	
Asn-43														
HN	R42	HA	3.7	2.2†	HB1	E7	HB1	4.0	2.0†	HA	F4	HD	6.0	2.0
HN	R42	HB2	4.0	2.0†	HB1	P9	HA	4.5	2.0†	HA	F4	HE	6.0	2.0
HN	R42	HN	4.0	2.0	HB2	E7	HB1	4.0	2.0†	HA	F45	HD	7.0	2.0
HA	N44	HN	3.7	2.3	HB2	P9	HA	4.5	2.0†	HA	F45	HZ	4.0	2.0
					HN	F4	HB1	4.0	2.0†	HB1	F4	HD	7.0	2.0†
					HN	E7	HB1	4.0	2.0	HD1	C5	HB2	3.0	2.0†
										HD1	E7	HB1	4.0	2.0†
										HD1	F22	HA	4.5	2.0
										HD1	F22	HB2	3.0	2.0†
										HD1	F22	HD	6.0	2.0
										HD1	F22	HE	7.0	2.0†
										HD1	Y23	HA	3.0	2.0†
										HD1	Y23	HB2	4.5	2.0†
										HD1	Y23	HN	3.0	2.0
										HD1	A25	HN	4.8	2.0†
										HD2	C5	HB2	4.0	2.0†
										HD2	P9	HD2	4.8	2.0†
										HD2	F22	HD	6.0	2.0
										HD2	Y23	HN	4.0	2.0
										HD2	A25	HN	4.0	2.0†
										HN	K41	HD2	4.0	2.0†
Asn-44														
HN	N43	HA	3.7	2.3	HA	R20	HB2	4.5	2.0†	HA	F33	HE	5.5	2.0†
HA	F45	HN	4.5	2.0	HA	F22	HA	3.0	2.0†	HA	F33	HZ	3.0	2.0†
					HB1	Y10	HB	6.0	2.0†	HB2	F33	HE	7.0	2.0
					HB1	A40	HB	5.5	2.0†	HD1	Y35	HB1	4.0	2.0†
					HB2	A40	HB	5.5	2.0†	HD1	Y35	HB2	4.0	2.0†
										HD1	Y35	HD2	4.5	2.0†

Appendix *(continued)*

	Sequential backbone				Medium-range backbone and long-range backbone					Constraints with sidechains				
			UP	LO				UP	LO				UP	LO
										HD1	Y35	HE1	4.5	2.0†
										HD1	A40	HB	4.5	2.0†
										HD2	Y35	HD1	3.0	2.0†
										HD2	Y35	HD2	4.5	2.0†
										HD2	Y35	HE2	4.5	2.0†
										HD2	A40	HB	4.5	2.0†
										HD2	K41	HN	4.0	2.0†
Phe-45														
HN	N44	HA	4.5	2.0	HA	S47	HN	4.0	2.0	HD	Y21	HB	7.5	2.0†
HA	L46	HN	3.6	2.2	HB1	S47	HN	4.0	2.0	HD	F22	HB1	7.0	2.0†
HB1	L46	HN	4.0	2.0	HB1	C51	HA	4.0	2.0†	HD	N43	HA	7.0	2.0
HB2	L46	HN	3.0	2.0	HB1	C51	HB2	4.0	2.0†	HD	K46	HN	6.7	2.0†
					HB2	S47	HN	4.0	2.0	HD	T54	HB	7.5	2.0
					HB2	D50	HB2	3.0	2.0†	HE	F4	HE	7.0	2.0
					HB2	C51	HB1	4.0	2.0†	HE	T54	HN	7.0	2.0†
					HN	R20	HB2	4.0	2.0	HZ	F4	HD	7.0	2.0
					HN	Y21	HB	5.5	2.0	HZ	F4	HE	6.0	2.0
					HN	Y21	HN	3.0	2.0	HZ	N43	HA	4.0	2.0
					HN	F22	HA	4.0	2.0†	HZ	C51	HA	4.0	2.0†
										HZ	T54	HB	4.0	2.0
										HZ	T54	HG	4.5	2.0
										HZ	C55	HN	4.5	2.0†
										QR	F22	HA	olap	
Lys-46														
HN	F45	HA	3.6	2.2	HB2	S47	HB1	4.5	2.0†	HA	R20	HE	4.0	2.0†
HN	F45	HB1	4.0	2.0	HN	Y21	HB	5.5	2.0†	HA	Y21	HD	7.0	2.0
HN	F45	HB2	3.0	2.0						HN	F45	HD	6.7	2.0†
HA	S47	HN	3.6	2.2										
HN	S47	HN	3.0	2.0						HA	R20	HD	olap	
										HB	Y21	QR	N	
Ser-47														
HN	K46	HA	3.6	2.2	HA	E49	HN	4.0	2.0	HA	Y21	HD	6.0	2.0
HN	K46	HN	3.0	2.0	HB1	K46	HB2	4.5	2.0†	HA	Y21	HE	6.0	2.0
					HB1	E49	HB2	3.0	2.0	HB2	Y21	HE	7.0	2.0
HA	A48	HN	3.6	2.2	HB1	E49	HN	3.5	2.0	HN	Y21	HD	6.5	2.0†
HB2	A48	HN	3.0	2.0	HB1	D50	HN	4.0	2.0†					
HN	A48	HN	4.0	2.0†	HB2	E49	HB1	4.0	2.0†					
					HB2	E49	HN	3.0	2.0†					
					HN	F45	HA	4.0	2.0					
					HN	F45	HB1	4.0	2.0					
					HN	F45	HB2	3.0	2.0					
					HN	C51	HB1	4.0	2.0†					
					HN	C51	HN	4.0	2.0†					
					HA	A48	HA	4.5	2.9†					
					HA	A48	HB	6.0	2.0†					
Ala-48														
					HA	C30	HB1	3.5	2.0†	HA	Y21	HD	7.0	2.0
HN	S47	HA	3.6	2.2	HA	C30	HB2	4.5	2.0†	HA	Y21	HE	7.0	2.0
HN	S47	HB2	3.0	2.0	HA	S47	HA	4.5	2.9†	HB	Y21	HD	5.5	2.0
HN	S47	HN	4.0	2.0†	HB	Y21	HB	6.0	2.0†	HB	Y21	HE	6.0	2.0
HA	E49	HN	3.6	2.0	HB	C30	HB1	4.0	2.0†	HB	M52	HG	5.5	2.0†
HB	E49	HN	4.5	2.0	HB	S47	HA	6.0	2.0†	HN	Y21	HD	7.0	2.0†
HN	E49	HN	3.0	2.0	HB	M52	HN	6.0	2.0†	HA	M52	HG	olap	

Appendix *(continued)*

	Sequential backbone				Medium-range backbone and long-range backbone					Constraints with sidechains				
			UP	LO				UP	LO				UP	LO
Glu-49														
HN	A48	HA	3.6	2.2	HA	C51	HN	4.5	2.0†	HA	M52	HG	5.5	2.0†
HN	A48	HB	4.5	2.0	HA	M52	HB2	3.8	2.0†					
HN	A48	HN	3.0	2.0	HA	R53	HB2	4.0	2.0†					
HA	D50	HN	3.6	2.2	HA	R53	HN	4.0	2.0					
HB1	D50	HN	4.0	2.0	HB1	S47	HB2	4.0	2.0†					
HB2	D50	HN	4.0	2.0	HB2	S47	HB1	3.0	2.0†					
HN	D50	HN	3.0	2.0	HN	S47	HA	4.0	2.0					
					HN	S47	HB1	3.5	2.0					
					HN	S47	HB2	3.0	2.0					
					HN	C51	HN	3.0	2.0†					
					HA	M52	HN	3.5	2.0†					
Asp-50														
HN	E49	HA	3.6	2.2	HB2	F45	HB2	3.0	2.0†					
HN	E49	HB1	4.0	2.0	HN	S47	HB1	4.0	2.0†					
HN	E49	HB2	4.0	2.0	HN	M52	HN	4.0	2.0†					
HN	E49	HN	3.0	2.0	HA	R53	HN	olap						
HA	C51	HN	3.7	2.3	HA	T54	HN	N						
HB1	C51	HN	4.0	2.0†										
HB2	C51	HN	3.5	2.0†										
HN	C51	HB1	4.0	2.0†										
HN	C51	HN	3.0	2.0										
Cys-51														
HN	D50	HA	3.7	2.3	HA	T54	HB	4.0	2.0†	HA	F45	HZ	4.0	2.0†
HN	D50	HB1	4.0	2.0†	HA	T54	HN	4.0	2.0	HB1	M52	HG	5.0	2.0†
HN	D50	HB2	3.5	2.0†	HA	C55	HB1	4.0	2.0†	HB2	M52	HG	4.5	2.0†
HB1	D50	HN	4.0	2.0†	HB1	Y21	HB	4.5	2.0†	HN	M52	HG	5.5	2.0†
HN	D50	HN	3.0	2.0	HB1	F45	HB2	4.0	2.0†	HE	Y23	HD	6.5	2.0†
HA	M52	HN	3.6	2.2†	HB1	S47	HN	4.0	2.0†	HE	Y23	HE	6.0	2.0†
HB1	M52	HN	4.0	2.0	HB2	C30	HB2	4.0	2.0†	HE	L29	HB1	6.0	2.0†
HB2	M52	HN	3.0	2.0	HB2	F45	HB1	4.0	2.0†					
HN	M52	HN	3.0	2.0	HN	F45	HB2	4.0	2.0†					
					HN	S47	HN	4.5	2.0†					
					HN	E49	HA	4.5	2.0†					
					HN	E49	HN	3.0	2.0†					
					HN	R53	HN	4.0	2.0†					
Met-52														
HN	C51	HA	3.6	2.2†	HA	C55	HB1	4.0	2.0†	HE	C30	HN	5.5	2.0†
HN	C51	HB1	4.0	2.0	HA	C55	HN	4.0	2.0†	HE	G56	HA2	4.5	2.0†
HN	C51	HB2	3.0	2.0	HA	G56	HN	3.0	2.0	HG	G56	HN	5.5	2.0†
HN	C51	HN	3.0	2.0	HB2	E49	HA	3.8	2.0†	HG	C30	HN	5.0	2.0†
HA	R53	HN	3.6	2.2	HN	A48	HB	6.0	2.0†	HG	A48	HB	5.5	2.0†
HB2	R53	HN	4.0	2.0	HN	E49	HA	3.5	2.0†	HG	E49	HA	5.5	2.0†
HN	R53	HN	3.0	2.0	HN	D50	HN	4.0	2.0†	HG	C51	HB1	5.0	2.0†
					HN	C55	HN	4.5	2.0†	HG	C51	HB2	4.5	2.0†
HB1	R53	HN	olap		HA	G56	HA	N		HG	C51	HN	5.5	2.0†
										HG	A48	HA	olap	
										HE	L29	HA	olap	
Arg-53														
HN	M52	HA	3.6	2.2	HB2	E49	HA	4.0	2.0†					

Appendix *(continued)*

Sequential backbone					Medium-range backbone and long-range backbone					Constraints with sidechains				
			UP	LO				UP	LO				UP	LO
HN	M52	HB2	4.0	2.0	HN	E49	HA	4.0	2.0					
HN	M52	HN	3.0	2.0	HN	C51	HN	4.0	2.0†					
HA	T54	HN	3.7	2.2										
HB1	T54	HN	4.0	2.0										
HB2	T54	HN	4.0	2.0										
HN	T54	HN	3.0	2.0										
Thr-54														
HN	R53	HA	3.7	2.2	HB	C51	HA	4.0	2.0†	HB	F4	HE	7.0	2.0†
HN	R53	HB1	4.0	2.0	HN	C51	HA	4.0	2.0	HB	F4	HZ	4.5	2.0†
HN	R53	HB2	4.0	2.0	HN	G56	HN	4.0	2.0†	HB	F45	HD	7.5	2.0
HN	R53	HN	3.0	2.0						HB	F45	HZ	4.0	2.0
HA	C55	HN	3.6	2.2†						HB	C51	HA	4.0	2.0
HB	C55	HN	3.0	2.0						HG	F4	HE	6.0	2.0
HN	C55	HN	4.0	2.0						HG	F4	HZ	5.3	2.0
										HG	F45	HZ	4.5	2.0
										HN	F45	HE	7.0	2.0†
										HG	C55	HN	olap	
Cys-55														
HN	T54	HA	3.6	2.2†	HB1	Y23	HB2	4.5	2.0†	HA	R1	HD1	4.0	2.0†
HN	T54	HB	3.0	2.0	HB1	C51	HA	4.0	2.0†	HB1	Y23	HD	7.0	2.0†
HN	T54	HN	4.0	2.2	HB1	M52	HA	4.0	2.0†	HN	F45	HZ	4.5	2.0†
HA	G56	HN	3.7	2.2†	HB2	Y23	HB1	4.5	2.0†					
HN	G56	HA2	4.5	2.0†	HB2	Y23	HB2	4.0	2.0†					
HN	G56	HN	3.0	2.0	HN	M52	HA	4.0	2.0†					
					HN	M52	HN	4.5	2.0†					
					HA	P2	HD2	3.5	2.0					
					HA	G57	HN	W		HA	P2	HG	N	
										HN	T54	HG	olap	
Gly-56														
HA2	C55	HN	4.5	2.0†	HA2	G57	HA2	4.8	2.9†					
HN	C55	HA	3.7	2.2†	HN	M52	HA	3.0	2.0	HN	M52	HE	5.5	2.0†
HN	C55	HN	3.0	2.0	HN	T54	HA	4.0	2.0†					
HA2	G57	HA2	4.8	2.9†										
HN	G57	HN	3.0	2.0†										
HA	G57	HN	olap		HA	M52	HA	N						
Gly-57														
HA2	G56	HA2	4.8	2.9†	HA1	A58	HA	4.5	2.9†					
HN	G56	HN	3.0	2.0†	HA2	G56	HA2	4.8	2.9†					
HA1	A58	HA	4.5	2.9†	HA2	A58	HB	5.5	2.0†					
HN	A58	HN	3.0	2.0										
HA	A58	HN	olap		HN	C55	HA	W						
Ala-58														
HA	G57	HA1	4.5	2.9†	HA	G57	HA1	4.5	2.9†					
HN	G57	HN	3.0	2.0	HB	G57	HA2	5.5	2.0†					

† These cross peaks are not reported for the wild-type BPTI.

olap These cross peaks are reported for the wild-type BPTI but we could not report them due to the overlap with other peaks.

W These cross peaks were located under the H_2O peak.

N These cross peaks were reported for the wild-type BPTI but were not observed in Ala 14/Ala38 BPTI.

8

The Tryptophan Synthase $\alpha_2\beta_2$ Multienzyme Complex

Relationship of the Amino Acid Sequence and Folding Domains to the Three-Dimensional Structure

Edith Wilson Miles

Introduction

Tryptophan synthase (EC 4.2.1.20) has been the subject of many important genetic and biochemical studies *(1)*. In the late 1940s, Yanofsky initiated studies of tryptophan synthase, attempting to reveal the structural relationship between gene and protein *(2)*. His group isolated a number of mutants from *Neurospora crassa* and from *Escherichia coli* that required tryptophan for growth. These studies led to the discovery that the enzyme from *E. coli* is an $\alpha_2\beta_2$ complex composed of two nonidentical dissociable subunits, the α and β subunits *(3)*. The separate α and β subunits have low activities in two half reactions, termed here the α and β reactions, respectively. The $\alpha_2\beta_2$ complex is much more active than the separate subunits in the α and β reactions. The $\alpha_2\beta_2$ complex also catalyzes the overall reaction, termed here the $\alpha\beta$ reaction.

α reaction: indole 3-glycerol phosphate → indole + D-glyceraldehyde 3-phosphate

β reaction: L-serine + indole → L-tryptophan + H_2O

$\alpha\beta$ reaction: L-serine + indole 3-glycerol phosphate → L-tryptophan + D-glyceraldehyde 3-phosphate + H_2O

Tryptophan synthase from *E. coli* was the first multienzyme complex consisting of nonidentical subunits to be discovered and characterized *(3)*. It has subsequently served as a model for understanding protein-protein interaction in many other important multienzyme complexes, including complexes involved in the synthesis of DNA and protein. Tryptophan synthase from *N. crassa* was later found to be a single polypeptide chain containing two regions functionally and structurally equivalent to the α and β subunits from *E. coli (4)*. This type of enzyme is termed a multifunctional enzyme. Multienzyme complexes and multifunctional enzymes that catalyze sequential reactions in a metabolic pathway may facilitate the intramolecular transfer of a metabolic intermediate. Early experiments showed that indole does not appear as a free intermediate in solution in the $\alpha\beta$ reaction *(5)*. These results indicate that indole produced at the active site of the α subunit is transferred intramolecularly to the active site of the β subunit where it reacts with L-serine to yield L-tryptophan.

The Three-Dimensional Structure of the Tryptophan Synthase $\alpha_2\beta_2$ Multienzyme Complex

The three-dimensional structure of the tryptophan synthase $\alpha_2\beta_2$ multienzyme complex from *Salmonella typhimurium* has been determined at the National Institutes of Health in a collaboration between my group and David Davies's crystallographic group *(6)*. The complex has an extended $\alpha\beta\beta\alpha$ subunit arrangement with an overall length of about 150 Å (Color Plate II). The active centers of the α and β subunits in each α/β pair are separated by about 25 Å and are connected by a "tunnel." The tunnel is wide enough to allow the passage of indole and long enough to accommodate four molecules of indole. The tunnel probably provides a pathway for the internal diffusion of indole between the two active sites and prevents the escape of indole to the solvent.

In this chapter, I will describe the tertiary fold of each subunit and will correlate this new structural information with information obtained previously from studies of amino acid sequences, mutants, and folding domains.

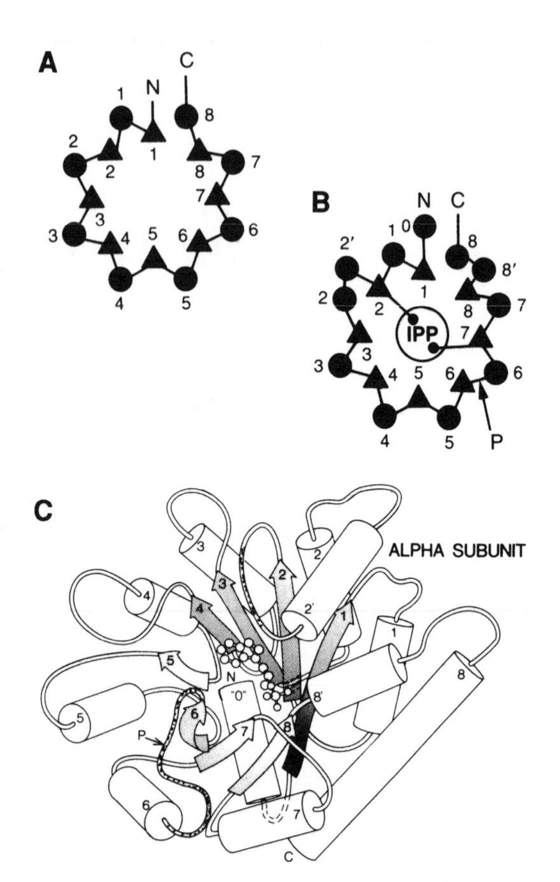

Fig. 1. Schematic representations of a canonical eight-fold α/β barrel protein (A) and of the α subunit of tryptophan synthase (B) and (C). (A) The eight alternating β strands (▲) and α helices (●) of a canonical eight-fold α/β barrel protein are numbered sequentially from the amino terminus (N) to the carboxyl terminus (C). (B) The α subunit, represented as in (A), contains three other helices labeled 0, 2′, and 8′. P indicates a known site of proteolysis at Arg-188 in a disordered loop between strand 6 and helix 6. Cleavage at this site yields an N-terminal fragment (α-1) and a C-terminal fragment (α-2). The active site is represented by a circle around IPP, the bound inhibitor, indole propanol phosphate. Two active site residues (Glu-49 and Gly-211) are indicated by—● (C). Schematic view of the overall fold of the α subunit based on the x-ray data. β strands are shown as flattened arrows with arrowheads at their C-termini. α helices are represented as cylinders and are labeled on their N-termini. In addition to the eight strands and helices found in a typical α/β barrel structure, the α subunit contains at least three other helices (labeled 0, 2′, and 8′). N and C mark the polypeptide amino- and carboxyl-termini. The darkened loops following strand 2 and strand 6 represent two polypeptide segments that are disordered in the crystal and are not currently part of the model. A known site of proteolysis (P) occurs in one of these disordered loops. The active site is centrally located near the C-terminal ends of the eight β strands. Indole propanol phosphate has been observed to bind in the active site as indicated by the ball-and-stick model. [Reprinted from *Adv. Enzymol. Relat. Areas Mol. Biol.* **64**, 93 (1991).]

The α Subunit

We were interested to find that the folding structure of the α subunit is a common one that has now been observed in triose phosphate isomerase and in at least 15 other enzymes *(6)*. This structure, which is termed an eightfold α/β-barrel structure, is built up from eight alternating α helices and β strands. The eight β strands form a parallel β barrel. The eight α helices form a larger cylinder of parallel helices concentric with the β barrel. The canonical eight-fold α/β-barrel structure is shown schematically in Fig. 1A where the α helices are shown by circles and the β strands are shown by triangles. The α subunit of tryp-

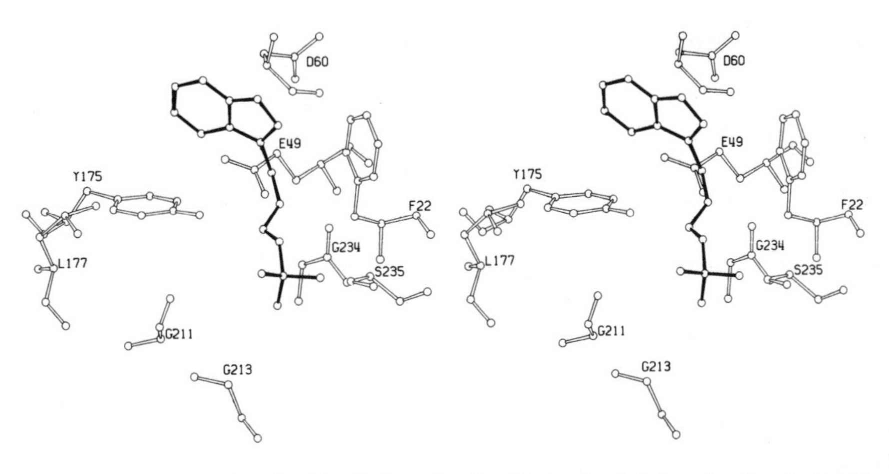

Fig. 2. Stereoview of the conformation of residues in the active site of the α subunit that are the sites of missense mutations or of second-site reversions as determined by x-ray crystallography. The binding site and conformation of the bound competitive inhibitor indole propanol phosphate determined from a difference electron-density map is shown near the center of the figure. I thank C. C. Hyde for this figure. [Reprinted from *Adv. Enzymol. Relat. Areas Mol. Biol.* **64,** 93 (1991).]

tophan synthase contains three extra helical segments, helices 0, 2′, and 8′, as shown schematically in Fig. 1B. The tertiary fold of the α subunit, based on the x-ray data, is shown in Fig. 1C. A substrate analogue, indole 3-propanol phosphate, binds in the hydrophobic center of the central β barrel. A close-up view of the active site in Fig. 2 shows the location of the bound inhibitor and of amino acid residues that have been identified as the sites of missense mutations or of second-site reversions by Yanofsky's group *(2)*. Figure 3 shows the location of these sites of mutations in the polypeptide chain of the α subunit.

Yanofsky found that the order of these sites of amino acid substitution in the polypeptide chain was the same as the order of the mutations in the trpA gene. This important finding that the gene and polypeptide were colinear was strong evidence that the gene sequence specifies the protein sequence. This fundamental molecular process was only fully understood after the genetic code was deciphered. Yanofsky's group found several cases in which multiple mutations at the same position on the trpA gene produced different amino acid changes at the corresponding position in the protein. These locations are shown in Fig. 3 at positions 49, 211, and 234. We now know that these amino acid changes result from different base changes in the nucleotide sequence of the parental codon. These mutant forms of the α subunit, with several different amino acid substitutions at a single site, have been very useful for a number of studies of the effects of single amino acid replacements on

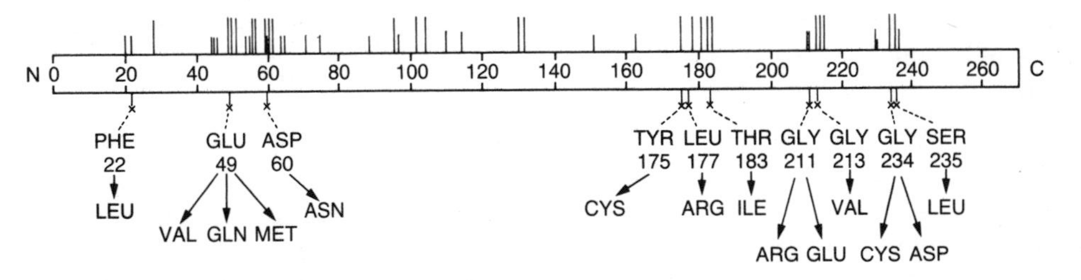

Fig. 3. The α subunit sequence: sites of missense mutations and conserved residues. The locations of mutations in the α subunit from *E. coli* and of amino acid residues that are highly conserved or invariant in homologous sequences. The 269 amino acids of the α subunit from *E. coli* are represented by the horizontal bar. Marks below the bar (–X) identify sites of missense mutations and of second-site reversions. These sites are identified by the amino acid change and the residue number (9–10). Marks above the bar identify the locations of amino acid residues that are highly conserved (short dash -) or invariant (long dash —) in many species of bacteria, *Neurospora crassa,* and *Saccharomyces cerevisiae (11)*. [Reprinted from *Adv. Enzymol. Relat. Areas Mol. Biol.* **64,** 93 (1991).]

the stability and folding of the α subunit *(7)*.

The crystallographic results show that all but one of the sites of mutation that were originally identified (Fig. 3) are clustered in the active site of the α subunit and are located close to the bound inhibitor (Fig. 2). Two active site residues (Glu-49 and Asp-60) are located in positions suitable to be catalytic residues. We have assigned catalytic roles to these residues on the basis of studies using site-directed mutagenesis *(8)*. Although Asp-60 was not one of the sites of mutation located in the early studies, this site has recently been identified by DNA sequence analysis of the only mutant that could not previously be analyzed. The site at position 60 was refractory to protein sequence analysis because it was located in a large, insoluble cyanogen bromide fragment. Our finding that these nine sites of mutation are located in the active site indicates that the α subunit contains only a small number of crucial positions at which a single amino acid replacement can completely inactivate the enzyme. Early studies of second-site revertants at positions 175, 177, 211, and 213 led to the suggestion that these residues are in close proximity in the folded α subunit *(10)*. That prediction is confirmed by the crystallographic data *(6)*. We thus conclude that the studies with mutants gave important clues to the relationship between the amino acid sequence and the structure and function of the α subunit.

Comparisons of homologous amino acid sequences also give important information about the structure and function of the α subunit *(11)*. An alignment of the amino acid sequences of the α subunit from many species of bacteria and of the corresponding N-terminal α domain of the multifunctional enzymes from *Saccharomyces cerevisiae* and *N. crassa* shows that relatively few (about 9%) of the residues in the α subunit are completely invariant. Most of these invariant residues and highly conserved residues are clustered near positions that were previously identified as the sites of missense mutations or of second-site reversions (Fig. 3) and are located close to the substrate binding site in the three-dimensional structure of the α subunit *(6)*. The sequence between residues 44 and 65 contains a high incidence of conserved amino acids and the two catalytic residues described above (Glu-49 and Asp-60). This sequence also contains part of a region (residues 53–78) which is inserted between strand 2 and helix 2 in the canonical α/β barrel. Residues 55–58 are located at the interface of the α and β subunits. These residues may be highly mobile due to their very poor electron density features. Since this region is highly conserved, it may be very important for function or for interaction with the β subunit.

Structure Prediction by Evolutionary Comparison

Comparison of the conserved amino acids in the aligned sequences of the α subunit from many species was used to facilitate the prediction of the secondary structure of the α subunit before the x-ray structure was available *(12, 13)*. The predictions assume that essential structural and functional features are conserved during divergent evolution while those of lesser significance will vary. One study predicted an eight-fold α/β-barrel secondary structure *(12)*, which agreed very well with the results of x-ray crystallography *(6)*. The second study initially predicted a β-sheet/α-helix structure *(13)*. The results were re-evaluated after the crystallographic results became available and found to be consistent with an α/β-barrel structure *(13)*.

Evidence for Folding Domains in the α Subunit

Several types of experiments provide evidence that the α subunit contains domains that fold independently upon chemical denaturation. The evidence comes from complementation experiments and from studies of limited proteolysis and of folding. The complementation experiments used dimers of the α subunit, which were formed in low yield after treatment of the α monomer with a high concentration of urea and removal of the urea by dialysis

(14). When dimers were formed from different mutant α chains, it was found that certain combinations of mutant α chains regained the enzymatic activity that was absent in the mutant monomers. For example, an α subunit with a mutational alteration at residue 49, which is in the N-terminal part of the polypeptide chain, complemented another α subunit with a mutational alteration at position 211, which is in the C-terminal part of the polypeptide chain. The results were rationalized by a model that proposed that dimer formation is caused by exchange of terminal portions of the contributing monomer chains (Fig. 4A). The reciprocal exchange of mutant α-chain termini results formally in the construction of one functional active-site region containing the wild-type residues (Glu-49 and Gly-211) and one nonfunctional active site containing the two altered residues (see circles in Fig. 4A). The crystallographic results confirm that Glu-49 and Gly-211 are combined in the active

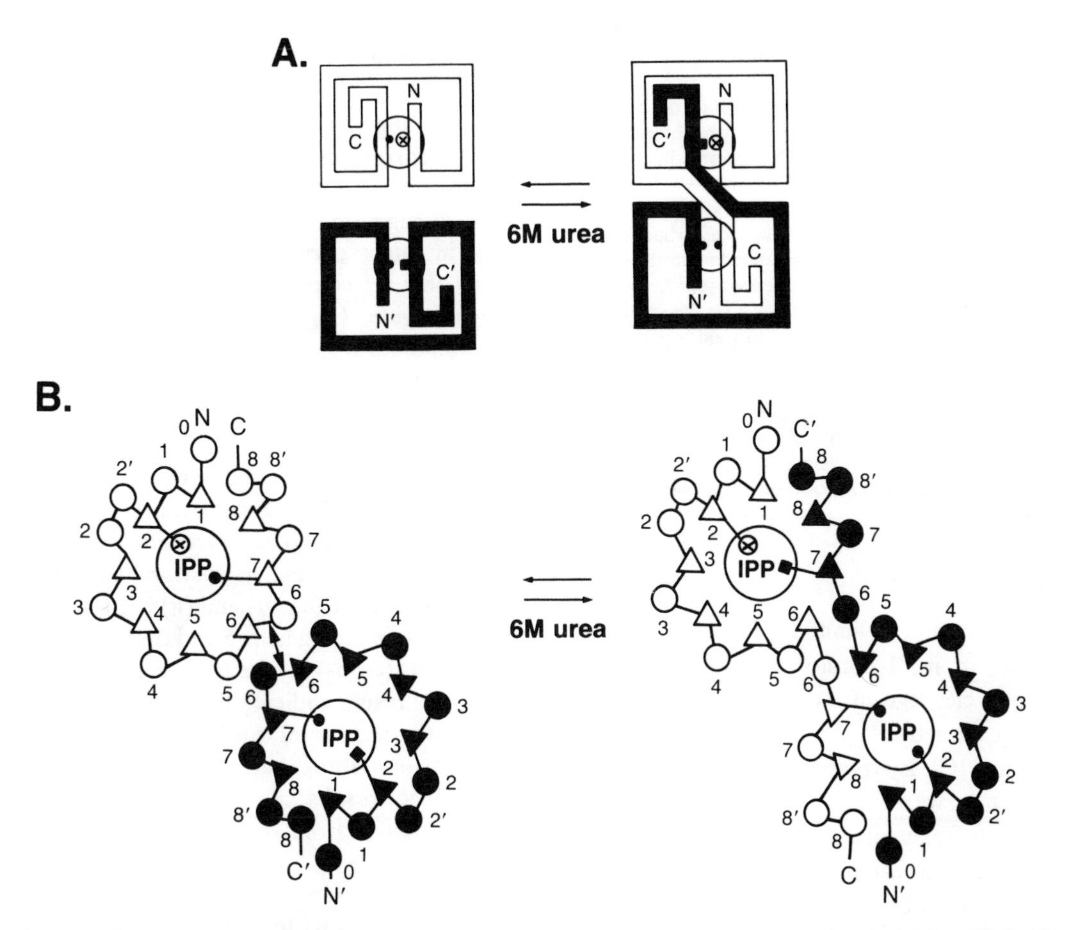

Fig. 4. Models for formation of dimers by the α subunit and for complementation by mutant forms of the α subunit. **(A)** Model proposed in *(14)*. The open bar (=) represents one polypeptide chain with an N-terminal mutation (X) such as one at residue 49. The solid bar (▬) represents a second polypeptide chain with a C-terminal mutation (■) at position 211. Unaltered residues at these positions are indicated by (●). The active site regions are indicated by the circled areas. The reciprocal exchange of mutant α-chain termini results formally in the the construction of one functional active site region and one doubly altered one. **(B)** Model based on x-ray structure and limited proteolysis experiments. The two mutant forms of the α subunit with amino acid substitutions at Glu-49 (X) or at Gly-211 (■) which are also shown in (A) are represented on the left by the schematic method described in Figs. 1A and 1B. The arrows point to the flexible loops between strand 6 and helix 6 which contain a site (Arg-188) that is susceptible to limited proteolysis. Cleavage at this site yields two fragments that correspond to folding domains (see text and Fig. 5). The model assumes that, following exposure to 6 M urea and removal of urea by dialysis, there is reciprocal exchange of the C-terminal folding domains as proposed in (A). This exchange may be facilitated by the long, flexible loop between strand 6 and helix 6.

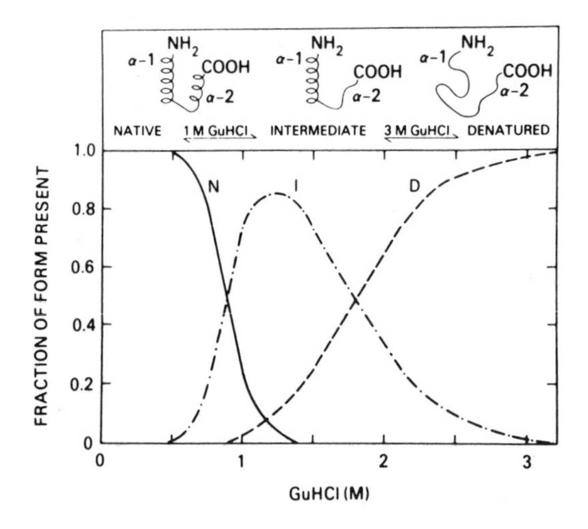

Fig. 5. Stepwise unfolding of the α subunit of tryptophan synthase from *E. coli* by guanidine hydrochloride. **(Bottom)** The fractions of native (N), intermediate (I), and denatured (D) states of the α subunit as a function of guanidine hydrochloride concentration at pH 7.0 and 26°C. **(Top)** Model of the denaturation process where α-1 and α-2 represent the domains corresponding to the α-1 and α-2 fragments obtained by tryptic cleavage at Arg-188. The α-2 domain of the intact α subunit (N) is shown to unfold at 1 M guanidine hydrochloride to yield a partially unfolded α intermediate (I); the α-1 domain is shown to unfold at 3 M guanidine hydrochloride to yield the fully denatured form (D). The model is based on the finding that the guanidine hydrochloride-induced unfolding of the α-2 and α-1 fragments, respectively, parallel these two steps. [Reprinted from *Biochemistry* **21**, 2586 (1982).]

site (Fig. 2).

Subsequent limited proteolysis experiments in my laboratory demonstrated that tryptic cleavage of the α subunit at Arg-188 yields an active "nicked" enzyme *(15)* (Figs. 1B and 1C). By treating the nicked α subunit with urea, we were able to separate and isolate the two fragments, termed α-1 and α-2. We showed that each fragment refolded independently after removal of urea. Our finding that the N-terminal residues 1–188 (α-1) refolded completely whereas the C-terminal residues 189–268 (α-2) refolded partially is evidence that the isolated fragments correspond to independent folding domains. Studies of the effect of guanidine hydrochloride *(16)* and of urea *(17)* concentration on the extent of unfolding of the α subunit under equilibrium conditions demonstrate that the unfolding process involves at least one stable intermediate (Fig. 5). Our finding that the stepwise unfolding of the α subunit by guanidine hydrochloride parallels the unfolding of the two proteolytic fragments supports the proposal that the principal folding intermediate has a folded N-terminal domain and an unfolded C-terminal domain *(18)* (Fig. 5). The conclusion that the α subunit has two folding domains is supported by subsequent hydrogen exchange experiments *(19)*.

Relationship between folding domains and the three-dimensional structure

Our finding that the α subunit contains regions corresponding to folding domains that are connected by a site susceptible to proteolysis suggested that these regions of the protein might correspond to structural domains that are connected by an easily cleaved hinge region (Fig. 5). The x-ray crystallographic results are thus surprising, since they show that the α subunit has a single structural domain (Fig. 1C). Arg-188, the site of proteolysis, is in a highly mobile surface loop that connects strand 6 and helix 6 (see arrow in Figs. 1B and 1C). Proteolytic cleavage in this loop results in (i) an N-terminal fragment containing the first five helix/strand structural units and strand 6 (α_5/β_6) and (ii) a C-terminal fragment containing helix 6 and the last two of these units (α_3/β_2). These folding and crystallographic results indicate that the N-terminal part of the α/β barrel can fold independently and that partial α/β barrels are stable structures. The results also show that folding domains differ from structural domains and that a protein with a single structural domain can have two or more folding domains.

We can now try to correlate the studies of complementary dimer formation described above and modeled in Fig. 4A with the crystal structure and the studies of folding domains. The complementary dimer is probably formed by the exchange of the terminal parts of two α chains near the loop that contains the proteolytic site at Arg-188 (Fig. 4B). This exchange and refolding is presumably facilitated by the flexibility of this loop and by the ability of each terminal part to refold independently.

The x-ray crystallographic structure now makes it easier to interpret previous studies on the kinetics of folding and of folding intermediates *(7, 20)*. Matthews is using some of the original missense mutants shown in Fig. 3 to determine which amino acid residues are involved in the "docking" of the folding domains *(21)*.

The β Subunit

Our representation of the three-dimensional structure of the $\alpha_2\beta_2$ complex of tryptophan synthase in Color Plate I shows the N-terminal residues 1–204 of the β subunit colored in yellow and the C-terminal residues 205–397 colored in red *(6)*. It is apparent that each β monomer contains two structural domains of nearly equal size *(6)*. One of these two domains is termed the N-domain, since it is largely composed of the yellow, N-terminal residues. The other domain is termed the C-domain, since it is largely composed of the red, C-terminal residues. Residues 53–85 in the N-terminal sequence "cross over" into the C-domain. The active site of the β subunit, which has been identified by the presence of the coenzyme pyridoxal phosphate, is located at the interface between these two domains. The tunnel passes through the interface between the N-domain and the C-domain to the active site of the β subunit and beyond. We think that the tunnel serves to transfer indole from the active site of the α subunit to the active site of the β subunit.

Figure 6 shows the folding patterns of each of the domains of the β subunit *(6)*. The core region of each domain contains four parallel strands with three helices packed on the interior side of the sheet and a fourth helix packed on the exterior side. Closer examination reveals that the structures of these two core regions are nearly superimposable. The presence of this high level of structural homology suggests that gene duplication followed by gene fusion may have occurred during the evolution of the β subunit. Figure 6 shows each domain from a point of view of the opposing domain. Since pyridoxal phosphate binds at

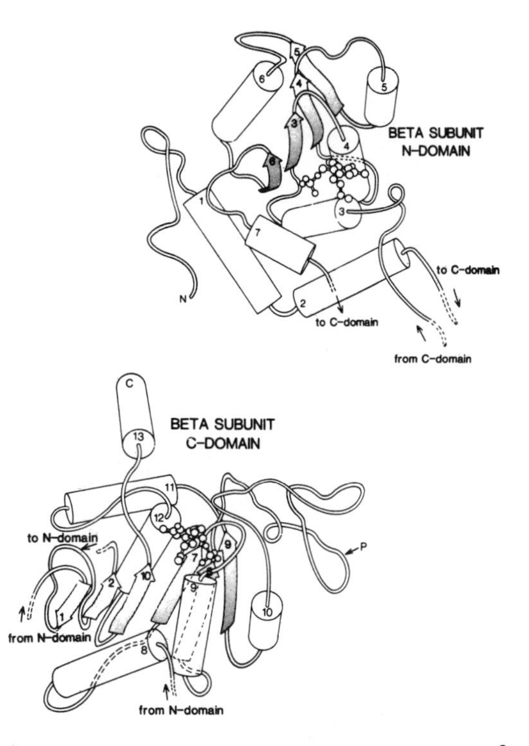

Fig. 6. Folding patterns of the two domains of the β subunit. **(Top)** Schematic view of the β subunit N-domain. The "core" of this domain is formed by a four-strand parallel β sheet (strands 6, 3, 4, and 5) packed on one side with three helices (3, 4, and 5) and by one helix on the opposite side (helix 6). The N-terminal helices 1 and 2 wrap around the core. The coenzyme pyridoxal phosphate (ball-and-stick model) binds covalently through a Schiff base linkage to Lys-87. A stretch of residues between helix 2 and helix 3 crosses over to and closely associates with the C-domain. Residues 1–8 at the N-terminus are disordered in the crystal. **(Bottom)** Schematic view of the β subunit C-domain showing the six-strand β sheet at its center. Strands 1 and 2 are formed from residues 53–85 of the N-terminal half of the chain. The "core" of the C-terminal domain, defined by helices 8, 9, 10, and 12 and strands 10, 7, 8, and 9 is topologically equivalent to the "core" of the N-domain. The pyridoxal phosphate:Lys-87 Schiff bases complex is shown in a ball-and-stick model with the phosphate group located toward the lower right. P shows a site susceptible to proteolytic cleavage. The cleavage site is within a stretch of residues (260–310), which contains three long β-hairpin loops and apparently lacks any other well-defined secondary structure. Each domain is shown here from a point of view from the opposing domain. [Reprinted from *Biotechnology* **8**, 27 (1990).]

the interface between the two domains, this coenzyme is on the surface of each domain. Pyridoxal phosphate is located near the C-terminal ends of the parallel strands in the core of each domain. The C-domain also contains a C-terminal helix and 50 residues (260–310) that fold in a complicated way and lack well defined secondary structural elements. The arrow in Fig. 6 indicates the location of three sites of tryptic cleavage that will be discussed below *(22)*.

Evidence for folding domains in the β subunit

When the β subunit is treated with various proteases, it is rapidly inactivated and yields a "nicked" protein consisting of two large polypeptide fragments termed the F1 and F2 fragments *(23)*. By treating the nicked protein with urea or guanidine hydrochloride, it is possible to dissociate and isolate the F1 and F2 fragments. Upon removal of the denaturing agent, each fragment spontaneously refolds to a conformation similar to that of the corresponding domain in the β subunit or nicked β subunit. These results are evidence that the isolated fragments correspond to independent folding domains. Michel Goldberg and his colleagues have used these fragments in a number of kinetic and immunological studies of intermediates on the pathway of folding of the β subunit *(24)*. The results provide evidence that the fragments correspond to regions of the protein that serve as intermediates in the pathway of unfolding and refolding of the intact protein.

Relationship between folding domains in the β subunit and the three-dimensional structure

Since the folding domains generated by limited proteolysis of the β subunit are derived from residues 1–272 and 284–397, they clearly differ from the two structural domains shown in Color Plate II and Fig. 6 which are largely derived from residues 1–204 and 205–397. The "hinge" between the two proteolytic domains, which is indicated by the arrow in Fig. 6, is located near the interface between the α subunit and the β subunit *(6)*. We previously predicted that the site of proteolysis was located in the subunit interaction site since we found that the rate of proteolysis of the β subunit is much slower in the presence of the α subunit *(15)* and since the nicked β subunit cannot form a complex with the α subunit *(23)*. The crystallographic results show that the site of proteolysis in the β subunit is in a flexible loop that is not a hinge between two structural domains. Nevertheless, the proteolytic fragments generated appear to correspond to folding domains and give useful information on the folding mechanism. The hinge between the two proteolytic domains also appears to be essential for enzyme activity *(23)*, for interaction with the α subunit *(23)*, and for conformational changes that occur upon substrate binding *(25)*.

Relationship between the amino acid sequence of the β subunit and the three-dimensional structure

The amino acid sequence homology of the β subunit is much higher than that of the α subunit *(11)*. The β subunit may have an unusually high sequence homology because it has several structural and functional features that must be preserved during evolution. In addition to the active site, the β subunit has large interaction sites between the two structural domains, between the two β monomers, and between the α and β subunits. It also contains the long, intramolecular tunnel for the passage of indole. The invariant amino acids in the β subunit are widely distributed in the three-dimensional structure. Although the locations of the invariant residues have not yet been thoroughly analyzed, certain of them are in the active site. Lys-87 is known to form a Schiff base with pyridoxal phosphate. The conserved residues 106–118 include residues 109, 114, and 115, which may be near the substrate-binding site. Conserved residues 229–237 contain residues 232–237, which are ligated to the phosphate of the coenzyme. The conserved residues 343–351 include Glu-350, which is located near the pyridine nitrogen of the coenzyme.

Conclusions and Future Directions

Further analysis of the three-dimensional structure of tryptophan synthase should help our group and others to identify residues in the structure that are potentially important for protein stability, protein folding, and protein-protein interaction. The roles of these residues can then be investigated by making a series of single amino acid replacements by site-directed mutagenesis and by investigating the properties of the mutant forms of tryptophan synthase and its subunits.

Acknowledgments

I am greatly indebted to Charles Yanofsky and Irving Crawford for sharing their interests in the tryptophan synthase problem and for sharing unpublished data. I am deeply saddened by Irving Crawford's untimely death October 8, 1989. I thank Craig Hyde for his excellent contributions to the crystallographic studies and for sharing unpublished data.

References

1. Yanofsky, C., and I. P. Crawford, *The Enzymes* **7,** 1 (1972); Miles, E. W., *Adv. Enzymol. Relat. Areas Mol. Biol.* **49,** 127 (1979); Miles, E. W., *Pyridoxal Phosphate and Derivatives*, vol. 1 of *Coenzymes and Derivatives*, D. Dolphin, R. Poulson, O. Avramovic, Eds. (John Wiley and Sons, New York, 1986), pp. 253–310; Miles, E. W., R. Bauerle, S. A. Ahmed, *Methods Enzymol.* **142,** 398 (1987); Miles, E. W., *Adv. Enzymol. Relat. Areas Mol. Biol.* **64,** 93 (1991).
2. Yanofsky, C., *Bioessays* **6,** 133 (1987).
3. Crawford, I. P., and C. Yanofsky, *Proc. Natl. Acad. Sci. U.S.A.* **44,** 1161 (1958).
4. Matchett, W. H., and J. A. DeMoss, *J. Biol. Chem.* **250,** 2941 (1975).
5. Yanofsky, C., and M. Rachmeler, Biochim. Biophys. Acta 28, 640 (1958); Matchett, W. H., and J. A. DeMoss, J. Biol. Chem. 250, 2941 (1975).
6. Hyde, C. C., S. A. Ahmed, E. A. Padlan, E. W. Miles, D. R. Davies, *J. Biol. Chem.* **263,** 17857 (1988); Hyde, C. C., and E. W. Miles, *Biotechnology* **8,** 27 (1990).
7. Yutani, K., K. Ogasahara, Y. Sugino, A. Matsushiro, *Nature* **267,** 274 (1977); Yutani, K., K. Ogasahara, A. Kimura, Y. Sugino, *J. Mol. Biol.* **160,** 387 (1982); Matthews, C. R., M. M. Crisanti, J. T. Manz, G. L. Gepner, *Biochemistry* **22,** 1445 (1983); Yutani, K., K. Ogasahara, K. Aoki, T. Kakuno, Y. Sugino, *J. Biol. Chem.* **259,** 14076 (1984); Hurle, M. R., N., B. Tweedy, C. R. Matthews, *Biochemistry* **25,** 6356 (1986); Beasty, A. M., M. R. Hurle, J. T. Manz, T. Stackhouse, J. J. Onuffer, C. R. Matthews, *Biochemistry* **25,** 2965 (1986); Yutani, K., K. Ogasahara, T. Tsujita, Y. Sugino, *Proc. Natl. Acad. Sci. U.S.A.* **84,** 4441 (1987).
8. Yutani, K., *et al., J. Biol. Chem.* **262,** 13429 (1987); Miles, E. W., P. McPhie, K. Yutani, *J. Biol. Chem.* **263,** 8611 (1988); Nagata, S., C. C. Hyde, E. W. Miles, *J. Biol. Chem.* **264,** 6288 (1989).
9. Shirvanee, L., V. Horn, C. Yanofsky, *J. Biol. Chem.* **265** 6624 (1990).
10. Helinski, D. R., and Yanofsky, C., *J. Biol. Chem.,* **238,** 1043 (1963); Yanofsky, C., V. Horn, D. Thorpe, *Science* **146,** 1593 (1964); Yanofsky, C., *Harvey Lect.* **61,** 145 (1967); Yanofsky, C., *J. Am. Med. Assoc.* **218,** 1026 (1971).
11. Hadero, A., and I. P. Crawford, *Mol. Biol. Evol.* **3,** 191 (1986); Crawford, I. P., *Annu. Rev. Microbiol.* **43,** 567 (1989).
12. Crawford, I. P., T. Niermann, K. Kirschner, *Proteins* **2,** 118 (1987).
13. Hurle, M. R., *et al., Proteins* **2,** 210 (1987).
14. Jackson, N. A., and C. Yanofsky, *J. Biol. Chem.* **244,** 4526 and 4539 (1969).
15. Higgins, W., T. Fairwell, E. W. Miles, *Biochemistry* **18,** 4827 (1979).
16. Yutani, K., K. Ogasahara, M. Suzuki, Y. Sugino, *J. Biochem. (Tokyo)* **85,** 915 (1979).
17. Matthews, C. R., and M. M. Crisanti, *Biochemistry* **20,** 784 (1981).
18. Miles, E. W., K. Yutani, K. Ogasahara, *Biochemistry* **21,** 2586 (1982).
19. Beasty, A. M., and C. R. Matthews, *Biochemistry* **24,** 3547 (1985).
20. Crisanti, M. M., and C. R. Matthews, *Biochemistry* **20,** 2700 (1981).
21. *Biotechnology* **7,** 324 (1989).
22. Ahmed, S. A., T. Fairwell, S. Dunn, K. Kirschner, E. W. Miles, *Biochemistry* **25,** 3118 (1986).
23. Hogberg-Raibaud, A., and M. E. Goldberg, *Biochemistry* **16,** 4014 (1977).
24. Zakin, M. M., G. Boulot, M. E. Goldberg, *Eur. J. Immunol.* **10,** 16 (1980); Zetina, C. R., and M. E. Goldberg, *J. Biol. Chem.* **255,** 4381 (1980); Zetina, C. R., and M. E. Goldberg, *J. Mol. Biol.*

157, 133 (1982); Chaffotte, A. E., and M. E. Goldberg, *Biochemistry* **22**, 2708 (1983); Blond, S., and M. E. Goldberg, *Proteins* **1**, 247 (1986); Blond, S., and M. E. Goldberg *Proc. Natl. Acad. Sci. U.S.A.* **84**, 1147 (1987); Murry-Brelier, A., and M. E. Goldberg, *Biochemistry* **27**, 7633 (1988); Friguet, B., L. Djavadi-Ohaniance, and M. E. Goldberg, *Res. Immunol.* **140**, 355 (1989).

25. Tschopp, J., and K. Kirschner, *Biochemistry* **19**, 4521 (1980); Chaffotte, A. F., and M. E. Goldberg, *Eur. J. Biochem.* **139**, 47 (1984); Djavadi-Ohaniance, L., B. Friguet, M. E. Goldberg, *Biochemistry* **25**, 2502 (1986).

9

Protein Structure Determination in Solution by Two-Dimensional and Three-Dimensional NMR[1]

Angela M. Gronenborn, G. Marius Clore

Introduction

Over the last few years we have witnessed a renewed interest in protein studies, especially aimed at understanding their structures, functions, and physiological roles. Some of this increased enthusiasm can be attributed to a variety of technological advances in the area of modern molecular biology, particularly molecular cloning of protein-encoding genes. The resulting wealth of data is clearly impressive; however, amino acid sequences per se are of limited value in understanding protein function. We need to know the three-dimensional (3-D) structure before we can begin to make progress in analyzing the intricate reactions carried out by proteins such as catalysis, ligand binding, gene regulation, and assembly. The design of modified proteins, rational design of drugs, as well as attempts at de novo design of protein molecules, all have to be based on concepts at the atomic level in three-dimensional space, thereby creating an increasing need for detailed structural analysis.

Until recently, the only experimental technique available for determining three-dimensional structures has been single crystal x-ray diffraction, and most of our structural knowledge about proteins is based on those crystal structures. There are approximately 400 coordinate sets available to date, comprising about 120 different protein folds. Analyzing protein structures by crystallography can be a slow and difficult undertaking since the first, and possibly hardest task, involves growing x-ray quality grade crystals, which have to be well ordered to give rise to good diffraction spots. Even if this task is accomplished, a second hurdle still needs to be overcome, as the phases have to be solved – commonly achieved by collecting data on heavy atom derivatives. Thus, despite spectacular advances in protein crystallography, we are faced with an enormous gap between the available primary sequence data and the tertiary structure data on proteins.

Over the last ten years, a second method for determining protein structures has been developed and is by now well established. This method makes use of nuclear magnetic resonance (NMR) spectroscopy. Unlike crystallography, NMR measurements are carried

1 An earlier version of this chapter appeared in *Analytical Chemistry* 62, 2 (1990), published by the American Chemical Society.

out in solution under potentially physiological conditions and are therefore not hampered by the ability or inability of a protein to crystallize.

The principal source of information used to solve three-dimensional protein structures by NMR spectroscopy resides in short interproton distances supplemented by torsion angles. The distances are derived from nuclear Overhauser effect (NOE) measurements. The size of the NOE between two protons is proportional to r^{-6}, where r is the distance between them. Torsion angles are obtained from an analysis of three-bond coupling constants which are related to dihedral angles. An essential prerequisite for obtaining interproton distance restraints and torsion angle restraints is the assignment of the NMR spectrum; that is to say, the identity of every proton resonance has to be determined. This is not a trivial task, considering that the proton spectrum of even a small protein, comprising only 80 amino acids, contains approximately 650 resonances. All of these exhibit several cross-peaks in the two-dimensional (2-D) spectrum such that the number of cross-peaks can easily reach several thousand whose identity has to be ascertained. Complete spectral assignment is, therefore, an integral part of the structure determination.

Although it was appreciated relatively early on that NMR could, in principle, provide the necessary information to obtain three-dimensional structures, it is only fairly recently that this goal has been realized. The reasons for this are threefold: (i) The development of 2-D NMR experiments *(1–3)* alleviated problems associated with resonance overlap, which for macromolecules prevents any analysis of the traditional one-dimensional spectrum. This is achieved by spreading all the information out in a plane, thereby permitting a detailed interpretation of the pertinent spectral features. This conceptual idea has been extended more recently to 3-D NMR *(4–7)*, again relieving problems arising from spectral crowding, and it is the 3-D approach, in particular, that will extend the present limits with respect to the size of the proteins that can be studied. (ii) The availability of high field magnets (500 and 600 MHz) has resulted in spectrometers with both a significant increase in signal-to-noise ratio and greater spectral resolution, and continuing development in this area again will extend the limits even further. (iii) Suitable mathematical algorithms and computational approaches that convert the NMR-derived restraints into three-dimensional structures have been developed *(8–14)*, and several robust and efficient methods are now available.

This chapter reviews the various stages involved in the determination of a three-dimensional protein structure by NMR, and the flow chart in Fig. 1 illustrates the individual steps. The general methodology is outlined; however, no attempt is made to provide an in-depth description of either the general NMR theory or the details of the mathematical algorithms. Emphasis is placed on the application of NMR to structural studies, and several examples illustrating various points are presented.

Basis of Two- and Three-Dimensional NMR

The principles of 2-D NMR have been discussed in depth [see *(15)* for a comprehensive review], and only a very basic and brief description will be given here. Each proton (spin) possesses a property known as magnetization. When a molecule is placed in a magnetic field, the magnetization lies parallel to it. Rotation of this magnetization away from its parallel orientation, either by a radio frequency pulse or a combination of pulses, allows one to follow the return of the magnetization to its equilibrium state. In any 2-D NMR experiment this is called the preparation period. This is followed by an evolution period in which this transient state of the spins is allowed to evolve for varying time periods t_1, a mixing period during which the spins are correlated with each other, and finally the detection period t_2. A number of experiments are recorded with increasing values for t_1 to generate a data matrix $s(t_1,t_2)$. Two-dimensional Fourier transformation of $s(t_1t_2)$ yields

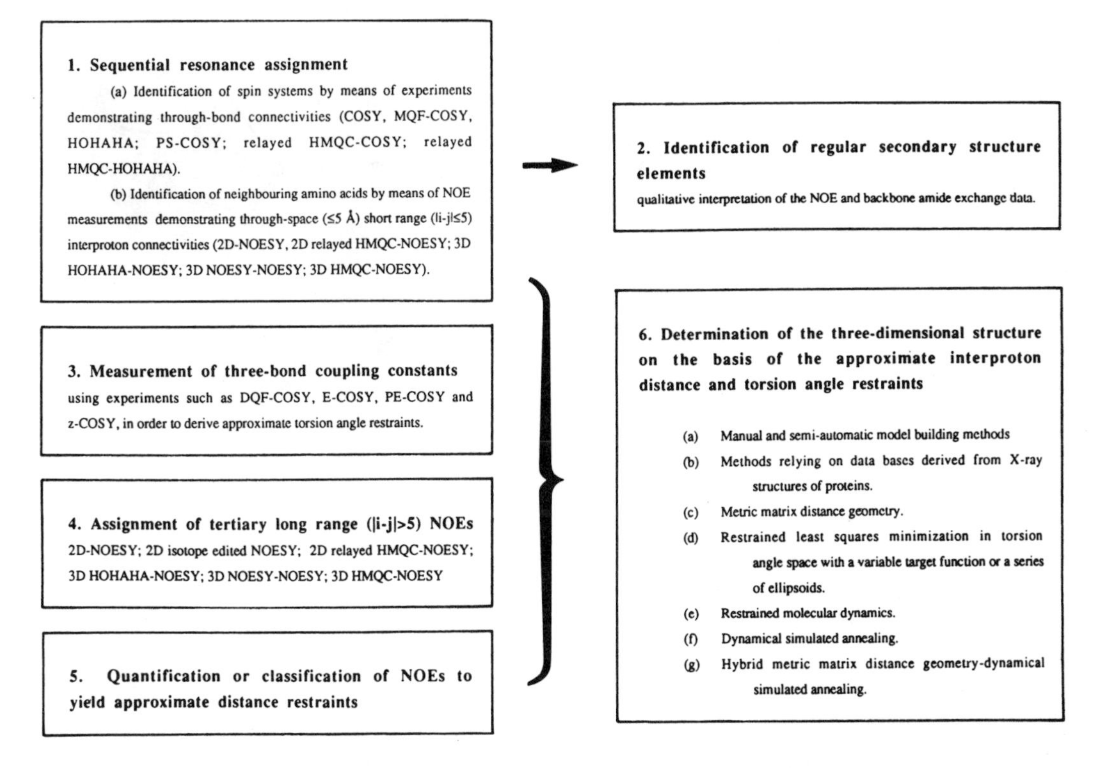

Fig. 1. Flow chart of the various steps involved in determining the three-dimensional structure of a protein in solution by NMR.

the 2-D spectrum $S(\omega_1,\omega_2)$. The second frequency dimension in the 2-D spectrum originates from the Fourier transformation of the t_1 modulation patterns in the 1-D spectra. The two frequency coordinates ω_1 and ω_2 of a particular resonance thus correspond to the t_1 and t_2 frequencies associated with the observed magnetization. This is illustrated schematically in Fig. 2. Most homonuclear 2-D experiments contain the 1-D spectrum on the diagonal with symmetrically placed cross-peaks on either side of the diagonal representing different kinds of interactions between the spins. The nature of the interaction depends on the type of experiment, with cross-peaks arising from through-bond scalar interactions in a COSY (correlated spectroscopy) *(2)* or HOHAHA (homonuclear Hartman-Hahn) *(16)* experiment and from through-space correlations in a NOESY (nuclear Overhauser and exchange spectroscopy) *(17)* experiment.

Since every 2-D NMR experiment consists of the basic scheme

Preparation – Evolution (t_1) – Mixing – Detection (t_2)

it is conceptually very simple to construct a 3-D experiment out of two 2-D experiments by omitting the detection period of the first 2-D experiment and the preparation period of the second one, then combining both into a single pulse train as illustrated in Fig. 3. The new detection period is now called t_3 and for every time variable t_2, a complete (t_1,t_3) 2-D data set is acquired. Fourier transformation of t_2 sections taken through these 2-D data sets converts the set of 2-D spectra into the 3-D spectrum. Since any two 2-D experiments can be combined into a 3-D experiment, one can envisage an enormous number of such experiments, starting from homonuclear versions and progressing to various heteronuclear 3-D experiments. A recent overview over a large number of combinations of 2-D experiments for 3-D versions can be consulted for further reading *(18)*. The power of the 3-D NMR experiments lies in overcoming resonance overlap, which for larger proteins becomes again a

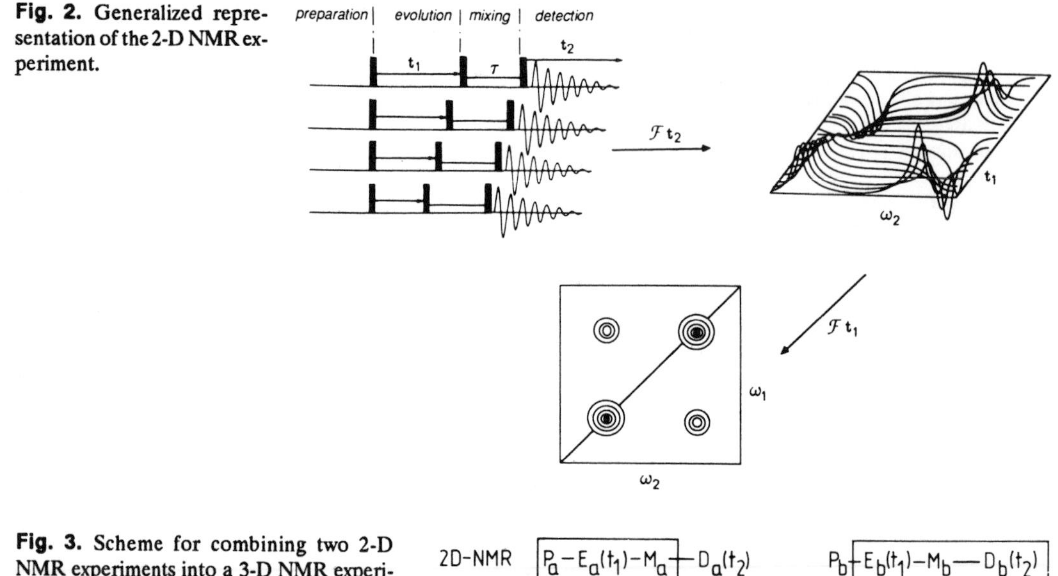

Fig. 2. Generalized representation of the 2-D NMR experiment.

Fig. 3. Scheme for combining two 2-D NMR experiments into a 3-D NMR experiment.

2D-NMR $P_a - E_a(t_1) - M_a - D_a(t_2)$ $P_b - E_b(t_1) - M_b - D_b(t_2)$

3D-NMR $P_a - E_a(t_1) - M_a - E_b(t_2) - M_b - D_b(t_3)$

major obstacle in the traditional 2-D ones. The first 3-D experiments on proteins were of the homonuclear type *(4, 5)*. While elegant and no doubt useful in certain cases, the applicability of these homonuclear 3-D experiments to larger proteins is limited, as the efficiency of magnetization transfer is severely reduced with increasing linewidths. This is not a problem for 3-D heteronuclear experiments *(6, 7)* that contain a heteronuclear shift correlation experiment such as 3-D NOESY-HMQC (heteronuclear multiple quantum coherence) or 3-D HOHAHA-HMQC. For those the heteronuclear magnetization transfer occurs via relatively large one-bond couplings and is therefore very effective. Such heteronuclear 3-D experiments represent, in essence, a series of 2-D NOESY or HOHAHA spectra, edited with respect to the chemical shift of the directly bonded heteronucleus such as ^{15}N or ^{13}C. Because these spectra can be regarded as stretched out heteronuclear edited 2-D spectra, their basic appearance and features resemble 2-D spectra, which allows for easy interpretation and data analysis. A schematic drawing of such a 3-D spectrum is presented in Fig. 4. Major advantages of these heteronuclear 3-D experiments are their high sensitivity (provided the protein is isotopically enriched) and ease of analysis.

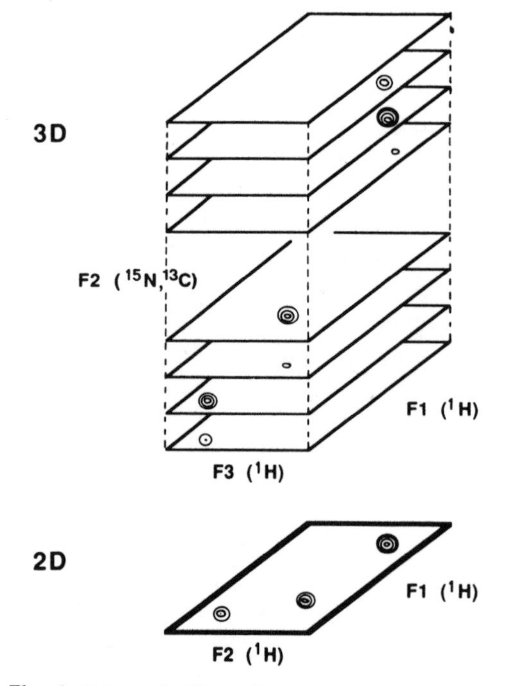

Fig. 4. Schematic illustration of a heteronuclear 3-D NMR experiment.

Nuclear Overhauser Effect

NMR-derived protein structures are mainly based on NOE measurements that can demonstrate the proximity of protons in space and allow determination of their approximate separation *(19–21)*. The principle of the NOE is relatively straightforward and is summarized in Fig. 5. Considering the simplest system with only two protons, each of which possesses a property known as magnetization, exchange of magnetization between the protons occurs by a process known as cross-relaxation. Because the cross-relaxation rates in both directions are equal, the magnetization of the two protons at equilibrium is equal. The approximate chemical analogy of such a system would be one with two interconverting species with an equilibrium constant of 1. The cross-relaxation rate is proportional to two variables: r^{-6}, where r is the distance between the two protons; and τ_{app}, the effective correlation time of the interproton vector. It follows that if the magnetization of one of the spins is perturbed, the magnetization of the second spin will change. In the case of macromolecules, the cross-relaxation rates are positive and the leakage rate from the system is very small, so that, in the limit, the magnetization of the two protons would be equalized. The change in magnetization of proton *i* upon perturbation of the magnetization of proton *j* is known as the nuclear Overhauser effect (NOE). The initial build-up rate of the NOE is equal to the cross-relaxation rate, and hence proportional to r^{-6}.

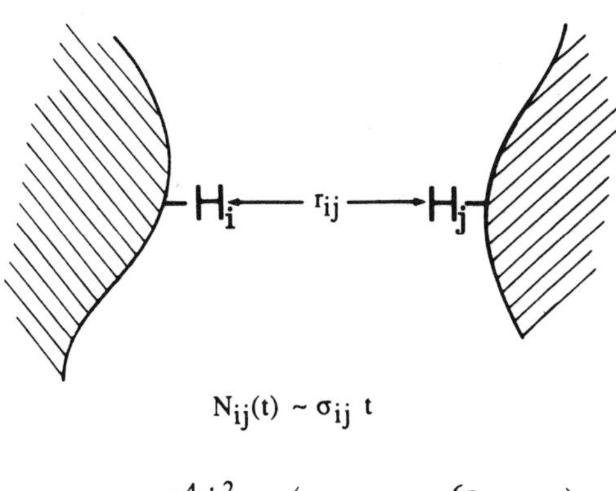

$$\sigma_{ij} = \frac{\gamma^4 \hbar^2}{10\, r_{ij}^6} \left(\tau_{app} - \frac{6\tau_{app}}{1 + 4\omega^2 \tau_{app}^2} \right)$$

Fig. 5. Basis of the NOE: r_{ij}, distance between the protons *i* and *j*; σ, cross-relaxation rate; *N*, NOE; τ_{app}, correlation time.

In 1-D NMR, the NOE can be observed in a number of ways, all of which involve the application of a selective radio frequency pulse at the position of one of the resonances. The simplest experiment involves the irradiation of resonance *i* for a time *t*, followed by acquisition of the spectrum. If proton *j* is close in space to proton *i*, its magnetization will be reduced. This is best observed in a difference spectrum, subtracting a spectrum without irradiation from one with selective irradiation. An alternative approach involves selective inversion of resonance *i* followed by acquisition after a time *t*. This particular experiment is the one-dimensional analogue of the two-dimensional experiment. In the two-dimensional experiment cross-peaks between proton resonances *i* and *j* are observed when the two protons are close in space, and thus exchange magnetization via cross-relaxation.

Sequential Resonance Assignment

Sequential resonance assignment of the ^{1}H-NMR spectra of proteins relies on two sorts of experiments: (i) those demonstrating through-bond scalar connectivities, and (ii) those demonstrating through-space (<5Å) connectivities *(22)*. The former, which are generally referred to as correlation experiments, serve to group together protons belonging to the same residue. The latter involve the detection of NOEs and serve to connect one residue with its immediate neighbors in the linear sequence of amino acids.

The first step in the assignment procedure lies in identifying spin systems, that is to say, protons belonging to one residue unit in the polypeptide chain. Such experiments have to be carried out both in H_2O and D_2O – the former to establish connectivities involving the exchangeable NH protons, and the latter to identify connectivities between nonexchangeable protons. Some spin systems are characteristic of single amino acids. This is the case for Gly, Ala, Thr, Leu, Ile, and Lys. Others are characteristic of several different amino acids. For example, Asp, Asn, Cys, Ser, and the aliphatic protons of all aromatic amino acids belong to the AMX spin system (i.e., they all have one $C^{\alpha}H$ and two $C^{\beta}H$ protons).

The simplest experiment used to delineate spin systems via scalar correlations is the COSY experiment, which was first described in 1976 by Aue *et al.* *(2)* and demonstrates direct through-bond connectivities. Thus, for a residue that has NH, $C^{\alpha}H$, $C^{\beta}H$, and $C^{\gamma}H$ protons, connectivities will only be manifested between the NH and $C^{\alpha}H$, $C^{\alpha}H$ and $C^{\beta}H$, and $C^{\beta}H$ and $C^{\gamma}H$ protons. This basic COSY experiment has now been superseded by slightly more sophisticated experiments such as DQF-COSY *(23)* and P.COSY *(24)*, which have the advantage of exhibiting pure phase absorption diagonals when the spectra are recorded in the pure phase absorption mode. This enables one to detect cross-peaks close to the diagonal.

Experiments that demonstrate only direct through-bond connectivities are of limited value if taken alone, owing to problems of spectral overlap. In a protein spectrum, the degree of spectral overlap tends to increase as one progresses from the NH and $C^{\alpha}H$ protons to the side-chain protons. For this reason, experiments that also demonstrate indirect or relayed through-bond connectivities, for example, between the NH and $C^{\beta}H$ protons, are invaluable. In this respect, the most useful experiment is the HOHAHA experiment *(16)*, also referred to as TOCSY for total correlated spectroscopy *(25)*. By adjusting the experimental mixing time, one can obtain successively direct, single, double, and multiple relayed connectivities. Further, the multiplet components of the cross-peaks are all in-phase in HOHAHA spectra, in contrast to COSY-type spectra where they are in antiphase. As a result, the HOHAHA experiment is, in general, more sensitive and affords better resolution than the COSY-type experiment. A schematic representation of cross-peak patterns observed in HOHAHA spectra for the various spin systems is illustrated in Fig. 6. Examples of protein HOHAHA spectra in H_2O and D_2O are shown in Fig. 7.

Once a few spin systems have been identified, one can then proceed to identify sequential through-space connectivities involving the NH, $C^{\alpha}H$, and $C^{\beta}H$ protons by means of 2-D NOESY. For the purpose of sequential assignment, the most important connectivities are the $C^{\alpha}H(i)$–$NH(i+1,2,3,4)$, $C^{\beta}H(i)$–$NH(i+1)$, $NH(i)$–$NH(i+1)$, and $C^{\alpha}H(i)$–$C^{\beta}H(i+3)$ NOEs. This is illustrated schematically in Fig. 8, and an example of a NOESY spectrum is shown in Fig. 9.

In the case of large molecules where the NH proton resonances are broad, the sensitivity of the conventional COSY spectrum can be improved by a factor of ~2 by recording a ^{15}N-filtered COSY spectrum *(26)* (also known as PS-COSY for pseudo single quantum COSY). Labeling the protein uniformly with ^{15}N allows for efficient generation of heteronuclear zero and double quantum

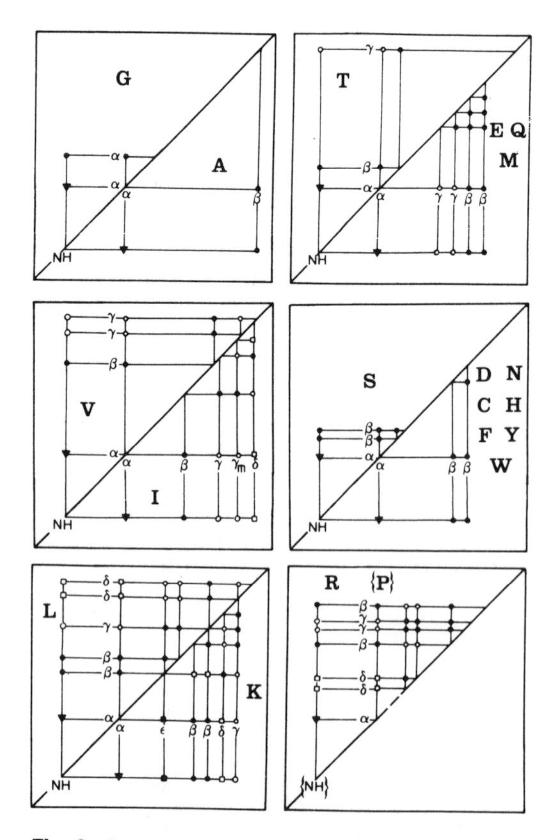

Fig. 6. Schematic representation of cross-peak patterns observed in HOHAHA spectra for various spin systems. In a COSY spectrum, only direct connectivities are observed.

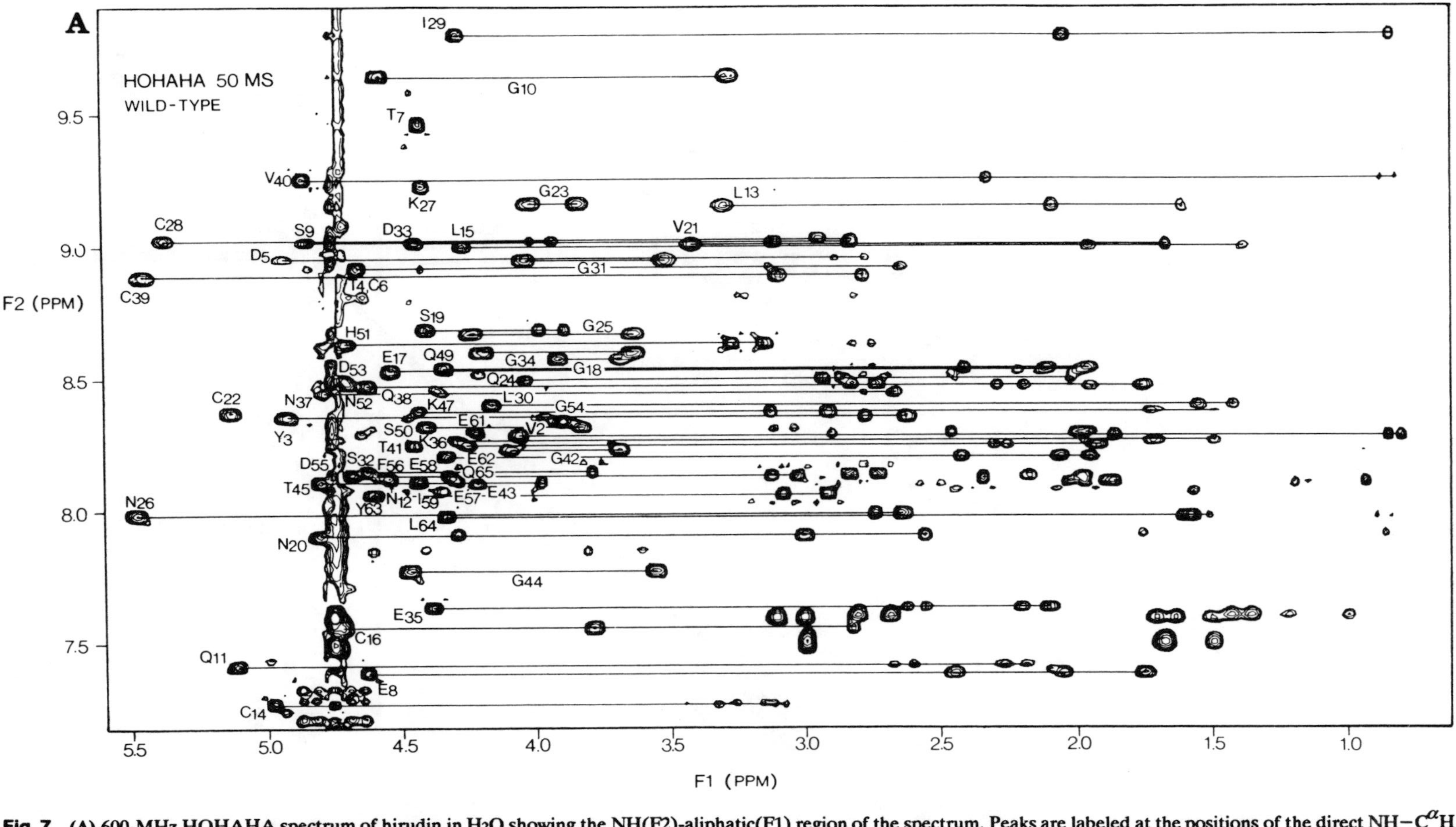

Fig. 7. (A) 600-MHz HOHAHA spectrum of hirudin in H_2O showing the NH(F2)-aliphatic(F1) region of the spectrum. Peaks are labeled at the positions of the direct NH–C^{α}H connectivities.

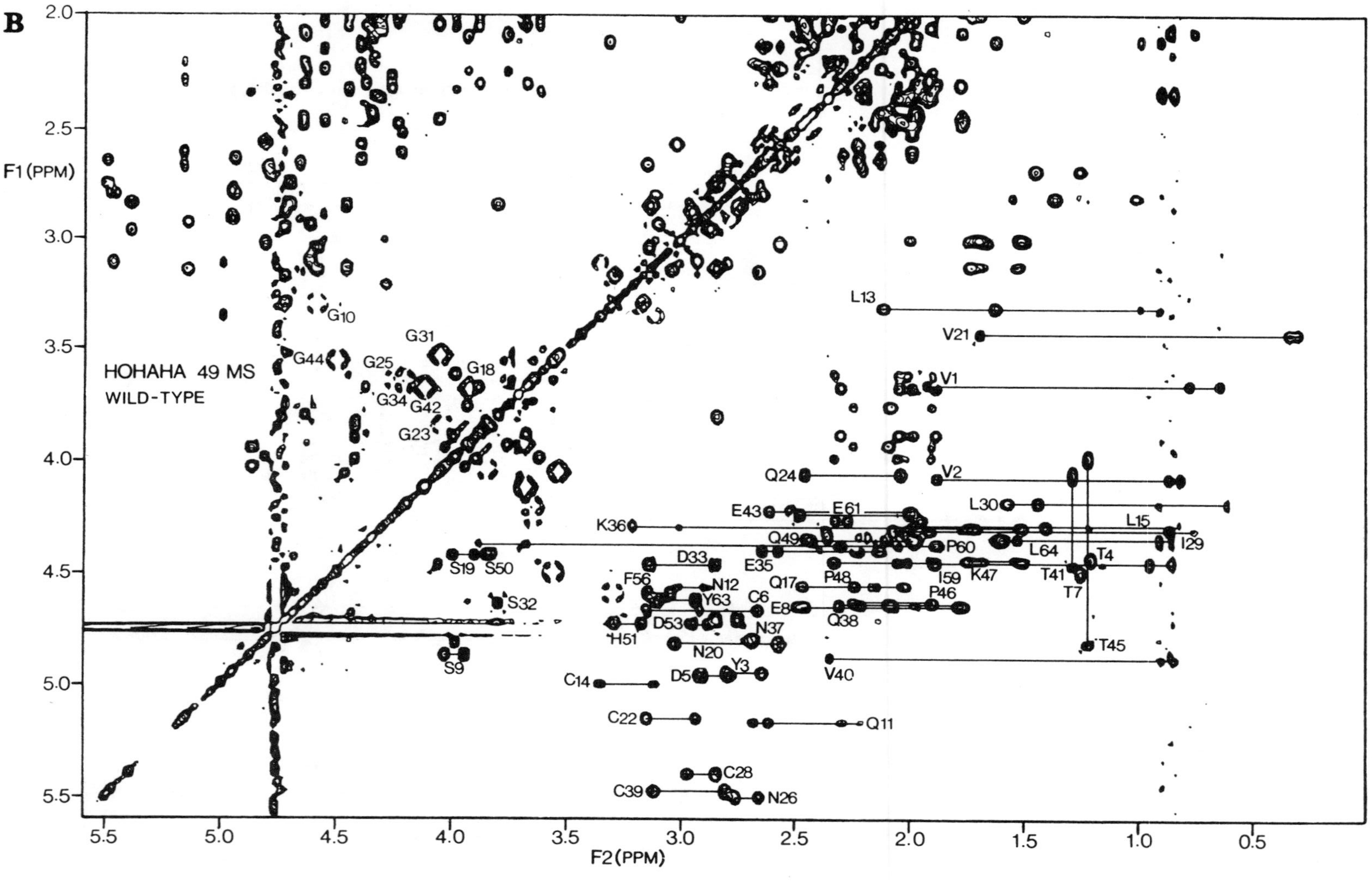

Fig.7. (B) 600-MHz HOHAHA spectrum of hirudin in D_2O showing the $C^{\alpha}H$(F1)-aliphatic(F2) region of the spectrum. A number of spin systems is indicated *(50)*.

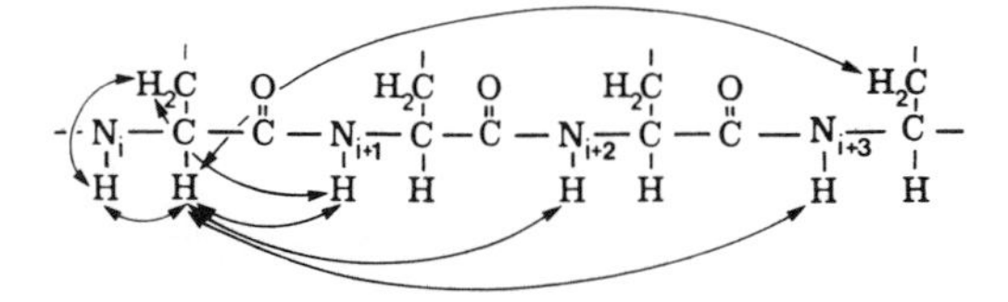

Fig. 8. Schematic illustration of the connectivities used for sequential resonance assignment of protein spectra.

coherences whose relaxation rates, to a first order approximation, are not affected by heteronuclear dipolar coupling. This permits one to eliminate one of the major line-broadening mechanisms for amide protons in proteins, namely, heteronuclear dipolar coupling to the nitrogen nucleus. As a result the multiple quantum resonances are significantly narrower than the corresponding NH resonances. The ^{15}N chemical shift contribution is easily removed from the multiple quantum frequency, yielding spectra that are similar in appearance to a regular COSY spectrum, apart from the line narrowing of the NH resonances.

As proteins get larger, problems associated with chemical shift dispersion become increasingly severe. One approach for alleviating such problems in the sequential assignment of proteins involves correlating proton-proton through-space and through-bond connectivities with the chemical shift of a directly bonded NMR active nucleus such as ^{15}N or ^{13}C. In the case of completely ^{15}N labeled protein, two sorts of experiments are particularly useful. The first are relayed experiments combining the heteronuclear multiple quantum coherence scheme *(27–31)* with experiments such as NOESY, COSY, or HOHAHA (Fig. 10) *(32)*. These experiments yield the same information as the homonuclear NOESY, COSY, or HOHAHA experiments, but the NH proton chemical shift axis is replaced by that of the ^{15}N chemical shift. Because it is very rare to find that both the ^{1}H and ^{15}N chemical shifts of two NH groups are degenerate, NOEs and through-bond correlations involving NH protons with the same chemical shift can readily be resolved in this manner. The second type of experiment involves the detection of long-range correlations between ^{15}N and CH atoms using ^{1}H-detected heteronuclear multiple-bond correlation (HMBC) spectroscopy *(33)*. In particular, the observation of two-bond

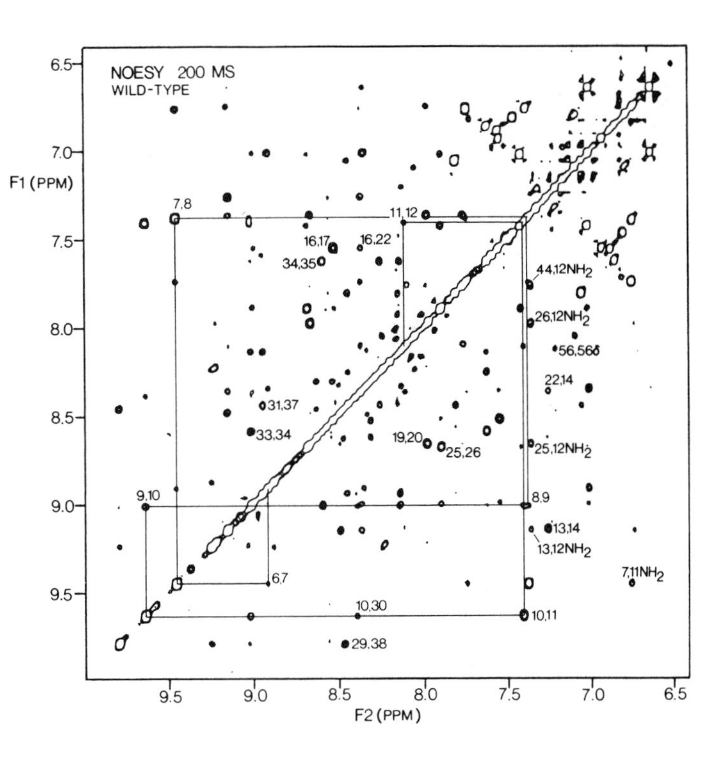

Fig. 9. 600-MHz NOESY spectrum hirudin in H_2O showing the NH(F1)–NH(F2) region of the spectrum. A stretch of NH(*i*)–NH(*i*+1) connectivities extending from residues 6 to 12 is indicated. In addition several long range NOEs are marked *(50)*.

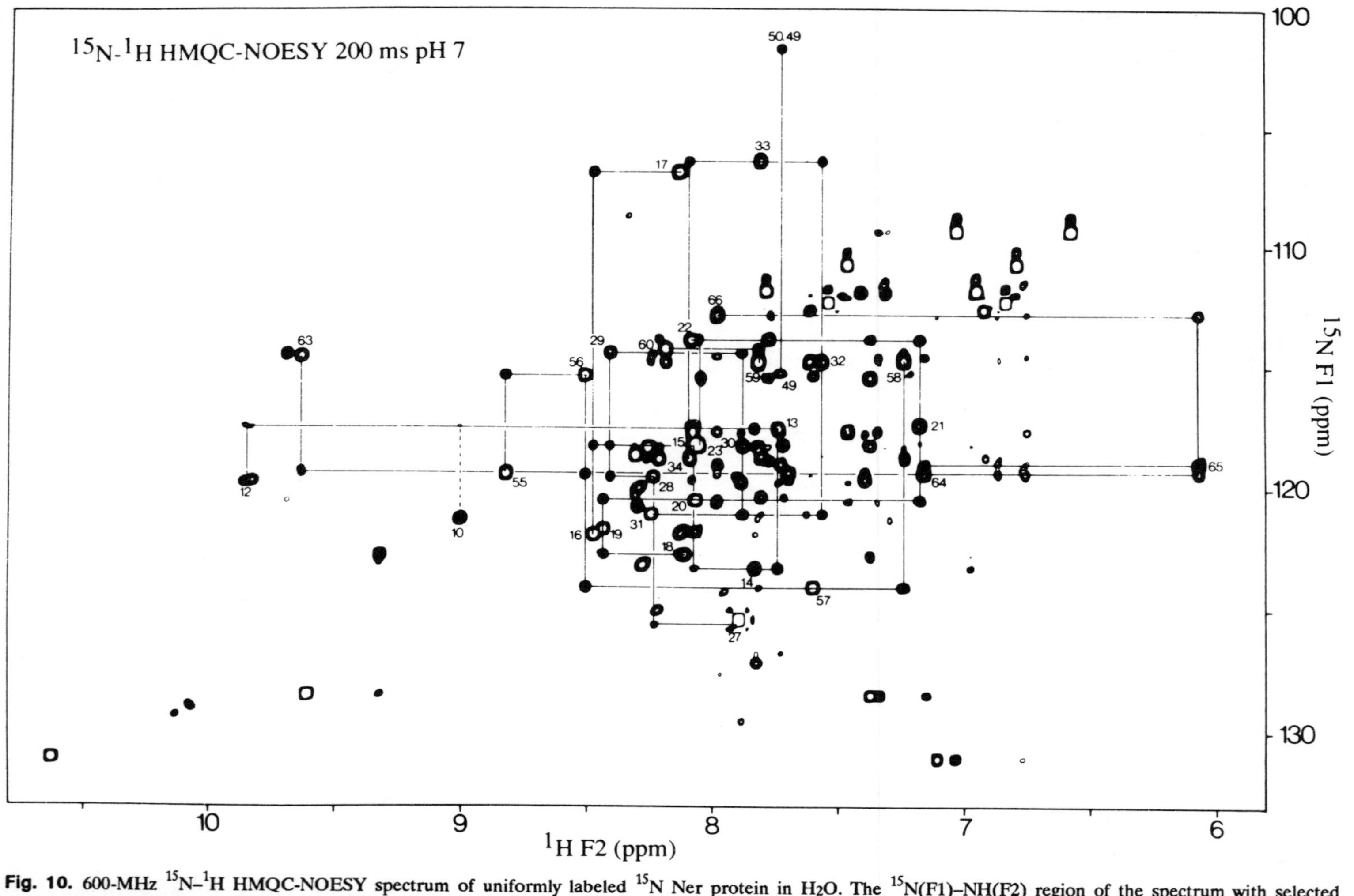

Fig. 10. 600-MHz ^{15}N–^{1}H HMQC-NOESY spectrum of uniformly labeled ^{15}N Ner protein in H_2O. The ^{15}N(F1)–NH(F2) region of the spectrum with selected NH(i)–NH(i+1) NOE connectivities is shown *(33)*.

$^{15}N(i)$–$C^{\alpha}H(i)$ and three-bond $^{15}N(i)$–$C^{\alpha}H(i - 1)$ correlations enables one to connect one residue with the next *(34)*. Additionally, because the size of the $^{15}N(i)$–$C^{\alpha}H(i - 1)$ coupling is very sensitive to the ψ backbone torsion angle, qualitative structural information can readily be derived.

A further avenue for large proteins involves the application of 3-D NMR. For ^{15}N 3-D spectra, the normal rules for making sequence-specific assignments can be readily applied. The only difference to the normal 2-D case is that connections between one residue and the next must be made not only between different sets of peaks but also between different planes of the spectrum. The major advantage in going from 2-D to 3-D is that the distribution of overlapping or closely spaced cross-peaks throughout the entire cube removes most of the ambiguities present in the 2-D spectrum. A comparison of the ^{15}N edited 2-D spectra with slices from the corresponding 3-D spectra is presented in Fig. 11, demonstrating the dramatic reduction in the number of cross-peaks. Combining such slices from a 3-D ^{15}N NOESY-HMQC spectrum

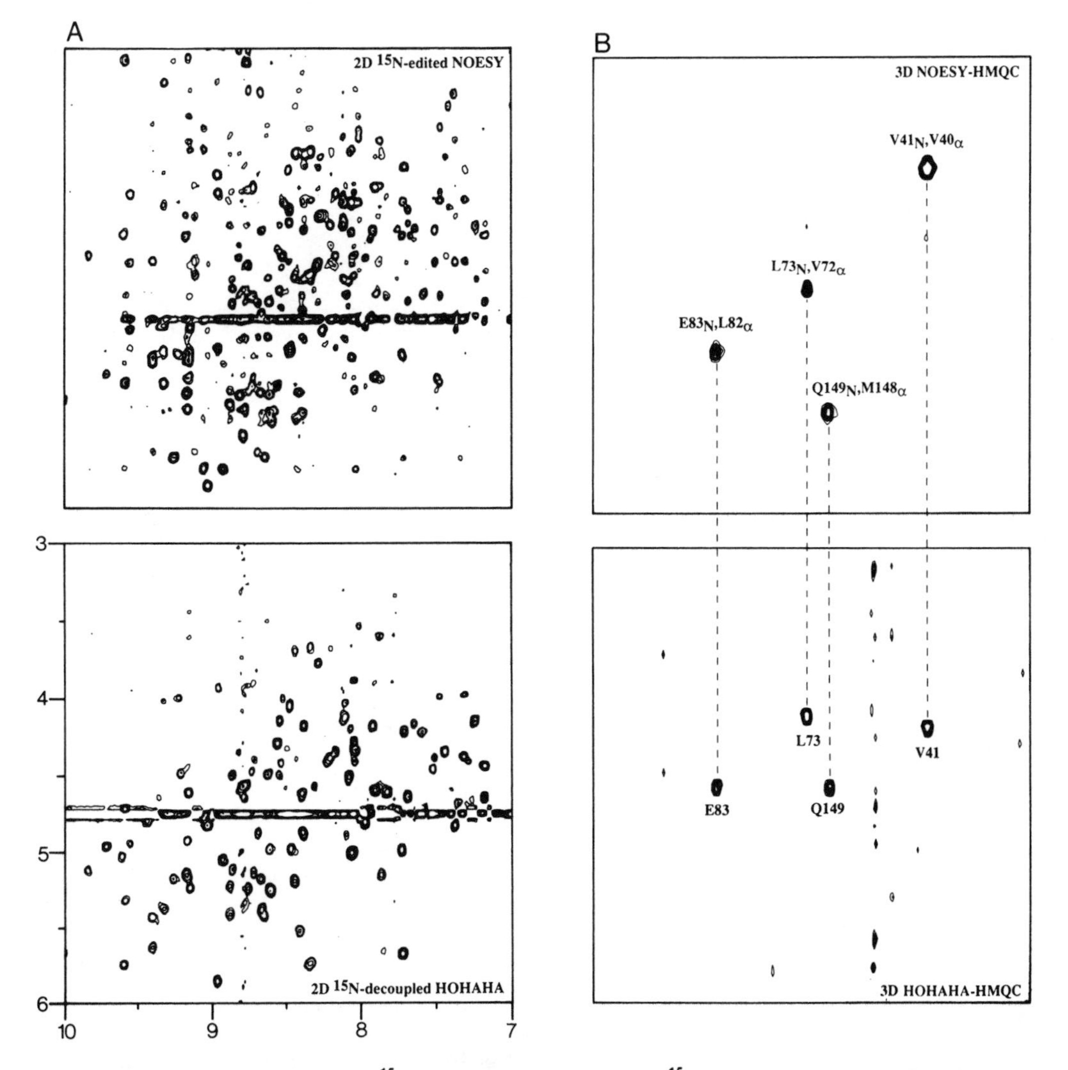

Fig. 11. (**A**) Fingerprint regions of the ^{15}N-edited NOESY (upper) and ^{15}N decoupled HOHAHA (lower) spectra of uniformly ^{15}N labeled interleukin-1β. (**B**) corresponding regions of a slice from the 3-D heteronuclear NOESY-HMQC (upper) and HOHAHA-HMQC (lower) spectra of the protein under identical conditions *(7b)*.

with those of a ^{15}N HOHAHA-HMQC spectrum allows the straightforward sequential assignment by hopping from one pair of HOHAHA/NOESY planes to another pair, connecting them either via $C^{\alpha}H(i)$–$NH(i+1)$ or $NH(i)$–$NH(i+1)$ NOEs in the same manner as for the 2-D case. This is illustrated for a stretch of residues for the protein interleukin-1β in Fig. 12.

Identification of Regular Secondary Structure Elements

Since each type of secondary structure element is characterized by a particular pattern of short range ($|i-j| \leq 5$) NOEs *(22, 35)*, qualitative interpretation of the sequential NOEs allows the identification of regular secondary structure elements. This is illustrated in Fig. 13. Thus, for example, helices are characterized by a stretch of strong or medium $NH(i)$–$NH(i+1)$ NOEs, and medium or weak $C^{\alpha}H(i)$–$NH(i+3)$, $C^{\alpha}H(i)$–$C^{\alpha}H(i+3)$ NOEs and $C^{\alpha}H(i)$–$NH(i+1)$ NOEs, sometimes supplemented by $NH(i)$–$NH(i+2)$ and $C^{\alpha}H(i)$–$NH(i+4)$ NOEs. Strands, on the other hand, are characterized by very strong $C^{\alpha}H(i)$–$NH(i+1)$ NOEs and by the absence of other short-range NOEs involving the NH and $C^{\alpha}H$ protons. β sheets can be identified and aligned from interstrand NOEs involving the NH, $C^{\alpha}H$, and $C^{\beta}H$ protons. It should also be pointed out that the identification of secondary structure elements

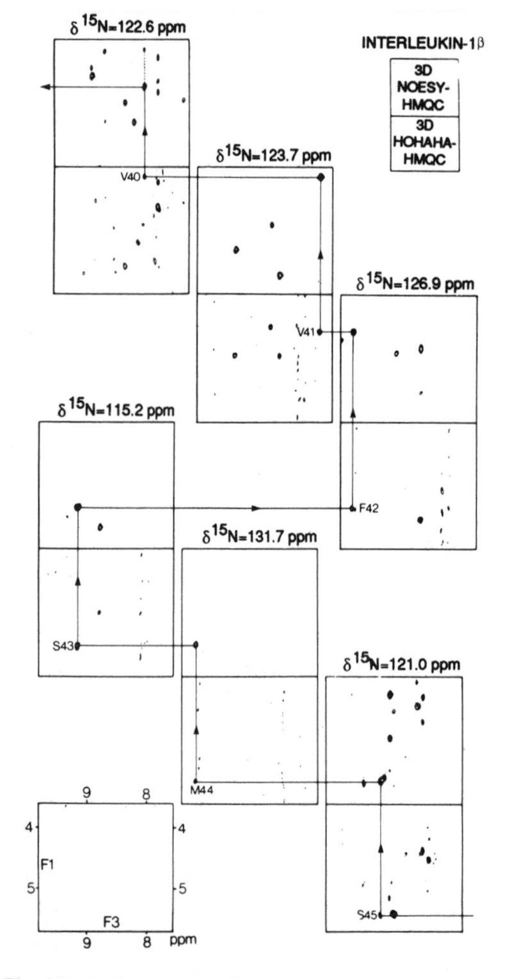

Fig. 12. Demonstration of the sequential assignment procedure for the 3-D spectra of interleukin 1β. Connectivities between neighboring amino acid spin systems is achieved by "plane hopping" through the 3-D cube *(7b)*.

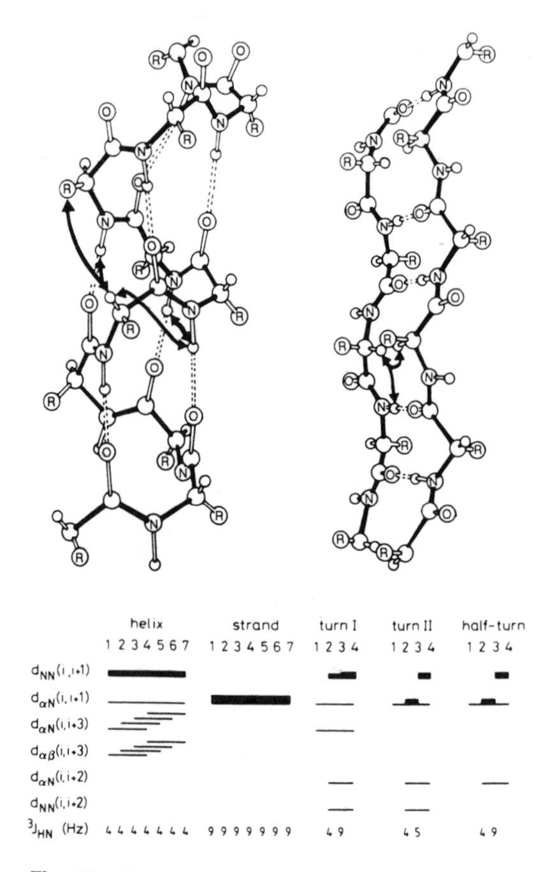

Fig. 13. Characteristic patterns of short range NOEs involving the NH, $C^{\alpha}H$ and $C^{\beta}H$ protons seen in various regular secondary structure elements. The NOEs are classified as strong, medium and weak, reflected in the thickness of the lines.

is aided by NH exchange data in that slowly exchanging NH protons are usually involved in hydrogen bonding, and by $^3J_{HN\alpha}$ coupling constant data. Figure 14 illustrates he application of this method to the protein interleukin-8 *(36)*. Inspection of the short- range NOE data immediately enables one to identify three β strands, several turns, and a long helix at the carboxy terminus.

In assessing the accuracy of secondary structure elements deduced using this approach, several factors should be borne in mind. Essentially it is a data based approach in so far that the expected patterns of short-range NOE connectivities for different secondary structure elements have been derived by examining the values of all the short-range distances involving the NH, $C^{\alpha}H$, and $C^{\beta}H$ protons in regular secondary structure elements present in protein x-ray structures. Thus, it tends to perform relatively poorly in regions of irregular structure such as loops. In addition, the exact start and end of helices tend to be rather ill-defined, particularly as the pattern of NOEs for turns is not all too dissimilar from that present in helices. Thus, a turn at the end of a helix could be misinterpreted as still being part of the helix. In the case of β sheets, the definition of the start and end is more accurate as the alignment is accomplished from the interstrand NOEs involving the NH and $C^{\alpha}H$ protons. Therefore, although this secondary structure delineation is a very easy and straightforward procedure, it can only be used in a qualitative fashion, and accurate positioning of the identified secondary structure elements can be only accomplished after the complete 3-D protein structure has been determined.

Interproton Distance Restraints

The initial slope of the time-dependent NOE, $N_{ij}(t)$, between two protons i and j is equal to the cross-relaxation rate σ_{ij} between the two protons *(37)*

$$\left.\frac{dN_{ij}}{dt}\right|_{t=0} = \sigma_{ij} \qquad (1)$$

σ_{ij} is simply the rate constant for exchange of magnetization between the two protons. σ_{ij}, in turn, is proportional to $\langle r_{ij}^{-6}\rangle$ and $\tau_{eff}(ij)$, where r_{ij} is the distance between the two protons, and $\tau_{eff}(ij)$ is the effective correlation time of the i–j vector:

$$\sigma_{ij} = \frac{\hbar^2\gamma^4}{10\,r_{ij}^{\,6}}\left(\tau_{eff}(ij) - \frac{6\tau_{eff}(ij)}{1 + 4\omega^2\tau_{eff}(ij)^2}\right) \qquad (2)$$

(γ is the gyromagnetic ratio of the proton, $\hbar$ is Planck's constant divided by 2π, and ω is the spectrometer frequency). It therefore follows that at short mixing times τ_m, ratios of NOEs can yield either ratios of distances or actual distances, if one distance is already known, through the relationship

$$r_{ij}/r_{kl} = (\sigma_{kl}/\sigma_{ij})^{1/6} \qquad (3)$$

providing the effective correlation times for the two interproton vectors are approximately the same.

In practice, initial slope measurements are not entirely trivial. First, the magnitude of the NOEs at very short mixing times are small, inevitably posing a signal-to-noise problem. Second, the measured NOE at short mixing times may not reflect the true magnitude of the NOE due to the particularities of the experimental set up. In addition, problems arising from spin diffusion in large multispin systems introduce errors into the initial rate approximation. These can, to some degree, be minimized by carrying out a full relaxation matrix analysis *(38)*, although certain problems still exist *(39)*. For proteins, however, variations in effective correlation time will always present a problem if one wants to extract accurate interproton distances, and it is therefore advisable to settle for only approximate interproton distance restraints. Because of the $\langle r^{-6}\rangle$ dependence of the NOE, such approximate interproton distance restraints can clearly be derived, even in the presence of large variations in effective correlation times. Empirically, the type of classification generally used is one in which strong, medium, and weak NOEs correspond to distance ranges of approximately 1.8–2.7 Å, 1.8–3.3 Å, and 1.8–5.0 Å, where the lower limit of 1.8 Å corresponds to the sum of the van der Waals radii of two protons. By using such a scheme, varia-

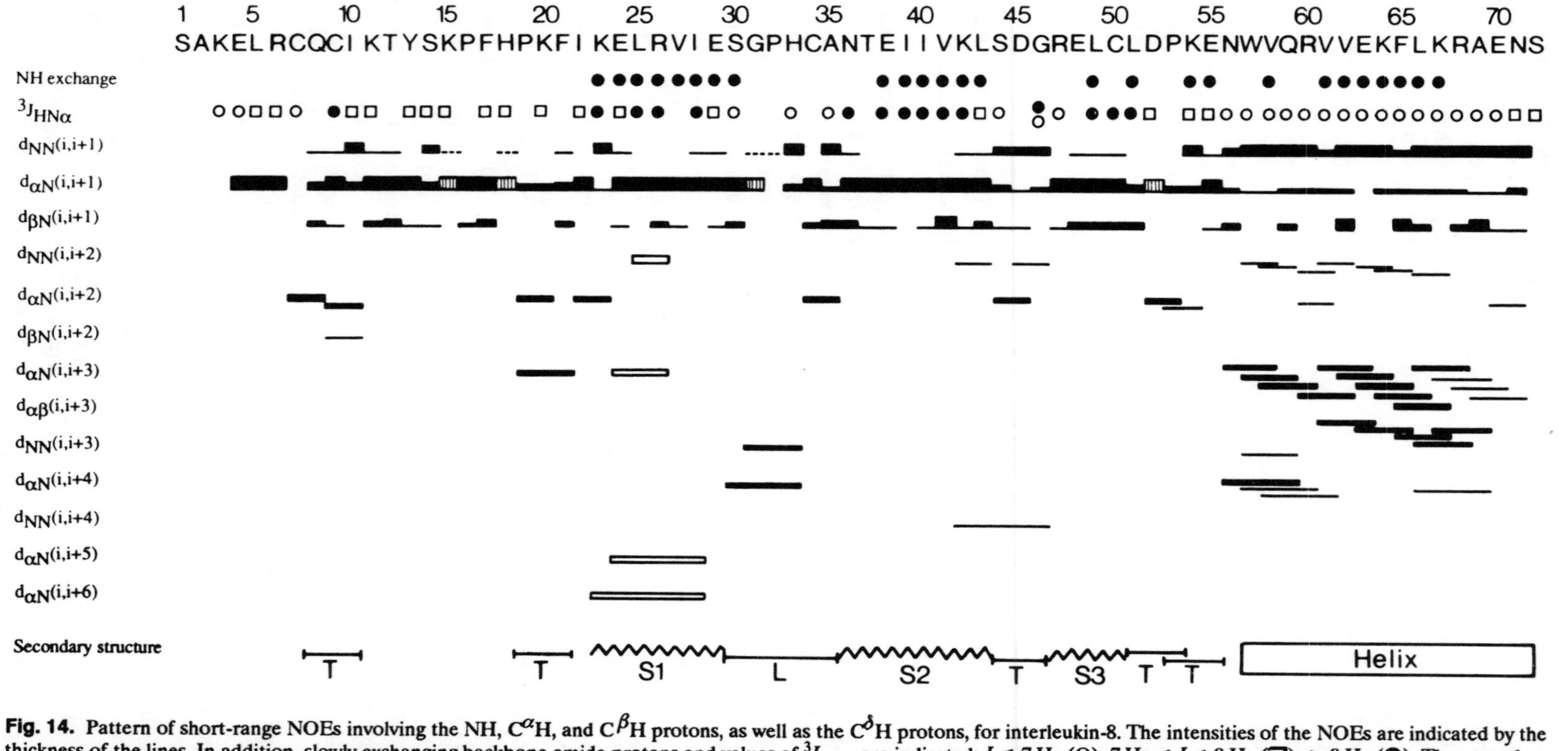

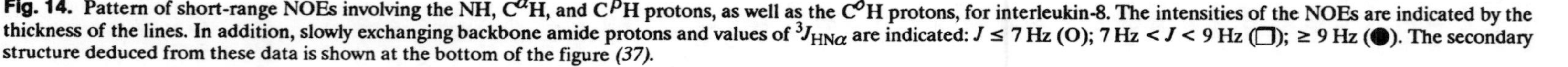

Fig. 14. Pattern of short-range NOEs involving the NH, $C^{\alpha}H$, and $C^{\beta}H$ protons, as well as the $C^{\delta}H$ protons, for interleukin-8. The intensities of the NOEs are indicated by the thickness of the lines. In addition, slowly exchanging backbone amide protons and values of $^3J_{HN\alpha}$ are indicated: $J \leq 7$ Hz (O); 7 Hz $< J <$ 9 Hz (□); ≥ 9 Hz (●). The secondary structure deduced from these data is shown at the bottom of the figure *(37)*.

tions in effective correlation times do not introduce errors into the distance restraints. Rather, they only result in an increase in the estimated range for a particular interproton distance.

Torsion Angle Restraints

Vicinal spin-spin coupling constants can provide useful information supplementing the interproton distance restraints derived from NOE data. In particular, ranges of torsion angles can be estimated from the size of the coupling constants. The latter may be obtained by analyzing the multiplet patterns in COSY and COSY-like (e.g., DQF-COSY, E-COSY, PE-COSY, z-COSY) spectra.

The easiest coupling constants to determine in proteins are the $^3J_{HN\alpha}$ coupling constants, which can be obtained by simply measuring the peak-to-peak separation of the antiphase components of the C^αH–NH COSY cross-peaks. The size of the $^3J_{HN\alpha}$ coupling constant is related to the ϕ backbone torsion angle through a Karplus-type relationship *(40)*. Consequently, values of $^3J_{HN\alpha} < 6$ Hz and > 8 Hz correspond to ranges of $-10°$ to $-90°$ and $-80°$ to $-180°$, respectively, for the ϕ backbone torsion angles. Considerable care, however, has to be taken in deriving ϕ backbone torsion angle ranges from apparent values of $^3J_{HN\alpha}$ coupling constants measured in this way, as the minimum separation between the antiphase components of a COSY cross-peak is equal to approximately one-half of the NH linewidth *(41)*. That is to say that small coupling constants can only be determined for relatively sharp resonances. This limitation can be overcome by measuring $^3J_{HN\alpha}$ from ^{15}N HMQC-COSY/HMQC-J spectra *(42)* because of the significantly narrower multiple quantum linewidths in these experiments.

χ_1 side-chain torsion angle restraints and stereospecific assignments can be obtained by analyzing the pattern of $^3J_{\alpha\beta}$ coupling constants and the relative intensities of the intraresidue NOEs from the NH and C^αH protons on the one hand to the two C^βH protons on the other, and in the case of Valine to the $C^\gamma H_3$ protons. The $^3J_{\alpha\beta}$ coupling constants are related to the χ_1 torsion angle and are best measured from correlation spectra that yield reduced multiplets such as β-COSY, E-COSY, P.E.COSY or z-COSY *(43, 44)*. In addition, under suitable conditions, they can be qualitatively assessed from the C^αH–C^βH cross-peak shapes in HOHAHA spectra *(45)*.

If both $^3J_{\alpha\beta}$ couplings are small (~3 Hz), then χ_1 must lie in the range $60 \pm 60°$. If, on the other hand, one of the $^3J_{\alpha\beta}$ couplings is large and the other small, χ_1 can lie either in the range $180 \pm 60°$ or $-60 \pm 60°$. These two possibilities are easily distinguished on the basis of short mixing time NOESY experiments which yield simultaneously stereospecific assignments of the β-methylene protons. Clearly, this approach may fail if a side chain has a mixture of conformations or the χ_1 angle deviates by more than ~40° from the staggered rotamer conformations (60°, 180,° and $-60°$). In the former case, the coupling and NOE data will be mutually inconsistent, while in the latter, it may not be possible to make an unambiguous distinction between two rotamer conformations. Fortunately, analysis of high resolution x-ray structures in the protein data bank has shown that 95% of all χ_1 angles lie within $\pm 15°$ of the staggered rotamer conformations and that there is a very clear correlation between the values of χ_1 and the degree of refinement: the better refined the structures, the closer the χ_1 angles to the ideal staggered rotamer conformations. These results suggest that stereospecific assignments can be obtained for up to 80% of β-methylene protons using this simple approach. In practice, of course, the percentage of stereospecific assignments will be lower due to either spectral overlap or large linewidths, preventing the determination of $^3J_{\alpha\beta}$ coupling constants.

A more rigorous approach for stereospecific assignment involves matching the observed $^3J_{HN\alpha}$ and $^3J_{\alpha\beta}$ coupling constants, together with approximate distances from the intraresidue C^αH–C^βH and NH–C^βH NOEs and the interresidue C^αH(i–1)–NH(i), C^αH(i)–NH($i+1$), C^βH($i-1$)–NH(i) and

$C^{\beta}H(i)$–$NH(i+1)$ NOEs to theoretical values held in a data base which are either derived for all combinations of ϕ, ψ, and χ_1 torsion angles (varied by 10°) in a model tripeptide segment or for tripeptide segments taken from high resolution x-ray structures *(46, 53)*. The data base search is carried out for both possible stereospecific assignments, and in those cases where only one of the two assignments satisfies the information in the data base, the correct stereospecific assignment together with ranges for the ϕ, ψ, and χ_1 torsion angles are obtained. A key advantage of this method is that it allows one to obtain much narrower limits for the ϕ, ψ, and χ_1 torsion angle restraints than would otherwise be possible.

Assignment of Long Range ($|i-j| > 5$) NOEs in Proteins

In globular proteins the linear amino acid chain is folded into a tertiary structure such that protons far apart in the sequence may be close together in space. These protons give rise to tertiary NOEs whose identification is essential for determining the polypeptide fold. Once complete assignments have been made, many such long range NOEs can be identified in a straightforward manner. It is usually the case, however, that the assignment of a number of long-range NOE cross-peaks remains ambiguous due to resonance overlap. In some cases, this ambiguity can be removed by recording additional spectra. Where ambiguities still remain, it is often possible to resolve them by deriving a low resolution structure on the basis of the available data (i.e., the secondary structure and the assignment of a subset of all the long range NOEs) either by model building or by distance geometry calculations. This low resolution structure can then be used to test possible assignments of certain long range NOEs. For larger proteins it becomes increasingly difficult to assign tertiary NOEs because of the associated overlap problems. In these cases the 3-D approach will become a necessity, in particular ^{13}C 3-D experiments, since they allow editing with respect to particular side-chain positions for which the individual NOEs can then be extracted. In Fig. 15 all experimental short- and intermediate-range NOE restraints (A) as well as all long range NOE restraints (B) that were measured from the NOESY spectra are shown as dotted lines superimposed on the framework of the finally determined structure of hirudin, illustrating the dense network of distances throughout the protein core.

Tertiary Structure Determination

Several different approaches can be used to determine the three-dimensional structure of a protein from experimental NMR data. The simplest approach, at least conceptually, is model building. This can be carried out either with real models or by means of interactive molecular graphics. It suffers, however, from the disadvantage that no unbiased measure of the size of the conformational space consistent with the NMR data can be obtained. Consequently, there is no guarantee that the modelled structure is the only one consistent with the experimental data. Furthermore, in this way nothing more than a very low resolution structure can be obtained. Nevertheless, model building can play an important role in the early stages of a structure determination, particularly with respect to resolving ambiguities in the assignments of some of the long-range NOEs.

The main computational methods for generating structures from NMR data comprise as common feature a conformational search to locate the global minimum of a target function, which is made up of stereochemical and experimental NMR restraints. The descent to the global minimum region is not a simple straightforward path, as the target function is characterized by many false local minima which have to be avoided or surmounted by all the methods. There are essentially two general classes of methods. The first can be termed real space methods. These include restrained least-squares minimization in torsion angle space with either a variable target function *(9)* or a sequence of ellipsoids of

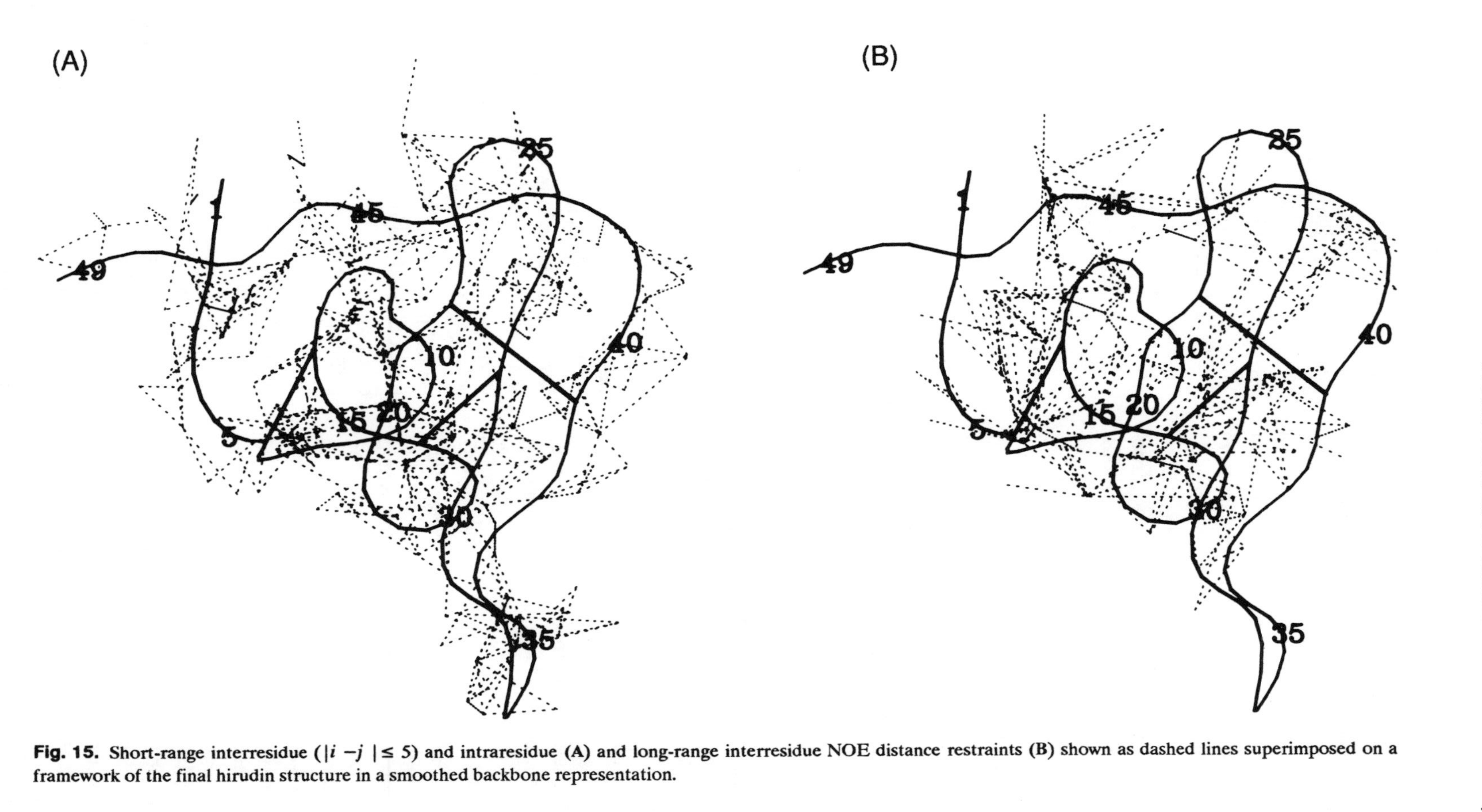

Fig. 15. Short-range interresidue ($|i - j| \leq 5$) and intraresidue (A) and long-range interresidue NOE distance restraints (B) shown as dashed lines superimposed on a framework of the final hirudin structure in a smoothed backbone representation.

constantly decreasing volume, each of which contains the minimum of the target function *(11)*, and restrained molecular dynamics *(10)* and dynamical simulated annealing *(12, 13)* in Cartesian coordinate space. All real space methods require initial structures. These can be (i) random structures with correct covalent geometry; (ii) structures that are very far from the final structure (e.g., a completely extended strand); (iii) structures made up of a completely random array of atoms; and (iv) structures generated by distance space methods. They should not, however, comprise structures derived by model building as this inevitably biases the final outcome. Because these methods operate in real space, great care generally has to be taken to ensure that incorrect folding of the polypeptide chain does not occur. A new real space approach involving the use of dynamical simulated annealing, however, has succeeded in circumventing this problem *(13)*. In contrast to the real space methods, the folding problem does not exist in the second class of methods which operates in distance space and is generally referred to as metric matrix distance geometry *(47)*. Here the coordinates of the calculated structure are generated by a projection from $N(N-1)/2$ dimensional distance space (where N is the number of atoms) into three-dimensional Cartesian coordinate space by a procedure known as embedding [see *(48)* for a comprehensive review].

A flow chart of the calculational strategy that is generally used to solve protein structures is shown in Fig. 16. Since a detailed description of all the various methods would go far beyond the scope of this review, the interested reader is referred to the references cited above. [A comparison between the different methods is given in *(49)*]. All the methods are comparable in convergence power. In general, however, the structures generated by dynamical simulated annealing or refined by restrained molecular dynamics tend to be better in energetic terms than the structures generated by the other methods, particularly with respect to nonbonded contacts and agreement with the experimental NMR data.

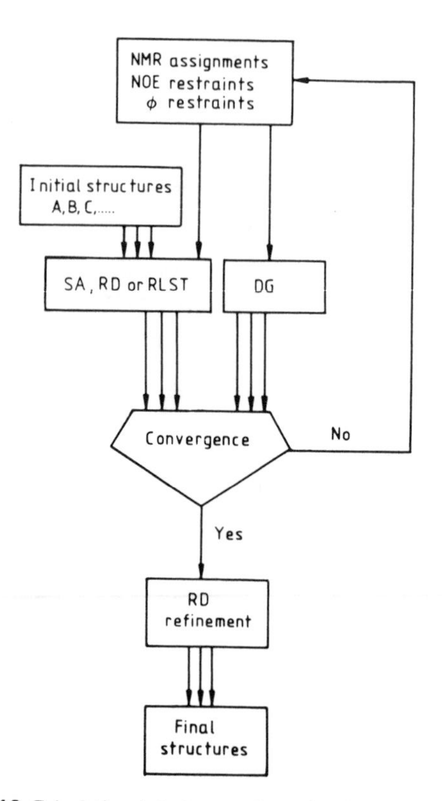

Fig. 16. Calculational strategy used to solve three-dimensional structures of macromolecules on the basis of NMR data.

In order to assess the uniqueness and the precision of the structures determined by any of the above methods, it is essential to calculate a reasonable number of structures with the same experimental data set, from different starting structures or conditions, and examine their atomic root-mean-square (rms) distribution. If these calculations result in several different folds of the protein while satisfying the experimental restraints, then the data is not sufficient to determine a unique structure and either more data has to be gathered, or the structure determination abandoned. If, however, convergence to a single fold is achieved with only small deviations from idealized covalent geometry, exhibiting good nonbonded contacts in addition to saisfying the experimental restraints, then one can be confident that a realistic and accurate picture of the solution structure of the protein has been obtained. The spread observed in the superposition within the family of structures or a plot of the rms distribution with respect to the mean allows one to assess the precision associated with different regions of the protein.

Examples

1. Hirudin. Hirudin is a small 65-residue protein from the leech and is the most potent natural inhibitor of coagulation known. It acts by interacting specifically with α-thrombin, thereby preventing the cleavage of fibrinogen. Two recombinant hirudin variants have been examined by NMR, namely, wild-type hirudin and the Lys-47→Glu mutant *(50)*. Analysis of the NMR data indicated that hirudin consists of a N-terminal compact domain (residues 1–49) held together by three disulfide bonds and a disordered C-terminal tail (residues 50–65). Evidence for the presence of a flexible C-terminal tail was provided by the absence of any intermediate-range or long-range NOEs beyond amino acid 49. Therefore, structure calculations were restricted to the N-terminal domain using the hybrid distance geometry–dynamical simulated annealing method. Experimental input data consisted of 701 and 677 approximate interproton distance restraints derived from NOE data for the wild-type and mutant hirudin, respectively, 26 ϕ backbone and 18 χ_1 torsion angle restraints derived from NOE and three-bond coupling constant data, and 8 backbone hydrogen bonds identified on the basis of NOE and amide exchange data were used as experimental input data. A total of 32 structures were computed for both the wild-type and mutant hirudins (Fig. 17). The structure of residues 2–30 and 37–48 constitute the core of the N-terminal domain formed by a triple stranded antiparallel β sheet and the atomic rms difference between the individual structures, and the mean structure is ~0.7 Å for the backbone atoms and ~ 1 Å for all atoms. The orientation of the exposed finger of antiparallel β sheet (residues 31–36) with respect to the core could not be determined as no long-range NOEs were observed between the exposed finger and the core. This is easily appreciated from Fig. 17, since the structures in that region exhibit a very large spread. Locally, however, the polypeptide fold of residues 31–36 is reasonably well defined.

The first five residues form an irregular strand which leads into a loop closed off at its base by the disulfide bridge between Cys-6 and Cys-14. This is followed by a mini-antiparallel β sheet formed by residues 14–16 (strand I) and 21–22 (strand I′) connected by a type II turn. This β sheet is distorted by a β

(A) Wild type

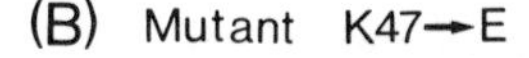

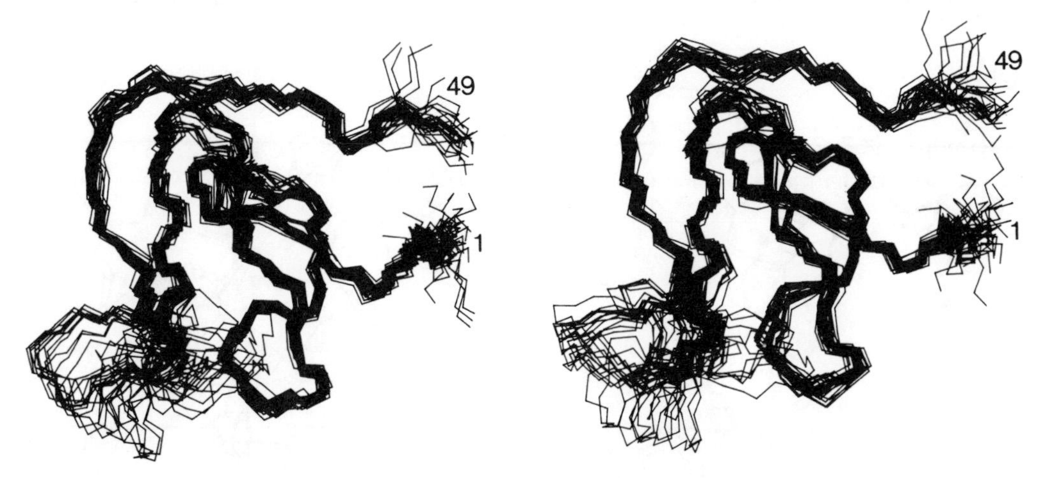

Fig. 17. Superposition of the backbone (N, C^{α}, C) atoms of 32 dynamical simulated annealing structures of wild-type hirudin (A) and the Lys- 47→Glu mutant (B) for the first 49 amino acids. The wild-type and mutant structures were calculated on the basis of 701 and 677 interproton distance restraints, respectively, 26 ϕ and 18 χ_1 torsion angle restraints *(50)*.

bulge at Cys-16. Strand I′ leads into a second antiparallel β sheet formed by residues 27–31 (strand II) and 36–40 (strand II′) connected by a β-turn (residues 32–35). Additionally, residues 10–11 exhibit features of a β bulge with the amide of Gly-10 and the carbonyl oxygen atom of Glu-11 hydrogen bonded to the carbonyl and amide groups, respectively, of Cys-28. Finally, strand II′ leads into an irregular strand which folds back onto the protein such that residue 47 (Lys in the wild type, Glu in the mutant) is in close proximity to residues in the loop closed off by the disulfide bridge between Cys-6 and Cys-14. Not only the backbone, but also many of the side-chain conformations are well-defined, especially those in the interior of the protein.

A superposition of the core (residues 1–30 and 37–49) of the restrained minimized mean structures of the wild-type and mutant hirudin provides a good representation of the differences between the two structures (Fig. 18A). Regions of noticeable difference can be identified where the atomic rms difference between the two mean structures is larger than the atomic rms distribution of the individual structures about their respective means. This analysis indicates the presence of clear differences for the backbone atoms of residues 3, 5, 8, 11 to 15, 22, 26, and 27. In the mutant, the backbone atoms of residues 3,4, and 11 to 15 are slightly closer to that of residues 45–47 than in the wild type, a change which can be rationalized in terms of the shorter length of the Glu side chain relative to that of Lys. Concomitantly, the backbone of residue 8 appears to be pushed away in the wild-type structure. Residues 22, 26, and 27 also move in the same direction as residues 11–15. This is secondary to the perturbation of residues 11–15 and can be attributed to the presence of numerous contacts between residues 8–11 on the one hand and 28–30 on the other, including hydrogen bonds between Cys-28 and Gly-10 and between Cys-28 and Glu-11. There seem to be no significant differences, however, with respect to the side-chain positions within the errors of the coordinates, even within the immediate vicinity of residue 47 (Fig. 18B).

At this point it may be worth mentioning that NMR can have two important applications with respect to genetically engineered proteins. The structural studies on hirudin were first initiated using natural protein extracted from the whole body of leeches *(51)*. All further work was subsequently carried out on recombinant products, either comprising the wild-type sequence or mutants thereof *(50)*. It was a fast and easy task to assess the structural identity of the recombinant product, since only a comparison of the appropriate 2-D spectra had to be carried out. These spectra can be regarded as a fingerprint of a particular protein, and hence a reflection of the 3-D structure in solution. Simple overlay often allows one to ascertain whether the

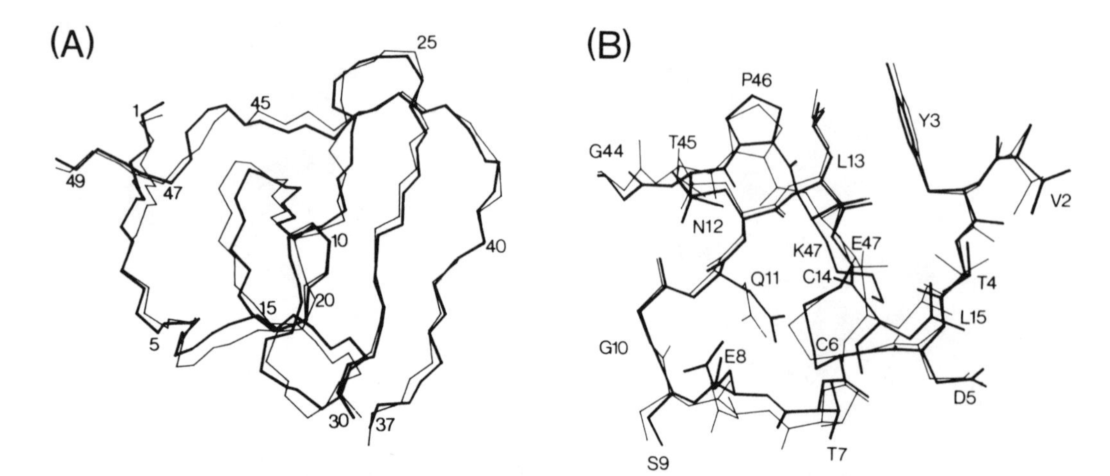

Fig. 18. (A) Superposition of the core (residues 1–30 and 37–49) of the restrained minimized mean structure of wild-type (thick lines) and Lys-47→Glu mutant (thin lines) hirudin. (B) View of all atoms around the site of mutation *(50)*.

recombinant product is folded indistinguishably from the natural counterpart. In the same way, NMR can establish whether the structure of a mutant is essentially unchanged from the wild-type one, since very similar spectra will arise from closely related structures. Analyses of this kind can be carried out without a full assignment of the spectra or a complete structure calculation, thus providing a useful tool for guiding genetic engineering projects.

2. The cellulose binding domain of cellobiohydrolase. Cellulases are enzymes involved in plant cell wall degradation which exhibit a common domain structure consisting of a catalytic domain (~400–500 amino acids), a highly conserved (~70% sequence identity) terminal domain (~40 amino acids) that is located either at the C-terminus or N-terminus, and a heavily glycosylated linker region (~30 amino acids) which connects the two domains. The domain architecture of the cellulases as well as the observation that the domains of CBH I retain their respective activities after cleavage suggested a dual approach to the problem of obtaining a three-dimensional structure of this cellulase, involving the application of both NMR and x-ray crystallography. To date there has been no success in the crystallization of an intact cellulase, whereas crystals of the catalytic domain of CBH II have been obtained *(52)*, opening the possibility for determining its x-ray structure. The cellulose binding domain being a very small polypeptide, on the other hand, seemed ideally suited for a structure determination by NMR. Because of its small size it was possible to chemically synthesize the 36 amino acid domain, and it was established that the biological properties were identical to the cleavage product. The subsequent NMR structure determination *(53)* was therefore carried out on the synthetic product and a large number of stereospecific assignments obtained using the data base approach resulted in numerous interproton distance and torsion angle restraints. Thus an exceptionally large experimental data set consisting of 554 interproton distance restraints, 33ϕ, 24ψ, and 25χ_1 torsion angle restraints, and 42 hydrogen bonding restraints, was used in the structure determination.

The converged set of 41 structures represents the best quality NMR structure to date, exhibiting an rms difference between the individual structures and the mean coordinate positions of 0.33 Å for the backbone atoms and 0.52 Å for all atoms. It was possible to determine the pairing of the two disulfide bridges, which previously was unknown. A backbone trace of the CBH I structure is shown in Fig. 19, and several regions including the amino acid side chains are displayed in Fig. 20. The protein has a wedge-like shape with an amphiphilic character, one face being predominantly hydrophobic and the other mainly hydrophilic. As can be readily appreciated from Figs. 19 and 20, the quality of this structure is comparable to a crystal structure at approximately 2 Å resolution, and such structures permit the detailed analysis of side chain–side chain interactions and other possibly interesting structural features.

Perspective and Concluding Remarks

It should be clear from the above discussion that NMR now stands side by side with x-ray crystallography as a powerful method for three-dimensional structure determination. What are the limitations of this approach? At present it is limited to proteins of molecular weight ≤ 20,000. Indeed the largest proteins

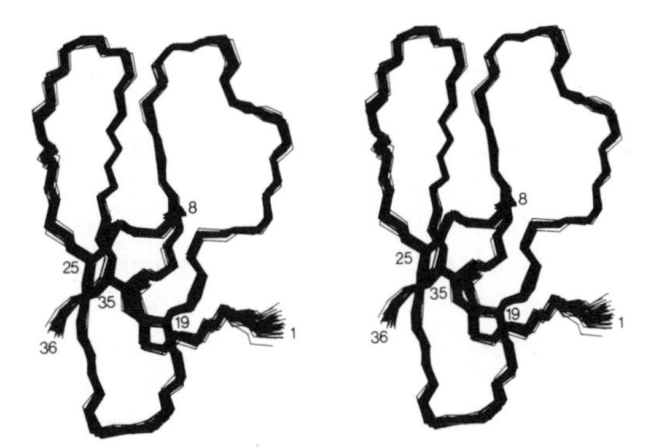

Fig. 19. Superposition of the backbone (N, C^{α}, C) atoms of 41 dynamical simulated annealing structures of the C-terminal domain of CBH I *(53)*.

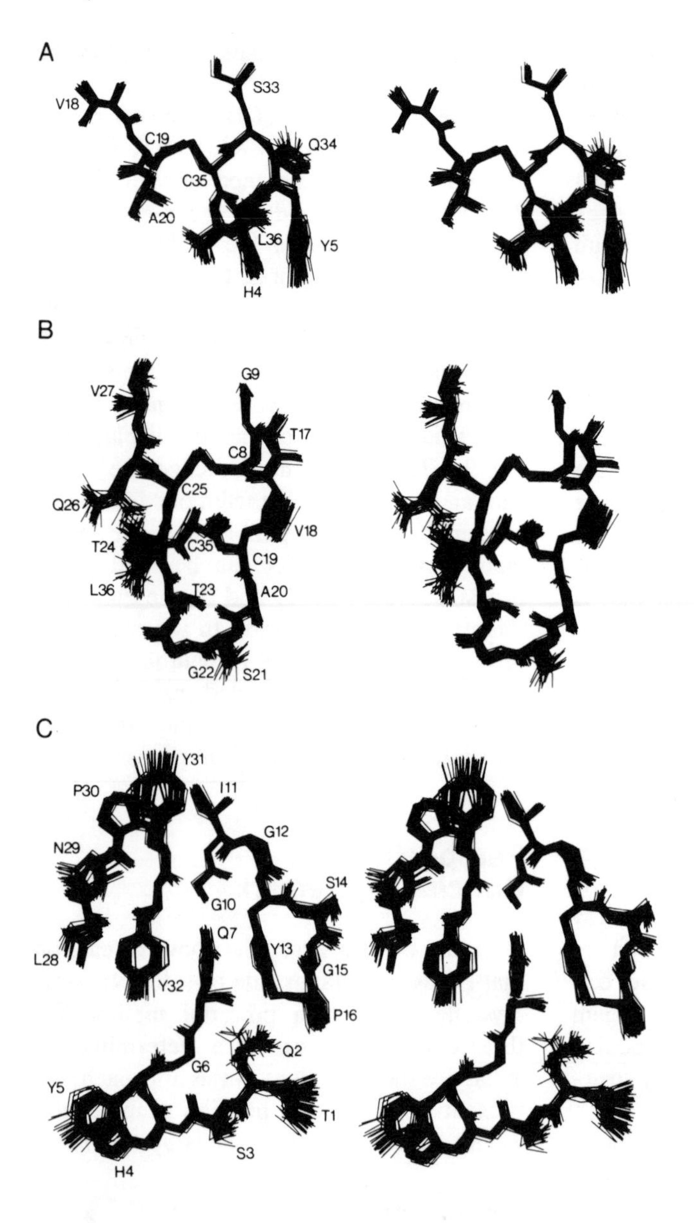

Fig. 20. Best fit superpositions of all atoms for three selected regions of the C-terminal domain of CBH I illustrating the excellent definition of side-chain positions *(53)*.

whose three-dimensional structures have been determined to date are plastocyanin [99 residues *(54)*], the globular domain of histone H5 [79 residues *(55)*], α-amylase inhibitor [74 residues *(56)*], and interleukin-8 [a dimer of 72 residues per monomer *(57)*]. Virtually complete assignments, however, have been made for a variety of larger systems, in particular hen egg white lysozyme [129 residues, *(58)*] and the lac repressor headpiece–operator complex [molecular weight ~15000 *(59)*], *Staphylococcal* nuclease, [148 residues *(60)*], and interleukin-1β [153 residues *(61)*]. Further development of novel techniques based on multidimensional NMR in combination with isotopic labeling and the introduction of yet more powerful magnets may make it possible to extend the molecular weight range up to proteins of molecular weight ~40,000 in the future. This, however, will probably present a fundamental limit as the large linewidths of such proteins significantly reduce even the sensitivity of ^{1}H-detected heteronuclear correlation experiments.

At this point it is appropriate to add a word of caution concerning the practical limits of structure determination by NMR. It is not always the size or the number of residues in a particular protein that determines the feasibility of an NMR structure determination. Other factors play equally important roles. For example, the protein should be soluble up to millimolar concentrations, nonaggregating, and preferably stable up to at least 40°C, particularly for large proteins. A further consideration is the chemical shift dispersion of the ^{1}H-NMR spectrum. This depends to a large extent on the structure of the protein under investigation. Proteins that are made up only of α-helices, loops, and turns, invariably exhibit fairly poor proton chemical shift dispersion, whilst the chemical shift dispersion in β-sheet proteins is usually very good. To a degree, such problems associated with chemical shift degeneracy can be overcome by heteronuclear 3-D experiments.

Another potential problem may arise from the fact that different regions or domains of a protein may be well-defined and, therefore, amenable to a NMR structure determination, while other parts of the same protein may not. This will be reflected in the absence of long-range tertiary NOEs for the ill-defined regions and leads to the inability to position these regions with respect to the rest of the protein satisfactorily (e.g., the case of hirudin discussed above). We therefore believe that it is necessary to calculate a reasonable number of structures (ca. 20) with the same experimental data set in order to obtain a good representation of the NMR structure. Only by analyzing such a family of structures can the local and global definition of a structure be assessed.

X-ray crystallography, of course, also has its limitations, the most obvious being the requirement for a protein to crystallize. Thus suitable crystals that diffract to high resolution have to be grown, and a successful search for heavy atom derivatives to solve the phase problem is necessary. Therefore, not every protein will be amenable to both NMR and x-ray crystallography. In those cases where this is feasible, the information afforded by NMR and crystallography is clearly complementary and may lead to a deeper understanding of the differences between the solution and crystalline state of the protein.

Finally, it should be stressed that, in addition to being able to determine three-dimensional structures of proteins, NMR has the potential to address other questions, in particular those concerning the dynamics of the system. This opens the possibility that a whole variety of different NMR studies can be initiated on the basis of an NMR structure such as the investigation of the dynamics of conformational changes upon ligand binding, unfolding kinetics, conformational equilibria between different conformational states, fast internal dynamics on the nanosecond time scale and below, and slow internal motions on the second and millisecond time scales. The results obtained from these kinetic studies can then be interpreted in the light of the previously determined structure, thus bringing together structure and dynamics of proteins in a unified picture.

Acknowledgments

The work in the authors' laboratory was supported in part by the AIDS targeted Anti-Viral Program of the Office of the Director of the National Institutes of Health.

References

1. Jeener, J., Ampere International Summer School, Basko Polje, Yugoslavia, unpublished lecture (1971).
2. Aue, W.P., E. Bartholdi, R. R. Ernst, *J. Chem. Phys.* **64**, 2229 (1976).
3. Jeener, J., B. H. Meier, P. Bachmann, R. R. Ernst, *J. Chem. Phys.* **71**, 4546 (1979).
4. Oschkinat, H., *et al.*, *Nature* (London) **332**, 374 (1988); Oschkinat, H., C. Cieslar, T. A. Holak, G. M. Clore, A. M. Gronenborn, *J. Magn. Reson.* **83**, 450 (1989).
5. Vuister, G. W., R. Boelens, R. J. Kaptein, *Magn. Reson.* **80**, 176 (1988).
6. Fesik, S. W., and E. R. P. Zuiderweg, *J. Magn. Reson.* **78**, 588 (1988); Zuiderweg, E.R.P., and S. W. Fesik, *Biochemistry* **28**, 2387 (1989).
7. Marion, D., L. E. Kay, S. W. Sparks, D. A. Torchia, A. Bax, *J. Am. Chem. Soc.* **111**, 1515

(1989); Marion, D., *et al., Biochemistry* **28,** 6150 (1989).
8. Havel, T. F., I. D. Kuntz, G. M. Crippen, *Math. Biol.* **45,** 665 (1983); Havel, T. F., and K. Wüthrich, *Bull. Math. Biol.* **46,** 673 (1984).
9. Braun, W., and N. Go, *J. Mol. Biol.* **186,** 611 (1985).
10. Clore, G. M., A. M. Gronenborn, A. T. Brünger, M. Karplus, *J. Mol. Biol.* **186,** 435 (1985); Kaptein, R., E. R. P. Zuiderweg, R. M. Scheek, R. Boelens, W. F. van Gunsteren, *J. Mol. Biol.* **182,** 179 (1985); Clore, G. M., A. T. Brünger, M. Karplus, A. M. Gronenborn, *J. Mol. Biol.* **191,** 523 (1986).
11. Billeter, M., T. F. Havel, K. Wüthrich, *J. Comput. Chem.* **8,** 132 (1987).
12. Nilges, M., A. M. Gronenborn, G. M. Clore, *FEBS Lett.* **229,** 317 (1988); Nilges, M., A. M. Gronenborn, A. T. Brünger, G. M. Clore, *Protein Engineering* **2,** 27 (1988).
13. Nilges, M., G. M. Clore, A. M. Gronenborn, *FEBS Lett.* **239,** 129 (1988).
14. Litharge, O., C. W. Cornelius, B. G. Buchanan, O. Jardetzky, *Proteins: Struct., Funct. & Genet.* **2,** 340 (1987).
15. Ernst, R. R., G. Bodenhausen, A. Wokaun, *Principles of Nuclear Magnetic Resonance* in *One and Two Dimensions* (Clarendon Press, Oxford, 1986).
16. Davis, D. G., and A. Bax, *J. Am. Chem. Soc.* **107,** 2821 (1985); Bax, A., and D. G. Davis, *J. Magn. Reson.* **65,** 355 (1985); Bax, A., in vol. 176 of *Methods in Enzymology*, James, T. L., and N. Oppenheimer, Eds. (Academic Press, New York, 1988), p. 151.
17. Macura, S., and R. R. Ernst, *Molec. Phys.* **41,** 95 (1980); Macura, S., Y. Huang, D. Suter, R. R. Ernst, *J. Magn. Reson.* **43,** 259 (1981).
18. Griesinger, C., O. W. Sørensen, R. R. Ernst, *J. Magn. Reson.* **84,** 14 (1989).
19. Overhauser, A., *Phys. Rev.* **89,** 689 (1953); Overhauser, A., *Phys. Rev.* **92,** 411 (1953).
20. Solomon, I., *Phys. Rev.* **99,** 559 (1955).
21. Noggle, J. H., and R. E. Schirmer, *The Nuclear Overhauser Effect – Chemical Applications* (Academic Press, New York, 1971).
22. Wüthrich, K., *NMR of Proteins and Nucleic Acids* (J. Wiley, New York, 1986).
23. Rance, M., O. W. Sørenson, G. Bodenhausen, G. Wagner, R. R. Ernst, K. Wüthrich, *Biochem. Biophys. Res. Commun.* **117,** 479 (1983).
24. Marion, D., and A. Bax, *J. Magn. Reson.* **79,** 352 (1988).
25. Braunschweiler, L., and R. R. Ernst, *J. Magn. Reson.* **53,** 521 (1983).
26. Bax, A., L. E. Kay, S. W. Sparks, D. L. Torchia, *J. Am. Chem. Soc.* **111,** 408 (1989).
27. Mueller, L., *J. Am. Chem. Soc.* **101,** 4481 (1979).
28. Redfield, A. G., *Chem. Phys. Lett.* **96,** 537 (1983); Griffey, R. H., and A. G. Redfield, *Q. Rev. Biophys.* **19,** 51 (1987).
29. Bax, A., R. H. Griffey, B. L. Hawkins, *J. Am. Chem. Soc.* **105,** 7188 (1983).
30. Bolton, P. H., and G. Bodenhausen, *Chem. Phys. Lett.* **89,** 139 (1982).
31. Sklenar, V., and A. Bax, *J. Magn. Reson.* **71,** 379 (1987).
32. Gronenborn, A. M., A. Bax, P. T. Wingfield, G. M. Clore, *FEBS Lett.* **243,** 93 (1989); Gronenborn, A.M., P. T. Wingfield, G. M. Clore, *Biochemistry* **28,** 5081 (1989).
33. Bax, A., and M. F. Summers, *J. Am. Chem. Soc.* **108,** 2093 (1986); Bax, A., and D. Marion, *J. Magn. Reson.* **78,** 186 (1988).
34. Clore, G. M., A. Bax, P. T. Wingfield, A. M. Gronenborn, *FEBS Lett.* **238,** 17 (1988).
35. Wüthrich, K., G. Wider, G. Wagner, W. Braun, *J. Mol. Biol.* **155,** 311 (1982); Billeter, M., W. Braun, W. Wüthrich, *J. Mol. Biol.* **155,** 321 (1982); Wagner, G., and K. Wüthrich, *J. Mol. Biol.* **155,** 347 (1982); Wüthrich, K., M. Billeter, W. Braun, *J. Mol. Biol.* **180,** 715 (1984).
36. Clore, G. M., E. Appella, M. Yamada, K. Matsushima, A. M. Gronenborn, *J. Biol. Chem.* **264,** 18907 (1989).
37. Wagner, G., and K. Wüthrich, *J. Magn. Reson.* **33,** 675 (1979); Dobson, C. M., E. T. Olejniczak, F. M. Poulsen, R. G. Ratcliffe, *J. Magn. Reson.* **48,** 97 (1982); Keepers, J.W., and T. L. James, *J. Magn. Reson.* **57,** 404 (1984); Clore, G.M., and A. M. Gronenborn, *J. Magn. Reson.* **61,** 158 (1985).
38. Borgias, B. A., and T. L. James, *J. Magn. Reson.* **79,** 493 (1988).
39. Clore, G. M., and A. M. Gronenborn, *J. Magn. Reson.* **84,** 398 (1989).
40. Karplus, M., *J. Am. Chem. Soc.* **85,** 2870 (1963); Pardi, A., M. Billeter, K. Wüthrich, *J. Mol. Biol.* **180,** 741 (1984).
41. Neuhaus, D., G. Wagner, M. Vask, J. H. R. Kgi, K. Wüthrich, *Eur. J. Biochem.* **151,** 257 (1985).
42. Kay, L. E., and A. Bax, *J. Magn. Reson.* **86,** 110 (1990); Forman-Kay, J. D., A. M. Gronenborn, L. E. Kay, P. T. Wingfield, G. M. Clore, *Biochemistry* **29,** 1566 (1990).
43. Mueller, L., *J. Magn. Reson.* **203,** 251 (1987).
44. Oschkinat, H., A. Pastore, P. Pfändler, G. Bodenhausen, *J. Magn. Reson.* **69,** 559 (1986); Oschkinat, H., G. M. Clore, M. Nilges, A. M. Gronenborn, *J. Magn. Reson.* **75,** 534 (1987).

45. Driscoll, P. C., G. M. Clore, L. Beress, A. M. Gronenborn, *Biochemistry* **28,** 2178 (1989).
46. Nilges, M., G. M. Clore, A. M. Gronenborn, *Biopolymers* **29,** 813 (1990).
47. Crippen, G. M., *J. Comp. Phys.* **24,** 96 (1977).
48. Crippen, G. M., and T. F. Havel, *Distance Geometry and Molecular Conformation* (John Wiley, New York, 1988).
49. Clore, G. M., and A. M. Gronenborn, *Crit. Rev. Biochem.* **24,** 479 (1989).
50. Folkers, P. J. M., *et al., Biochemistry* **28,** 2601 (1988).
51. Clore, G. M., D. K. Sukumaran, M. Nilges, J. Zarbock, A. M. Gronenborn, *EMBO J.* **6,** 529 (1987).
52. Bergfors, T., *et al., J. Mol. Biol.* **209,** 167 (1989).
53. Kraulis, P. J., *et al., Biochemistry* **28,** 7241 (1989).
54. Moore, J. M., *et al., Science* **240,** 314 (1988).
55. Clore, G. M., A. M. Gronenborn, M. Nilges, D. K. Sukumaran, J. Zarbock, *EMBO J.* **6,** 1833 (1987).
56. Kline, A. D., W. Braun, K. Wüthrich, *J. Mol. Biol.* **189,** 377 (1986); Kline, A. D., W. Braun, K. Wüthrich, *J. Mol. Biol.* **204,** 675 (1988).
57. Clore, G. M., E. Appella, M. Yamada, K. Matsushima, A. M. Gronenborn, *Biochemistry* **29,** 1689 (1990).
58. Redfield, C., and C. M. Dobson, *Biochemistry,* **27,** 122 (1988).
59. Boelens, R., *et al., J. Mol. Biol.* **193,** 213 (1987).
60. Torchia, D. A., S. W. Sparks, A. Bax, *Biochemistry* **27,** 5135 (1988).
61. Driscoll, P. C., G. M. Clore, D. Marion, P. T. Wingfield, A. M. Gronenborn, *Biochemistry* **29,** 3542 (1990).

Part III

Folding Mechanisms

C. N. Pace

The denatured states of proteins are becoming more complicated – a reflection of the fact that we are beginning to learn more about them. In the 1960s, Tanford's group published a series of careful studies aimed at better characterizing the denatured states of proteins *(1–4)*. One major conclusion was that proteins in 6 M guanidine hydrochloride (GdnHCl), with their disulfide bonds broken closely approach a randomly coiled conformation *(5)*. (Since proteins are not homopolymers, they can never be true random coils.) Less detailed studies suggested that the same was true in 8 M urea *(6)*, and that even with the disulfide bonds intact, proteins appeared to unfold as completely as possible in these two solvents, given the restraints imposed by the disulfide bonds *(5)*. This general view was probably accepted by most protein chemists for the next 20 years. However, there were some disquieting findings. For example, Tsong showed that cytochrome c retained structure in 9 M urea at neutral pH that could be unfolded by lowering the pH *(7)*. (Some of his results are remarkably similar to the results described in chapter 11 from the Fink group.)

In 1986, Shortle and Meeker rekindled interest in the denatured states produced by urea and GdnHCl by reaching ". . . the surprising conclusion that single amino acid substitutions, even quite conservative ones, can produce major changes in the "structure" of the denatured state of staphylococcal nuclease . . ." *(8)*. Further studies of these mutant proteins led them to suggest ". . . the existence of two alternative denatured states of staphylococcal nuclease . . . D1 representing a compact structured, but nonnative state, and D2 representing an expanded, less structured state" *(9)*. These unfolded states were observed in the absence of GdnHCl by using unstable fragments of staph nuclease missing from 13 to 46 amino acids from the C-terminus. By adding GdnHCl, the residual structure could be unfolded. The picture that emerges for staph nuclease, then, is that the states resulting from urea or GdnHCl denaturation may not be completely unfolded, but increasing the denaturant concentration further leads to a greater extent of unfolding. Thus, these results are not necessarily in conflict with those from the Tanford group.

Recent studies from our laboratory suggest that different proteins unfold to different extents in the presence of urea and GdnHCl and, therefore, must not all approach randomly coiled conformations *(10)*. For proteins that unfold by two-state folding mechanisms, the dependence of the free energy of unfolding on denaturant concentration, i.e., the m value, depends on the amount and composition of the polypeptide chain that is freshly exposed to solvent by unfolding *(11)*. Model compound data can be used to estimate the fraction of the polypeptide chain that must be freshly exposed to solvent, $\Delta \alpha$, in order to account for the measured m values *(12)*. A list of these $\Delta \alpha$ values for several proteins is given in Table 1. Note that substantial differences exist among the six proteins with no disulfide bonds. For example, the $\Delta \alpha$ value for barnase is ≈60% greater than the value for dihydrofolate re-

Table 1. Change in accessibility ($\Delta\alpha$) calculated from measured m values using Tanford's model.

Protein	MW	Disulfides	$\Delta\alpha$(urea)[a]	$\Delta\alpha$(GdnHCL)[a]
Lysozyme	14300	4	.23	.20
RNase A	13700	4	.35	.35
RNase T1	11100	2	.33	.35
RNase T1	11100	0	.38	
Barnase	12400	0	.53	.50
Nuclease	16800	0	.44	.43
Apo Mb	16900	0	.46	
DHFR	17700	0	.32	
T4 Lyso	18600	0		.39

[a] $\Delta\alpha$ was calculated from the measured m values as described in *(10)*.

ductase. The most reasonable interpretation of these differences is that the unfolded states of these proteins differ significantly in their ability to interact with denaturants, and this indicates that they unfold to different extents. Thus, we suggested that the conformations assumed by unfolded proteins may depend, to at least some extent, on the amino acid sequence of the protein, as does the conformation of the folded protein *(10)*.

Tanford's group next turned their attention to the denatured states produced by acid or thermal denaturation *(13)*. They concluded that ". . . these transitions retain regions of ordered structure, susceptible to disruption by guanidine hydrochloride." This and considerable other evidence suggested that in most cases, the products of acid or thermal denaturation appeared to be less completely unfolded than the products of GdnHCl or urea denaturation *(11)*. This view was questioned by Pfeil and Privalov because the thermodynamics of unfolding in water and GdnHCl solutions did not differ significantly *(14)*. The Privalov group now concedes that "... significant residual structure ... appears at room temperature in the denatured states of some globular proteins (e.g. myoglobin and lysozyme) at neutral pH . . ." *(15)*. However, they show that this residual structure makes only a small contribution to the thermodynamics of folding.

The nonnative state that has generated the most recent interest is called the "molten globule," a name suggested by Ohgushi and Wada *(16)*. Molten globules are compact and have a high content of secondary structure like native proteins, but have their tertiary structure disrupted like denatured proteins. It appears as if the native structure expands just enough to allow the hydrophobic residues to become exposed to solvent. Kuwajima has recently written an excellent review on the molten globule state *(17)*.

In addition to gaining a better understanding of the structure of nonnative states present at equilibrium, great progress has been made recently in characterizing the kinetic intermediates that proteins pass through during folding. Results with ribonuclease A *(18)*, cytochrome c *(19)*, and barnase *(20)*, have confirmed that protein folding follows a preferred pathway with identifiable intermediates and that elements of secondary structure are formed early in the folding process. It seems clear from these early results that this approach, based on nuclear magnetic resonance (NMR), is going to give us a much more detailed view of the mechanism of folding than seemed possible just a few years back *(21)*.

The three chapters in this section confirm the fact that we are gaining a better understanding of the nonnative conformations of proteins and that this information is relevant to the mechanism of protein folding.

Chapter 10 by Ptitsyn and Semisotnov provides an excellent summary of our current knowledge of the molten globule state and then argues convincingly that the folding of all proteins proceeds through a universal kinetic intermediate with properties that ". . . are similar (though not completely identical) to the properties of the equilibrium "molten globule" state.

The results from Fink's group, chapter 11, give us a much clearer understanding of the unfolded states resulting from acid unfolding of proteins and of the relationship of these states to the molten globule state.

The studies by Dobson's group, chapter 12, show some of the ways in which NMR is being used to better characterize the unfolded states of proteins, and they present an interesting comparison of the folded, unfolded, and molten globule states of proteins using lysozyme and α-lactalbumin as examples. With regard to unfolded states they conclude ". . . that local conformational preferences exist . . . and there is evidence for clustering of hydrophobic residues. Although the description of the unfolded state may differ significantly from that of a classical "random coil," there is no evidence for persistent secondary structure or, indeed, for any significant population of native-like structural features." In contrast, their experiments show that ". . . regions of native-like structure . . . persist in the molten globule state."

References

1. Tanford, C., K. Kawahara, S. Lapanje, *J. Am. Chem. Soc.* **88,** 729 (1967).
2. Nozaki, Y., and C. Tanford, *J. Am. Chem. Soc.* **89,** 742 (1967).
3. Tanford, C. *et al., J. Am. Chem. Soc.* **89,** 5023 (1967).
4. Lapanje, S., and C. Tanford, *J. Am. Chem. Soc.* **89,** 5030 (1967).
5. Tanford, C., *Adv. Prot. Chem.* **23,** 121 (1968).
6. Lapanje, S., *Croat. Chem. Acta* **41,** 115 (1969).
7. Tsong, T.Y., *Biochemistry* **14,** 1542 (1975).
8. Shortle, D., and A. K. Meeker, *Proteins: Struct. Funct. Genet.* **1,** 81 (1986).
9. Shortle, D., and A. K. Meeker, *Biochemistry* **28,** 936 (1989).
10. Pace, C. N., D. V. Laurents, J. A. Thomson, *Biochemistry* **29,** 2564 (1990).
11. Pace, C. N., *CRC Crit. Rev. Biochem.* **3,** 1 (1975).
12. Pace, C. N., *Methods Enzymol.* **131,** 266 (1986).
13. Aune, K. C., A. Salahuddin, M. H. Zarlengo, C. Tanford, *J. Biol. Chem.* **242,** 4486 (1967).
14. Pfiel, W., and P. L. Privalov, *Biophys. Chem.* **4,** 33 (1976).
15. Privalov, P., *et al., J. Mol. Biol.* **205,** 737 (1989).
16. Ohgushi, M., and A. Wada, *FEBS Lett.* **164,** 21 (1983).
17. Kuwajima, K., *Proteins: Struct. Funct. Genet.* **6,** 87 (1989).
18. Udgaonkar, J. B., and R. L. Baldwin, *Nature* **335,** 694 (1988).
19. Roder, H., G. A. Elove, S. W. Englander, *Nature* **335,** 700 (1988).
20. Matouschek, A., J. T. Kellis, L. Serrano, M. Bycroft, A. R. Fersht, *Nature* **346,** 440 (1990).
21. Kim, P. S., and R. L. Baldwin *Ann. Rev. Biochem.* **59,** 631 (1990).

10

The Mechanism of Protein Folding

Oleg B. Ptitsyn, Gennady V. Semisotnov

Introduction

Two main processes are involved in the creation of a protein in a living cell: the chemical and the physical. The chemical (biochemical) process is the enzymatic synthesis of the desirable amino acid sequence on a ribosome. The physical process is the folding of this sequence into a native (functioning) three-dimensional (3-D) structure.

It is well known *(1)* that many proteins can fold from the unordered state in the absence of ribosomes and other factors. This suggests that, in principle, all information that a protein needs for correct folding is ciphered in its amino acid sequence. This does not mean that ribosomes and other factors that are present in the living cell do not facilitate protein folding. Moreover, some factors (e.g., protein-disulfide isomerase, proline isomerase, molecular chaperones) almost certainly do so. However, the fact that the same result (creation of native protein) can be achieved without these factors suggests that they mainly accelerate and facilitate the "intrinsic" process of protein folding rather than essentially altering this process and its result.

These arguments serve to emphasize that the study of protein folding in vitro can substantially facilitate understanding the means by which proteins are created in a living cell. In this chapter, we shall outline the results of our experimental studies of protein folding in vitro (protein *renaturation*). The main result of these studies is that protein folding proceeds through a universal kinetic intermediate state that is similar to the equilibrium "molten globule" state of a protein molecule revealed in our laboratory in 1981 [*(2)*; see also *(3, 4)*].

The Molten Globule: An Equilibrium Intermediate State of Protein Molecules

A number of proteins at mild denaturating conditions (high temperatures, low or high pH, moderate concentrations of strong denaturants) are in a state that is intermediate between the native and the completely unfolded (unordered) ones. Ohgushi and Wada *(5)* called this state the *molten globule* state. The properties of the molten globule state have been described in two reviews *(4, 6)*. Therefore, we shall only briefly summarize their characteristics here.

(i) The molten globule is compact. Hydrodynamic properties and small angle x-ray scattering of molten globules show that their linear dimensions can be only ~10% larger (i.e., their volume is only ~30% larger) than that of native protein *(2, 3, 7–9)*. Since this increase of the molecular volume is accompanied by neither a corresponding increase of the partial specific volume nor of the compressibility *(69)*, this "excess" 30% of volume is mainly filled by water.

Nevertheless, large angle x-ray scattering shows that the core of a protein molecule also remains in the molten globule state, being packed more loosely than that in the native protein *(10, 11)*.

(ii) The molten globule has a pronounced secondary structure. Far-ultraviolet circular dichroism (UV CD) spectra show a high content of the secondary structure of the molten globule *(2, 3, 7, 12–14)*. Infrared spectra show that its content can be similar to that of the native protein *(4, 8, 15)*.

(iii) The molten globule is more flexible than the native state. Near-UV CD *(2, 3, 7, 12–14, 16)* and ^{1}H nuclear magnetic resonance (NMR) spectra *(3, 4, 14, 17–21)* show that the native rigid environment of side groups is greatly reduced in the molten globule state. Direct spin echo measurements show that the mobility of aliphatic side groups in the molten globule state is similar to that in the unfolded state, while the mobility of aromatic groups in the molten globule state is intermediate between that in the native and the unfolded states *(21)*. Moreover, the most bulky tryptophan groups can have practically the same mobility in the molten globule and in the native states *(2, 3, 14)*. Deuterium exchange shows that large-scale fluctuations in the molten globule state are substantially increased, compared with the native state *(2, 3, 8)*.

(iv) The molten globule state is already molten. The molten globule does not melt cooperatively at heating *(2, 3, 8, 15, 22)*, while temperature and guanidine hydrochloride (GuHCl)-induced transitions from the native to the molten globule state are highly cooperative *(2, 3, 8, 22)*.

All these properties of the molten globule state strongly imply that this state differs from the native one mainly by an essential decrease in van der Waals attraction between side groups and by a substantial increase of the intramolecular mobility *(2–4, 23, 24)*. This idea is the focus both of the theory of the molten globule state *(25)* and of transition from native to molten globule states *(23, 24)*. According to this theory, there are two types of equilibrium (thermodynamically stable) states: the native state—with a small conformational entropy but with large van der Waals attraction—and the family of denatured (less condensed) states—with large conformational entropy but smaller van der Waals attraction.

The molten globule state is the most condensed among all these denaturated states, having just enough free intramolecular space to allow the rotational isomerization of side groups. The values of volume lying between the native and the molten globule states are forbidden because they correspond to the decrease of van der Waals attraction, which is not compensated by a substantial increase of the side-chain movement. Therefore, the denaturation of proteins (including the native to molten globule state transition) is the phase transition of the first order. There is some evidence that the molten globule to unfolded state transitions are also highly cooperative *(14)*, but this problem awaits a more detailed investigation.

In summary, we conclude that among the three main types of interactions that stabilize the structure of native proteins (van der Waals and other specific attractions, peptide hydrogen bonds, and hydrophobic effect), the first (van der Waals attraction) is substantially reduced in the molten globule state, while the two others (hydrogen bonds and hydrophobic effect) remain mainly unaltered. Therefore, the existence of the molten globule state gives us a unique possibility of understanding the role of these interactions in the determination of the two levels of the structure of a protein molecule [for example, see *(26)*]: (i) the "protein fold" or the "folding pattern," i.e., the approximate positions of the α and β regions in a protein chain and in a space, and (ii) the exact tertiary structure of a protein (a set of atomic coordinates).

The second level is unique for each protein and, of course, is determined by all these types of interactions. The first level can be similar, or even the same, for a large number of related proteins despite the wide differences in their amino acid sequences (e.g., "globin fold," "immunoglobulin fold"). This suggests that the protein folds (folding patterns) are not determined by van der Waals interactions, which strongly depend on all details of amino acid sequence and, therefore, are different in each member of these protein families. If this is the case, protein folding pattern prediction becomes much simpler. The experimental approach to the problem outlined above is to determine whether the molten globule has the same crude 3-D structure as that of the native protein.

The study of guinea pig α-lactalbumin by NMR methods [*(20)*; see also chapter 12, in this volume] gives evidence that in the molten globule state this protein has some features of the native-like 3-D structure:

(i) In the acid (molten globule) state of this protein, some regions of the protein backbone are highly protected from deuterium exchange; all of them are also protected in the native state. It was even possible to assign NMR resonances to a number of the protected residues and to show that they correspond to native α-helices.

(ii) 2-D NMR spectra of this protein in both the molten globule and the native states show that there is a cluster in each consisting of one tyrosine, one phenylalanine, and two tryptophan residues, which is consistent with the cluster from TYR 103, PHE 31, TRP 26, and TRP 104, following from x-ray data of closely related proteins.

These data suggest a strong similarity between at least some features of the native and the molten globule state structures.

Accumulation of the Molten Globule State During Protein Folding

The first evidence that protein folding in vitro includes a compact intermediate came from urea gradient gel electrophoresis *(27, 28)*. The accumulation of the kinetic inactive intermediate characterized by compactness and a secondary structure, but without a fixed tertiary structure, has been shown by direct methods (viscosimetry and circular dichroism) for bovine carbonic anhydrase B *(29)*. This means that the kinetic intermediate in protein folding has properties similar to those of the molten globule state.

This conclusion was later confirmed by Zerovnik and Pain *(30)* by rapid size-exclusion chromatography with fast protein, peptide, and polynucleotide liquid chromatography (FPLC). The results published in our joint paper *(31)* show that compact kinetic intermediates accumulate upon refolding of bovine carbonic anhydrase B, *Staphylococcus aureus* β-lactamase, and recombinant human interleukin 1β.

The elution volumes of these kinetic intermediates are somewhat smaller than those of the native states (14.3 and 15.7 for β-lactamase, 14.1 and 15.0 for carbonic anhydrase, 14.8 and 15.3 for interleukin 1β), which corresponds to a small increase of the molecular volumes. It has been shown also that the elution volumes of the compact intermediates are even closer to the native elution volumes than those for the equilibrium molten globule states (Fig. 1). This means that the kinetic intermediates can be more compact than the equilibrium molten globule state, though less compact (more loosely packed) than the native state. In the course of time, the peak elution volume of a compact intermediate decreases, and the corresponding peak of a native (or native-like) state increases (Fig. 1), suggesting the all-or-none kinetic transition between these two states.

Our evidence *(29)* that the compact intermediate of carbonic anhydrase has a pronounced secondary structure but not a rigid

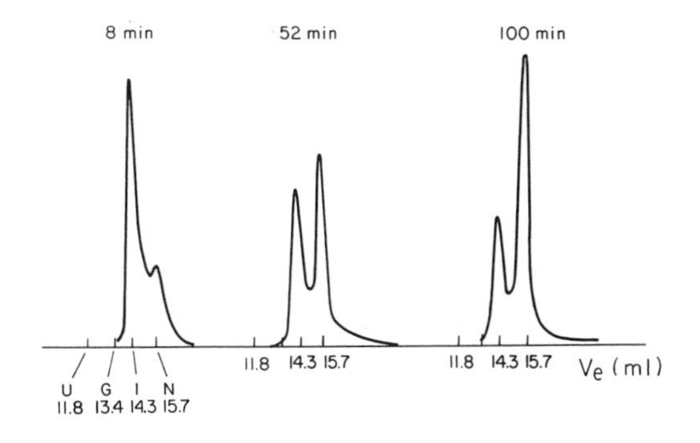

Fig. 1. Size exclusion chromatography (FPLC data) for *S. aureus* β-lactamase refolding from the molten globule state (2.0M urea) to the native state (0.5M urea) at 15°C and pH 7 at different times after the start of refolding *(30, 31)*. Identical data have been obtained for refolding from 4.0M urea (unfolded state). The smaller the molecular volume of a protein, the larger is its elution volume V_e. Marks show the elution volumes of β-lactamase in the equilibrium native (N = 0M urea, pH 7.0), molten globule (G = 2.0M urea) and unfolded (U = 4.0M urea) states as well as in the kinetic intermediate (I). Adapted from *(31)*.

tertiary structure has been also confirmed by R. H. Pain and his collaborators for interleukin 1β *(32)* and β-lactamase *(31)*. Moreover, ^{1}H NMR spectra of these kinetic intermediates for carbonic anhydrase *(33)* and β-lactamase *(31)* are much simpler than the spectra of native proteins (Fig. 2) and similar to the NMR spectra of the equilibrium molten globule state *(4, 21)*. The activity of the compact kinetic intermediates of carbonic anhydrase *(29, 34)* and β-lactamase *(35)* is completely absent.

Thus, the compact intermediates that accumulate upon refolding of carbonic anhydrase, β-lactamase, and interleukin 1β have properties very similar to those of the equilibrium molten globule state. The only difference detected to date is that the kinetic intermediates are even more compact than the equilibrium ones. This can be easily explained by the difference in the conditions: the kinetic intermediates were studied at native conditions (neutral pH, low concentrations of denaturants), while the equilibrium intermediates were studied either at acid pH or at rather high concentrations of strong denaturants. In both these cases, the effect of electrostatic repulsion or of strong denaturants can increase the dimensions of the rather flexible molten globule state.

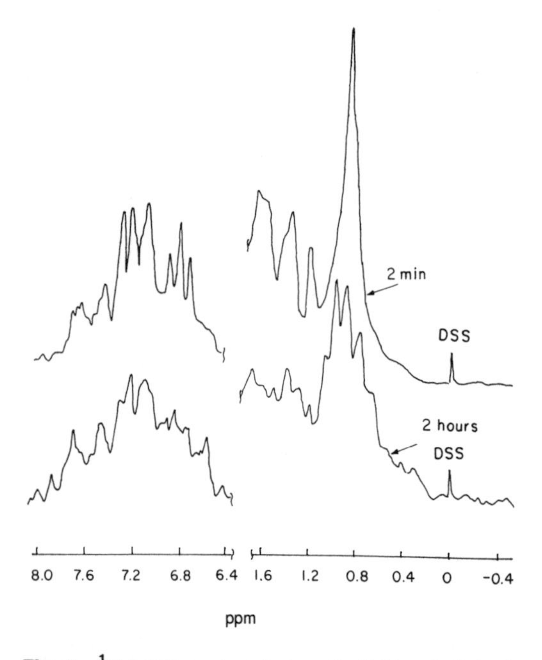

Fig. 2. ^{1}H NMR spectra of aromatic (**left**) and aliphatic (**right**) groups of bovine carbonic anhydrase B, 2 min and 2 hours after the start of refolding from 8.5M urea to 4.2M urea at 18°C. Spectra were obtained at 400 MHZ at protein concentration ~5 mg/ml *(72)*.

Kinetics of Protein Folding: Secondary Structure

Now, let us consider the *kinetics* of the formation of different protein structure features at protein folding. Long-standing evidence indicates *(29, 34, 36–41)* that far-UV ellipticity is restored during protein folding faster than near-UV ellipticity, i.e., that the formation of protein secondary structure precedes the formation of a rigid tertiary structure. In 1987–88, it was shown by the stop flow technique that the far-UV ellipticity of bovine α-lactalbumin *(42, 43)*, cytochrome c *(44)*, β-lactoglobulin *(44)*, and parvalbumin *(45)* is completely or essentially restored within 0.01 sec, suggesting that the secondary structure of these proteins is formed at a very early stage of their folding. Recently, the rates of far-UV ellipticity restoration have been systematically measured and compared with the rates of compact state formation *(46, 70)*.

Figures 3 and 4 illustrate the results of these measurements for two proteins, carbonic anhydrase B *(70)* and the β_2 subunit of tryptophan synthase *(46)*. Carbonic anhydrase (Fig. 3) is an example of proteins where practically all the secondary structure is restored within 0.01 sec. Cytochrome c *(44)* and human α-lactalbumin *(70)* also belong to this family. During refolding of β-lactoglobulin, the far-ultraviolet ellipticity in the earliest kinetic intermediate (after 0.01 sec) is even more negative than in the native state. Then it slowly relaxes to its native value with the same rate as near-ultraviolet ellipticity *(44)*. This can be explained by the appearance of a (positive) contribution of aromatic groups in the far-UV CD *(3, 47, 48)*; therefore, β-lactoglobulin can also be considered as a part of this group.

On the other hand, in the β_2 subunit of tryptophan synthase *(46)*, only 60% of far-UV ellipticity is restored within 0.01 sec, while the

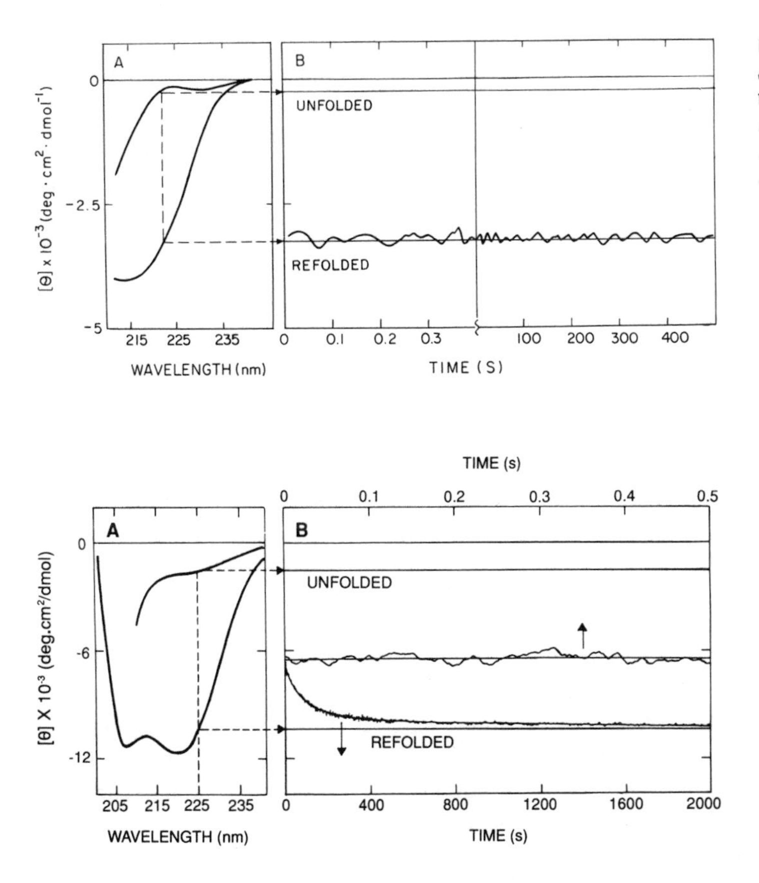

Fig. 3. (A) Far-ultraviolet circular dichroism spectra of bovine carbonic anhydrase B in the unfolded (8.5M urea, pH 8.0) and refolded (0.7M urea, pH 8.0) states. (B) Time dependence of molar ellipticity at 222 nm (marked by arrowheads) during protein refolding from 8.5M to 0.7M urea *(70)*.

Fig. 4. (A) Far-ultraviolet circular dichroism spectra of the β_2 subunit of *E. coli* tryptophan synthase in the unfolded (5.5M urea) and refolded (0.4M urea) states; (B) Time dependence of molar ellipticity at 225 nm (marked by arrowheads) during protein refolding from 5.5M to 0.4M urea. [Reprinted from *(46)* with permission.]

other 40% is restored with the rate comparable with that of near-UV CD (Figs. 4 and 5). Thus, unlike the case of lactoglobulin, both fast and slow phases of the change of far-UV ellipticity have the same sign. The effect of the contribution of aromatic groups (which would be negative in this case) also cannot be completely excluded. However, it seems more likely that the existence of two phases of restoration of far-ultraviolet ellipticity, in this case, reflects the two stages of restoration of a protein secondary structure. Similar results have been obtained also for parvalbumin *(45)* as well as for β-lactamase, apomyoglobins, and yeast phosphoglycerate kinase *(70)* for which 50–80% of far-UV ellipticity is restored within 0.01 sec.

Thus, the secondary structure of proteins at strong native conditions is either completely or substantially formed within 0.01 sec, i.e., as discussed below, long before all other stages of protein folding.

Kinetics of protein folding: Loosely packed compact intermediate

As the compact intermediate is formed much faster than in 1 min, we again must use the stop-flow technique to determine the rate of its formation. However, the problem is how to monitor the compactization of molecules if this process is very rapid. Very fast measurements of viscosity or elution volume are obviously impossible. We have turned to optical (or radio-optical) methods that provide at least indirect approaches to monitoring fast compactization. We have used three such methods for this purpose:

(i) Energy transfer from tryptophan residues to the fluorescent dansyl labels. The excitation spectrum of dansyl labels strongly overlaps with the emission spectrum of tryptophan residues. Therefore, the excitation energy can transfer from tryptophans to dansyl labels if they are close to one another in space. In this case (Fig. 6), the excitation

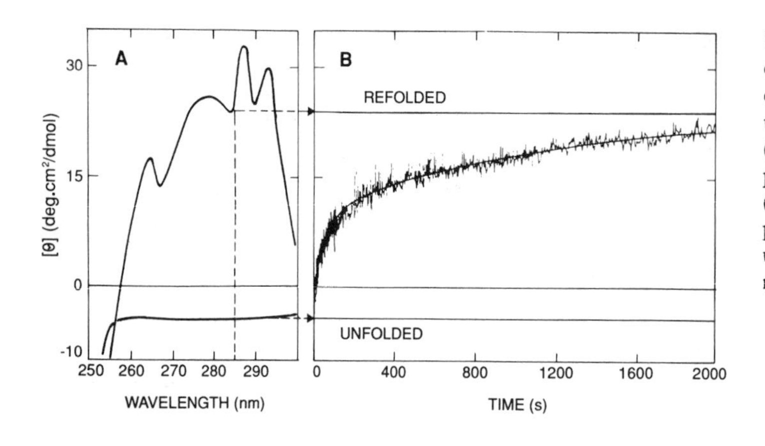

Fig. 5. (A) Near-ultraviolet circular dichroism spectra of the β_2 subunit of *E. coli* tryptophan synthase in the unfolded (5.5M urea) and refolded (0.4M urea) states. (B) Time dependence of molar ellipticity at 285 nm (marked by arrowheads) during protein refolding from 5.5M to 0.4M urea. [Reprinted from *(46)* with permission.]

spectrum of dansyl labels includes that of the tryptophan residues, and we must observe the increase of dansyl emission (~500 nm) under tryptophan excitation (~280 nm), compared with that in the absence of energy transfer. This effect obviously must be much stronger in the compact (globular) state of a protein molecule than in its unfolded state (Fig. 6) because the average distances between tryptophan residues and dansyl labels are smaller in the globular state. If a protein has several tryptophan residues, and if several dansyl labels have been covalently attached to it, the energy transfer is averaged over many tryptophan-dansyl pairs and so reflects the overall compactness of a protein molecule *(33, 49)*.

(ii) Immobilization of spin labels. The signal intensity of spin labels covalently attached to globular protein is usually lower than in the completely unfolded, coil-like one (Fig. 7). This is due to the more restricted motion (or stronger immobilization) of spin labels in globular protein than in the coil-like one, and results in broader and less intensive lines in the spin-label spectra in the compact state. This effect also can be used for monitoring protein compactization *(33, 49, 74)*.

Since both of these methods are indirect, it was necessary to compare them with a direct method, e.g., with viscosimetric data. Figure 8 represents the results of this comparison for the equilibrium unfolding of human carbonic anhydrase B by GuHCl. The figure shows that energy transfer and electron spin resonance (ESR) signal can be used for monitoring the change of molecular dimensions in the cases when direct methods cannot be applied.

(iii) Adsorption of a hydrophobic fluores-

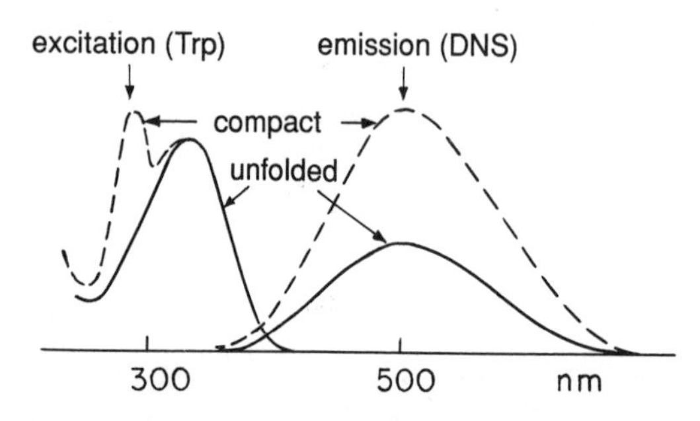

Fig. 6. Excitation and emission spectra of dansyl labeled bovine carbonic anhydrase B in the compact (molten globule) state (pH 3.8) and in the unfolded state (8.5M urea). [Reprinted from *(49)* with permission.]

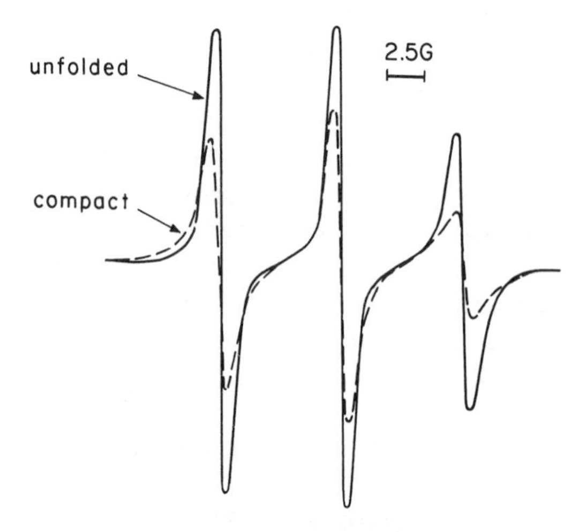

Fig. 7. Electron spin resonance spectra of spin labelled bovine carbonic anhydrase B in the compact (native) state (pH 8.0) and in the unfolded state (8.5M urea). [Reprinted from *(49)* with permission.]

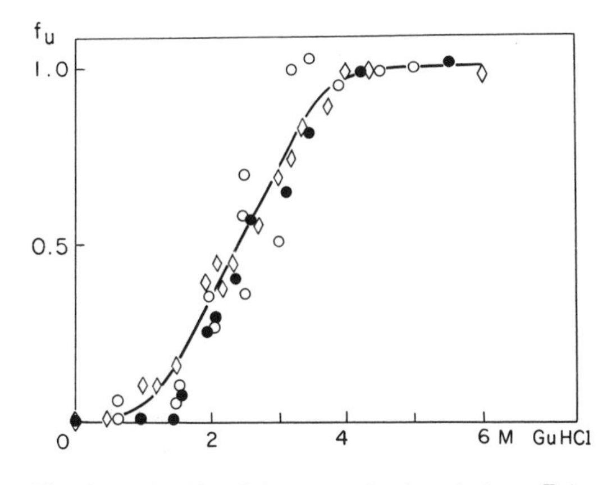

Fig. 8. Unfolding of human carbonic anhydrase B by GuHCl monitored by the direct method — intrinsic viscosity (◇) and by indirect methods — energy transfer from tryptophan residues to dansyl labels (•) and intensity of electron spin resonance spectra of spin label (o). All data are shown in relative units — "unfolding degrees" $f_U = (x - x_u)/(x_N - x_U)$ where x is the measured parameter while x_N and x_U are its values in the native and unfolded states. [Reprinted from *(49)* with permission.]

cent probe, 1-anilinonaphthalene, 8-sulfonate (ANS). Hydrophobic fluorescent probes have long been used to study exposed hydrophobic sites in proteins *(50, 51)*. It has been shown that the binding of a probe to a protein usually leads to a drastic increase of its fluorescence intensity and to a shift of a fluorescent spectrum to a shorter wavelength. We have shown *(14, 33, 52)* that ANS, as well as other hydrophobic probes, binds very strongly to proteins in the molten globule state. The affinity of ANS to molten globules is always much more than its affinity to the unfolded state; it is also usually much more than its affinity to the native state. (This is usually the case even for proteins that have remarkable surface hydrophobic clusters and can also bind ANS in the native state). The reason for this strong affinity is probably the presence of solvent-accessible hydrophobic clusters in the core of the molten globule, which can be penetrated by hydrophobic molecules. Unfolded proteins cannot strongly absorb ANS as they have no stable hydrophobic clusters, while native proteins very often cannot do so because the hydrophobic clusters in their cores are tightly packed and inaccessible for exogenic hydrophobic molecules *(50, 52)*. The binding of ANS to the protein can be easily monitored because it leads to a very large increase of the of ANS fluorescence intensity.

Thus, if the molten globule state accumulates upon protein folding or unfolding, the affinity of ANS to the protein (and, therefore, the intensity of ANS fluorescence) must have a maximum. This is illustrated by Fig. 9 for the equilibrium unfolding of human and bovine carbonic anhydrases B by GuHCl. This figure shows that the intensity of ANS fluorescence has a maximum at 1.5–2.1M GuHCl when the molten globule state is accumulated *(52)*. At a low concentration of GuHCl, both proteins are in the native states, while at higher concentrations of GuHCl, they are in the unfolded state *(12, 14)*. Thus, the increase of ANS affinity to a protein molecule is a specific tool for the accumulation of the molten globule state in protein folding.

Figure 10 shows the kinetics of the ANS fluorescence change for six different proteins upon their folding *(31, 46)*. It has been shown *(52)* that ANS binds to the molten globule state within the dead time of the stop-flow experiments (0.002 sec). Therefore, the increase of ANS fluorescence shown in Fig. 10 really reflects the *formation* of the ANS binding intermediates rather than the ANS binding to the *preexisting* intermediates. This is confirmed by Figs. 11 and 12, which show that the

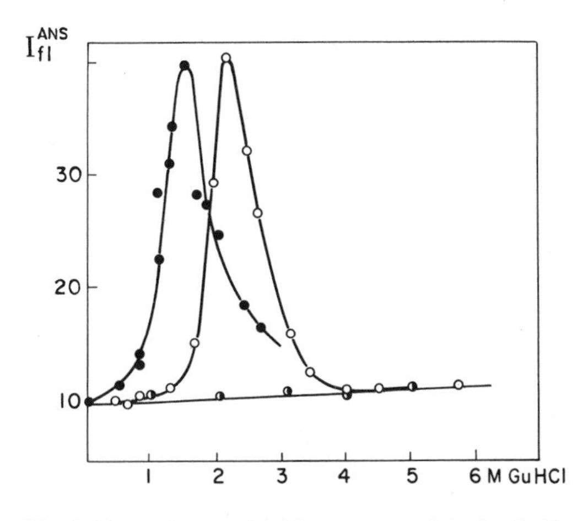

Fig. 9. Dependences of the fluorescence of a hydrophobic probe (ANS) in the presence of human (•) and bovine (o) carbonic anhydrase B on GuHCl concentrations. The increase of fluorescence reflects the increase of ANS affinity to the proteins. Data for "free" ANS (in the absence of proteins) are shown for comparison (◐). [Reprinted from *(49)* with permission.].

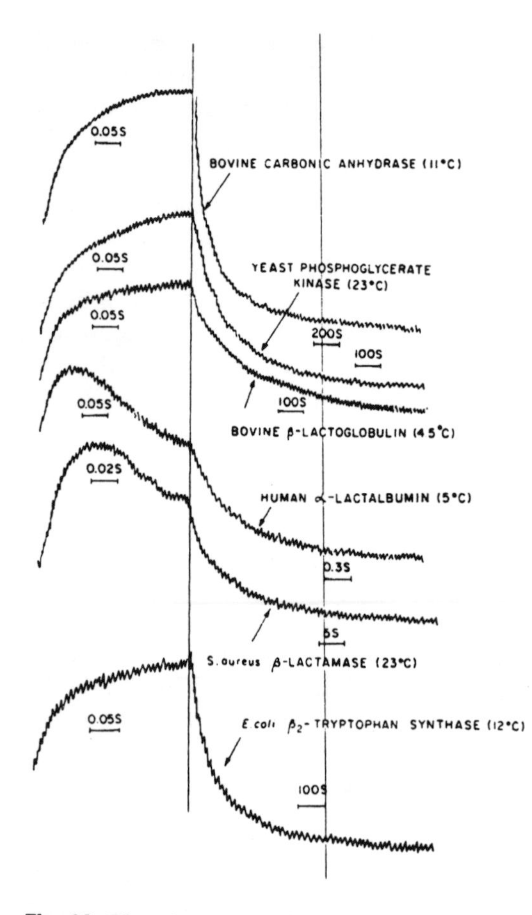

Fig. 10. Time dependence of ANS fluorescence during protein refolding. The maxima reflect the highest ANS affinity to the protein molecule in loosely packed kinetic intermediates. [From the data published in *(31)* and *(46)*.]

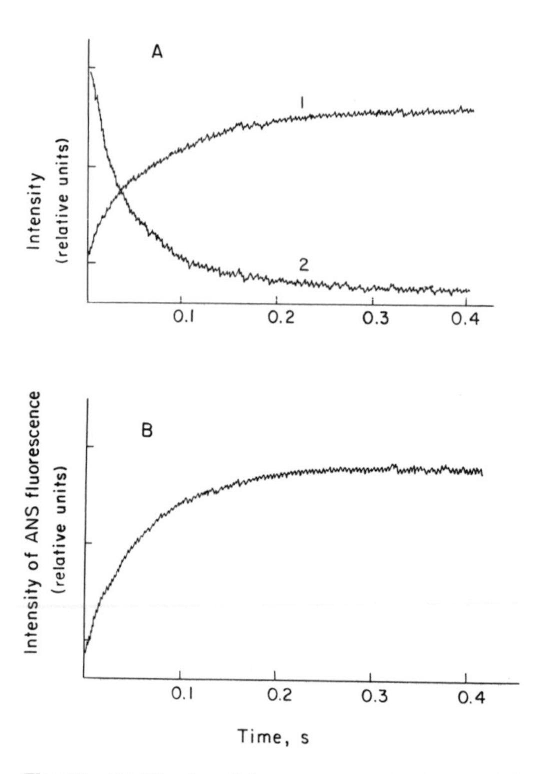

Fig. 11. **(A)** Kinetics of the energy transfer increase (1) and the decrease of intensity of spin labels (2) during carbonic anhydrase B refolding from 8.5M to 4.2M urea at 23°C. **(B)** Kinetics of ANS binding at the same conditions. [Adapted from *(33)*.]

rates of ANS fluorescence increase are practically the same as the rates of molecular compactization monitored by other methods. For carbonic anhydrase, it has been shown that the rate of ESR intensity decrease is also very near to the values for ANS fluorescence and energy transfer *(33)*.

Figures 10–12 show that at least half of the ANS fluorescence increase takes place within the dead time of our experiments (0.01 sec), while the remaining part needs much more time. Very fast increase of ANS fluorescence may reflect the formation of protein secondary structure as hydrophobic probes bind to β sheets *(52–54)* on which hydrophobic clusters are usually formed. (Note that all the studied proteins have a large amount of β structure.) On the other hand, we cannot exclude very fast formation of some elements of 3-D (globular) structure.

In any case, the maximum of ANS fluorescence reflects the formation of a loosely packed molten globule-like intermediate. This formation needs much more time than the overall or substantial formation of the secondary structure. On the other hand, a subsequent *decrease* of ANS fluorescence or the decrease of its affinity (Fig. 10) indicates a tight, rather than loose, packing of nonpolar clusters in the hydrophobic core of protein molecules (see below).

We can conclude from these experimental data that the loosely packed, compact kinetic intermediate is formed in a variety of proteins, including carbonic anhydrase, phosphoglycerate kinase, β-lactoglobulin, α-lactalbumin, β-lactamase, and the β_2 subunit of tryptophan synthase *(33, 31, 46)*, as well as in

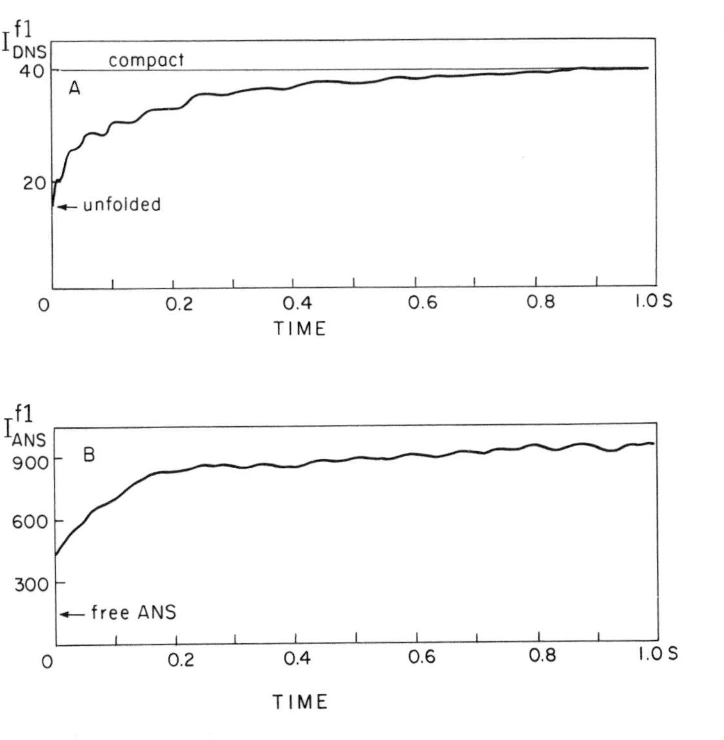

Fig. 12. Kinetics of the energy transfer increase (A) and ANS binding (B) during the refolding of β-lactoglobulin from 4M to 0.4M GuHCl at 4.5°C *(73)*.

interleukin-1β *(31)* and probably in lysozyme *(39)*. These proteins belong to different structural types, including β, $\alpha + \beta$, and α/β, with or without disulfide bridges. The times of complete refolding of these proteins can be as small as ~0.5 sec (α-lactalbumin), or as large as 2500 sec (carbonic anhydrase, β-lactamase). These proteins include not only proteins that have a loosely packed molten globule state at some equilibrium conditions (carbonic anhydrase, α-lactalbumin, β-lactoglobulin, β-lactamase) but also those that apparently do not have such a state (e.g., phosphoglycerate kinase or interleukin 1β). This has permitted us to suggest that *the loosely packed compact or molten globule state is a general kinetic intermediate in protein folding (31)*.

Kinetics of Protein Folding: Rigid Tertiary Structure

Figure 10 shows the kinetics of ANS desorption from six proteins, suggesting the transition from the loosely packed compact state to the tightly packed one. This interpretation is confirmed by the fact that high-field ^{1}H NMR signals of carbonic anhydrase appear simultaneously with ANS desorption *(33)*. As high-field ^{1}H NMR signals reflect specific rigid contacts between aliphatic and aromatic groups inside the nonpolar core of a native protein, their appearance directly shows the formation of a tightly packed hydrophobic core.

In the number of proteins, including the β_2- subunit of tryptophan synthase *(46)* as well as α-lactalbumin, β-lactoglobulin, and phosphoglycerate kinase *(70)* near-UV ellipticity is restored simultaneously with ANS desorption. This suggests the existence of a *single* process transforming a loosely packed, compact intermediate into a tightly packed native structure. On the other hand, in bovine carbonic anhydrase B, near-UV ellipticity (as well as enzymatic activity) is fully restored after ANS completely desorbs from a protein *(33)*. This suggests the existence of a *late* native-like, kinetic intermediate in which the hydrophobic core already has rigid structure, but the tryptophan residues and side groups involved in the enzymatic active center are still not fixed in their native positions.

The time required for complete refolding of a number of proteins at 23°C can be as long as 500–2500 sec. Slow phases of protein folding are often connected with cis-trans isomerization of prolines *(55, 56)*. Each proline residue in the native protein is in the fixed (cis or trans) conformation, while in the unfolded protein chain, the prolines can slowly jump (with $t_{1/2} \approx 100$ sec for trans-cis transition) from one state into another. Therefore, the unfolded protein chain is a mixture of various species, differing one from another by cis or trans isomers of proline residues. In these cases, only those molecules that occasionally have the "correct" proline isomers in the unfolded state can fold quickly, while the others must fold much more slowly, with a rate determined by the rate of proline isomerization.

A good experimental tool for proline-controlled protein folding is the double-jump technique *(57, 58)*. The native protein is quickly transferred into strong unfolding conditions (e.g., 8M urea or 6M GuHCl), stays under these conditions for a given time, and then is quickly transferred back into strong native conditions. If the existence of the slow phase is connected with proline isomerization (more generally, with the existence of slowly interconverted different species in the unfolded chain), the amplitude of this phase must be greater, the longer the protein has been incubated under unfolding conditions. In fact, for the short stay in the unfolded state, protein has no time to disorder its correct proline isomers and, therefore, can refold quickly. After the long stay, its proline isomers became completely disordered.

Figure 13 shows the results of double-jump experiments on the refolding of carbonic anhydrase *(59)*. It is clearly seen that the amplitude of the slow phase of protein folding increases with the increase of the incubation. This is very strong evidence for the proline-controlled limiting step of protein folding. Moreover, it has also been shown that not only the "slow" phase of carbonic anhydrase folding (with $t_{1/2} \approx 100$ sec) but also its "superslow" phase (with $t_{1/2} \approx 600$ sec) is proline controlled. The existence of proline-controlled stages with a half time much

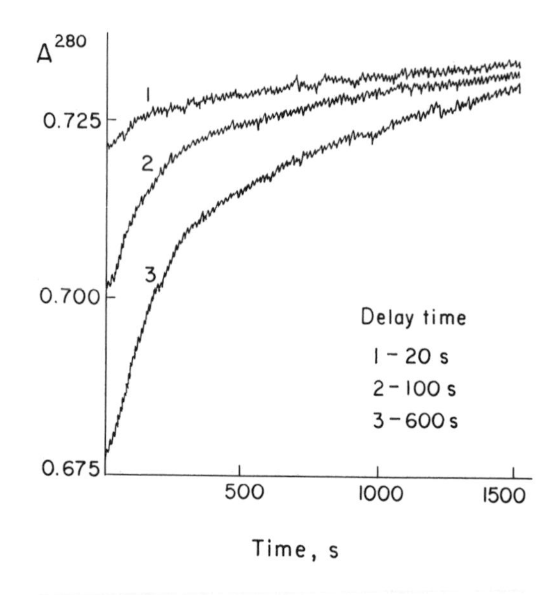

Fig. 13. Double-jump experiments for bovine carbonic anhydrase B. The protein has been dissolved in the native buffer, placed in unfolding conditions (6M GuHCl) for 20, 100, and 600 sec and then returned to the native conditions. The time dependence of absorbance restoration shown in the figure demonstrates the increase of amplitude of the slow refolding phase with the increase of incubation time under unfolding conditions. [Adapted from *(59)*.]

greater than the time of proline isomerization has been explained by the simultaneous transitions of several prolines into their correct isomers *(59, 60)*.

On the other hand, the slow refolding of yeast phosphoglycerate kinase is not proline controlled. It has been proved *(71)* by the independence of the amplitude of this slow stage from the time of protein incubation in the unfolded state. Similar results have been obtained also for horse *(61)* and pig *(71)* phosphoglycerate kinases.

It could be assumed that the slow, proline-independent folding of phosphoglycerate kinases is due to interdomain interaction in the two-domain protein. However, it was shown recently *(71)* that even one domain of pig or yeast phosphoglycerate kinase (produced by limited trypsinolysis) folds as slowly as the intact protein. It follows that proline isomerization is not the single reason for slow protein folding. There are also some slow *intradomain*, purely conformational processes that can limit the rate of protein folding. The nature of these processes and the reasons why their rates are so different for different proteins remain to be studied.

Kinetics of Protein Folding: First Features of Native Structure

Thus, we conclude that the formation of a loosely packed compact intermediate with a substantial secondary structure always precedes the formation of the native, tightly packed, rigid protein. However, the most intriguing point in the mystery of protein folding is probably neither the formation of secondary or compact structure nor the formation of the rigid tertiary structure, but the formation of the unique *tertiary fold* of protein molecules. The stage of protein folding at which protein achieves the first features of its crude native structure is, therefore, one of the central questions of the whole field.

Two approaches to this problem have been elaborated by M. Goldberg and his collaborators for the refolding of the β_2 subunit of tryptophan synthase. The first involves the energy transfer between two *definite* residues, which are close to one another in the native protein but well removed in the unfolded one. Blond and Goldberg *(62)* have measured the energy transfer from a single tryptophan residue of the β chain (Trp 177) to the pyridoxal 5′-phosphate attached to Lys 87 or to the dansyl group linked to Cys 170. The existence of a single tryptophan residue, together with the addressed labeling of two other residues, permits the measurement of the rates of bringing Trp 177 close to Cys 170 or Lys 87.

The second approach uses monoclonal antibodies specific to the native protein to capture the moment of formation of the native-like antigenic site that can be recognized by this antibody *(63)*. The results indicate that the half time of bringing together Trp 177 with Lys 87 is 35 sec, and that for Trp 177 with Cys 170 is 90 sec *(62)*; the half time of formation of the first detected antigenic site on the folding molecule is 12 sec *(64, 65)*.

We have combined these data on the formation of the first features of the protein native structure with the measurements of kinetics of ANS binding-desorption as well as of the restoration of far- and near-UV ellipticity *(46)*. The result is that the native-like antigenic determinant is formed *before* the complete formation of the native tertiary structure (Fig. 14), i.e., *within* the time of the existence of a loosely packed compact kinetic intermediate.

This was one of the first attempts to measure directly the rates of formation of some native-like structures in a folding protein. Of course, it is too early to generalize these first results and to speculate on the rate of this important step of protein folding. It is worthwhile to remember that the very fast formation of the N- and C-terminal α helix in cytochrome c is probably accompanied by the formation of their complex *(66)*, which suggests that the first feature of the native structure of this simple protein may form very early in the course of its folding. Much more work is needed before we can determine at which stage of its folding the protein acquires its native folding pattern or other important features of its native structure.

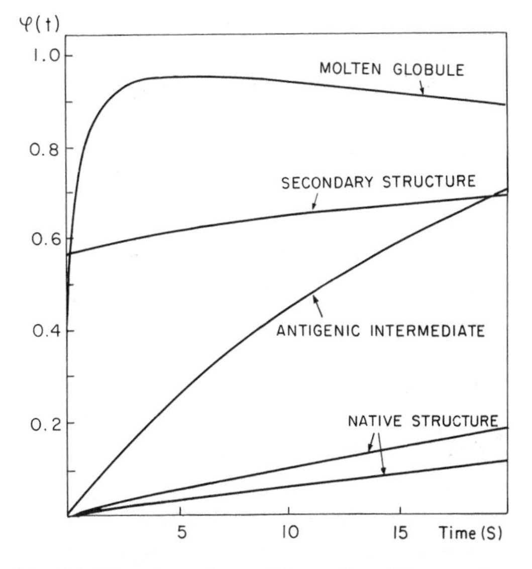

Fig. 14. Time dependence of formation of the secondary structure, molten globule state, antigenic determinant, and the complete native tertiary structure for the β_2 subunit of *E. coli* tryptophan synthase. Time dependences of formation of the native structure have been obtained by ANS desorption (**upper curve**) and near-ultraviolet circular dichroism (**lower curve**). [From the data published in *(46)* and *(65)*.]

Table 1. Main kinetic stages of protein folding.

Time	Experimental evidences	What happens?
< 0.01 s	Restoration of far-UV ellipticity (50–100%)	Complete or substantial formation of the secondary structure
< 1 s	Maximal affinity to a hydrophobic probe	Formation of a compact, loosely packed (molten globule) intermediate
	Energy transfer	
	Immobilitzation of spin labels	
1–1000 s	Decrease of the affinity to a hydrophobic probe	Formation of a rigid tertiary structure
	Restoration of high field NMR signals	
	Restoration of near–UV ellipticity	
	Restoration of activity	

Conclusions

Summarizing this brief review of our recent studies of the kinetics of protein folding *(31, 33, 46, 49, 52, 59, 70)*, we can come to the following conclusions:

(i) Protein folding proceeds through a universal kinetic intermediate, which is compact, loosely packed, and has either a whole or a substantial amount of native secondary structure but has no rigid tertiary structure. The properties of this kinetic intermediate are similar (though not completely identical) to the properties of the equilibrium molten globule state. This intermediate forms within ~1 sec.

(ii) The protein secondary structure, either completely or to a substantial extent, is formed much faster than the loosely packed kinetic intermediate (within 10^{-2} sec). The remaining part of the secondary structure can be formed as slowly as within 1000 sec (simultaneously with the formation of a rigid tertiary structure).

(iii) The native tertiary structure of a protein forms in a time period of 1 to 10^3 sec, depending on the protein. The slow (in $\geq 10^2$ sec) formation of the native tertiary structure can be either dependent or independent of the cis-trans isomerization of proline residues. Sometimes the formation of the rigid tertiary structure in a protein core can precede its formation on a protein surface.

(iv) The first features of a protein native structure (e.g., some antigenic intermediates) can be formed during the existence of the loosely packed compact intermediate, i.e., before formation of the rigid tertiary structure.

The main stages of protein folding that follow from these experiments are shown in Table 1. As early as 1973, one of us *(67)* suggested that protein folding involves at least three steps: (i) formation of embryos of the native structure in a chain; (ii) collapse of these embryos into an intermediate compact state, stabilized mainly by hydrophobic interactions; and (iii) the local adjustment of this intermediate state to the tightly packed native structure stabilized mainly by van der Waals interactions. This chapter summarizes our data, which basically confirm this hypothesis. In fact, the formation of the whole or of the substantial part of a secondary structure does precede the complete formation of a compact intermediate, and this intermediate always exists as a kinetic precursor of the native state.

However, to be really important for the mechanism of protein folding, this scheme must imply that the embryos of the secondary structure are formed in their "proper" (native) places in a protein chain and that the overall structure (folding pattern) of a kinetic intermediate must be similar to that of the native protein. Both these suggestions have been made in the above-mentioned hypothesis *(67)*. The data on far-UV molar ellipticity does not prove that the secondary structure is formed in the proper places. However, this very important point has been confirmed by 2-D NMR studies of the trapped deuterium exchange intermediates *(66, 68)*. It is much more difficult to check the second assumption. The first evidence that something native (in addition to the native-like secondary structure) may form before the formation of the rigid tertiary structure *(46, 66)* is far from sufficient to draw definite conclusions.

References

1. Creighton, T. E., *Proteins. Structure and Molecular Properties* (W. H. Freeman and Co., New York, 1983).
2. Dolgikh, D. A., *et al., FEBS Lett.* **136,** 311 (1981).
3. Dolgikh, D. A., *et al., Eur. Biophys.* **13,** 109 (1985).
4. Ptitsyn, O. B., *J. Protein Chem.* **6,** 273 (1987).
5. Ohgushi, M., and A. Wada, *FEBS Lett.* **164,** 21 (1983).
6. Kuwajima, K., *Proteins: Struct. Funct. & Genet.* **6,** 87 (1989).
7. Wong, K.-P., and L. M. Hamlin, *Biochemistry* **13,** 2678 (1974).
8. Dolgikh, D. A., *et al., Dokl. Akad. Nauk SSSR* **272,** 1481 (1983).
9. Gast, K., D. Zirwer, H. Welfle, V. E. Bychkova, O. B. Ptitsyn, *Int. J. Biol. Macromol.* **8,** 231 (1986).
10. Damaschun, G., Ch. Gernat, H. Damaschun, V. E. Bychkova, O. B. Ptitsyn, *Int. J. Biol. Macromol.* **8,** 226 (1986).
11. Ptitsyn, O. B., G. Damaschun, C. Gernat, H. Damaschun, V. E. Bychkova, *Stud. Biophysica* **112** 207 (1986).
12. Wong, K.-P., and C. Tanford, *J. Biol. Chem.* **248,** 8518 (1973).
13. Kuwajima, K., K. Nitta, M. Yoneyama, S. Sugai, *J. Mol. Biol.* **106,** 359 (1976).
14. Rodionova, N. A., *et al., Mol. Biol. (USSR)* **23,** 683 (1989).
15. Brazhnikov, E. V., Yu. N. Chirgadze, D. A. Dolgikh, O. B. Ptitsyn, *Biopolymers* **24,** 1899 (1985).
16. Nozaka, M., Kuwajima, K., K. Nitta, S. Sugai, *Biochemistry* **17,** 3753 (1978).
17. Ptitsyn, O. B., D. A. Dolgikh, R. I. Gilmanshin, E. I. Shakhnovich, A. V. Finkelstein, *Mol. Biol. (USSR)* **17,** 569 (1983).
18. Ikeguchi, M., K. Kuwajima, S. Sugai, *J. Biochem.* **99,** 1191 (1985).
19. Kuwajima, K., Y. Harushima, S. Sugai, *Int. J. Peptide Protein Res.* **27,** 18 (1986).
20. Baum, J., C. M. Dobson, P. A. Evans, C. Hanley, *Biochemistry* **28,** 7 (1989).
21. Semisotnov, G. V., V. P. Kutyshenko, O. B. Ptitsyn, *Mol. Biol. (USSR)* **23,** 808 (1989).
22. Pfeil, W., V. E. Bychkova, O. B. Ptitsyn, *FEBS Lett.* **198,** 287 (1986).
23. Shakhnovich, E. I., and A. V. Finkelstein, *Dokl. Akad. Nauk SSSR* **267,** 1247 (1982).
24. Shakhnovich, E. I., and A. V. Finkelstein, *Biopolymers* **28,** 1667 (1989).
25. Finkelstein, A. V., and E. I. Shakhnovich, *Biopolymers* **28,** 1681 (1989).
26. Richardson, J., *Adv. Prot. Chem.* **34,** 167 (1981).
27. Creighton, T. E., *J. Mol. Biol.* **137,** 61 (1980).
28. Creighton, T. E., and R. H. Pain, *J. Mol. Biol.* **137,** 431 (1980).
29. Dolgikh, D. A., A. R. Kolomietz, I. A. Bolotina, O. B. Ptitsyn, *FEBS Lett.* **165,** 88 (1984).
30. Zerovnik, E., and R. H. Pain, *Protein Engineering* **1,** 248 (1987).
31. Ptitsyn, O. B., R. H. Pain, G. V. Semisotnov, E. Zerovnik, O. I. Razgulyaev, *FEBS Lett.* **262,** 20 (1990).
32. Craig, S., V. Schmeissner, P. Wingfield, R. H. Pain, *Biochemistry* **26,** 3570 (1987).
33. Semisotnov, G. V., *et al., FEBS Lett.* **224,** 9 (1987).
34. McCoy, L. F., E. S. Rowe, K.-P.Wong, *Biochemistry* **19,** 4738 (1980).
35. Mitchinson, C., and R. H. Pain, *J. Mol. Biol.* **184,** 331 (1987).
36. Robson, B., and R. H. Pain, in *Conformation of Biological Molecules and Polymers,* E. D. Bergmann and A. Pullman, Eds. (Academic Press, New York, 1973), pp. 161 – 172.
37. Robson, B., and R. H. Pain, *Biochem. J.* **15,** 331 (1976).
38. Labhardt, A. M., *Proc. Natl. Acad. Sci. U. S. A.* **81,** 7674 (1984).
39. Kuwajima, K., Y. Hiraoka, M. Ikeguchi, S.

Sugai, *Biochemistry* **24,** 874 (1985).
40. Ikeguchi, M., K. Kuwajima, M. Mitani, S. Sugai, *Biochemistry* **25,** 6965 (1986).
41. Brems, D. N., S. M. Plaisted, J. J. Dougherty, Jr., T. E. Holzman, *J. Biol. Chem.* **262,** 2590 (1987).
42. Gilmanshin, R. J., and O. B. Ptitsyn, *FEBS Lett.* **223,** 327 (1987).
43. Gilmanshin, R. I., O. B. Ptitsyn, G. V. Semisotnov, *Biofizika (USSR)* **33,** 204 (1988).
44. Kuwajima, K., H. Yamaya, S. Miwa, S. Sugai, T. Nagamura, *FEBS Lett.* **221,** 115 (1987).
45. Kuwajima, K., A. Sakuruaka, S. Fueki, M. Yoneyama, S. Sugai, *Biochemistry* **27,** 7419 (1988).
46. Goldberg, M. E., G. V. Semisotnov, B. Friguet, K. Kuwajima, O. B. Ptitsyn, S. Sugai, *FEBS Lett.* **263,** 51 (1990).
47. Sears, D. W., and S. Beycock, in *Physical Properties and Techniques of Protein Chemistry,* part C, S. J. Leach, Ed. (Academic Press, New York, 1973), pp. 445–593.
48. Bolotina, I. A., *Mol. Biol. (USSR)* **21,** 1625 (1987).
49. Rodionova, N. A., Ph.D. thesis, Moscow Physico-Technical Institute (1990).
50. Stryer, L., *J. Mol. Biol.* **13,** 482 (1965).
51. Turner, D. C., and L. Brand, *Biochemistry* **7,** 3381 (1968).
52. Semisotnov, G. V., N. A. Rodionova, O. I. Razgulyaev, V. N. Uversky, A. F. Gripas, R. I. Gilmanshin, *Biopolymers*, in press (1990).
53. Lynn, J., and G. D. Fasman, *Biochem. Biophys. Res. Comm.* **33,** 327 (1968).
54. Witz, G., and B. L. Van Duren, *J. Phys. Chem.* **27,** 648 (1973).
55. Brandts, J. F., H. R. Halvorson, M. Brennan, *Biochemistry* **14,** 4953 (1975).
56. Kim, P. S., and R. L. Baldwin, *Ann. Rev. Biochem.* **51,** 459 (1982).
57. McPhie, P., *Biochemistry* **21,** 5509 (1982).
58. Goto, Y., and Hamaguchi, K., *J. Biochem.* **99,** 1501 (1986).
59. Semisotnov, G. V., *et al., J. Mol. Biol.* **213,** 561 (1990).
60. Creighton, T. E., *J. Mol. Biol.* **125,** 401 (1978).
61. Betton, J. M., M. Desmadril, A. Mitzaki, J. M. Yon, *Biochemistry* **24,** 4570 (1985).
62. Blond, S., and M. E. Goldberg, *Proteins: Struct. Funct. & Genet.* **1,** 247 (1986).
63. Blond, S., and M. E. Goldberg, *Proc. Natl. Acad. Sci. U. S. A.* **84,** 1147 (1987).
64. Murry-Brelier, A., and M. E. Goldberg, *Biochemistry* **27,** 7633 (1988).
65. Bloud-Elguindi, S., and M. E. Goldberg, *Biochemistry* **29,** 2409 (1990).
66. Roder, R. H., G. A. Elöve, S. W. Englander, *Nature* **335,** 700 (1988).
67. Ptitsyn, O., *Dokl. Akad. Nauk SSSR* **213,** 473 (1973).
68. Udgaonkar, J. B., and R. L. Baldwin, *Nature* **335,** 694 (1988).
69. Kharakoz, D. P., and V. E. Bychkova, unpublished data.
70. Semisotnov, G. V., and K. Kuwajima, unpublished data.
71. Semisotnov, G. V., and M. Vas, unpublished data.
72. Semisotnov, G. V., and V. P. Kutyshenko, unpublished data.
73. Semisotnov, G. V., unpublished data.
74. Ebert, B., G. V. Semisotnov, N. A. Rodionova, *Studia Biophysica* **137,** 125 (1990).

11

Conformation States in Acid-Denatured Proteins

Anthony L. Fink, Linda J. Calciano, Yuji Goto, Daniel Palleros

Years ago, Tanford and co-workers showed that several proteins, when denatured with high concentrations of urea or guanidine.HCl, had hydrodynamic radii equivalent to those calculated for a random coil conformation. Thus the notion arose that unfolded proteins were random coils. However, it has also been known for a long time that denaturation, especially by heat or extremes of pH, frequently does not lead to a fully unfolded protein, rather there is some residual structure *(1)*.

In connection with investigations of the kinetics of protein folding, it became apparent to us that better characterization of the "unfolded" state of proteins was needed, since several observations indicated that, in many cases, ostensibly unfolded proteins were in fact far from fully unfolded. (We use the term "fully unfolded" to imply complete lack of ordered structure, analogous to a random coil.) As a result, we have been pursuing a systematic study of the structure and properties of denatured proteins.

Here we report details concerning the acid denaturation of proteins. Perhaps the two most unexpected results are that (i) some proteins retain their native state to pH as low as 0.5, and (ii) as expected due to charge-charge repulsion, many other proteins initially unfold in the vicinity of pH 3, when the pH of a salt-free solution is lowered by the addition of hydrochloric acid (HCL). Then as the pH is lowered below 2, these same proteins begin to refold, in many cases attaining a compact structure with as much secondary structure as the native state! Our data suggest that there is a common pattern of behavior for all proteins, which is relevant to other forms of denaturation, as well as to the kinetics and pathway of the folding process.

The effect of decreasing pH on the conformation of apomyoglobin will be used to illustrate the type of phenomena frequently observed. If we adjust the pH to 7.0, dialyze to remove all salts, and then titrate with HCl, monitoring the conformational state by near- and far-ultraviolet circular dichroism (CD), tryptophan fluorescence, Fourier transform infrared spectroscopy (FTIR), and 1-anilino-naphthalene, 8-sulfonate (ANS) binding, and monitoring the hydrodynamic properties by fast protein liquid chromatography (FPLC) gel filtration, light scattering or viscometry, we observe that the native state predominates until about pH 5, at which time unfolding begins (Fig. 1). This is a rather broad (i.e., noncooperative) transition, culminating in unfolded protein at pH 2.5 (U_A). The apomyoglobin appears to be nearly as unfolded as by 6 M Gdn.HCl at pH 2, which gives a far-ultraviolet CD spectrum closely resembling that of a protease-digested pro-

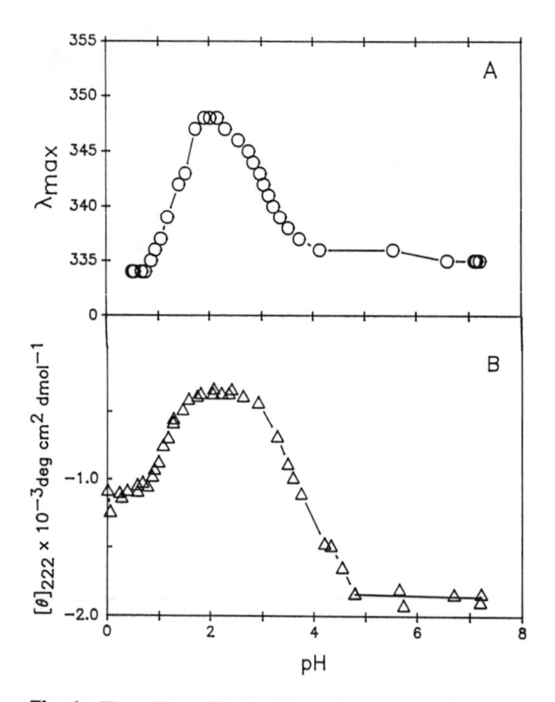

Fig. 1. The effect of increasing concentration of HCl on the conformation of apomyoglobin. The transition was monitored by tryptophan fluorescence emission in panel **A**, and by circular dichroism ellipticity at 222 nm in panel **B**. The sample was prepared at pH 7, 20°C, in a salt-free solution.

tein, indicating that apomyoglobin in HCl, pH 2, is close to fully unfolded.

Thus far there is nothing surprising; however, if we continue the HCl titration, the protein begins to *refold* and become compact as the pH decreases (Fig. 1). This is clearly an unexpected observation. We will refer to the resulting conformation as the A state. Unlike many proteins, apomyoglobin in the HCl-induced A state is only partly refolded, compared to the native (N) state. [In the presence of several other anions, the apomyoglobin A state shows almost as much secondary structure as the native state *(2)*.] In general, the A states have the properties of a molten globule (MG), that is, a structure almost as compact as the native state, with significant amounts of secondary structure, little or no native-like tertiary structure, significant amounts of exposed hydrophobic surface, and significant ANS binding *(3–5)*.

Other proteins such as cytochrome c and beta-lactamase *(5, 6)* show similar titration properties to apomyoglobin and have A states with similar amounts of secondary structure as the native state. That the A states are molten globules is apparent from examination of their structural properties. The compactness is shown from measurements of the hydrodynamic radius obtained from molecular exclusion on Superose 12 (Table 1). Measurements by far-ultraviolet CD show comparable amounts of secondary structure, whereas near-ultraviolet CD shows no tertiary structure (Fig. 2). FTIR measurements show characteristic changes between native, unfolded and A state, especially in the Amide I and III regions (Fig. 3). The tryptophan fluorescence

Table 1. Hydrodynamic radii for native, unfolded, and molten globule conformational states.[a]

Protein	State[b]	K_d	R_s	Relative volume
Apomyoglobin	N	0.50	21.5	1.0
	U	0.27	43	8.0
	A	0.36	35	4.3
Cytochrome c	N	0.54	17	1.0
	U	0.34	34	8.0
	A	0.51	20	1.6
β-Lactamase	N	0.46	24	1.0
	U	0.21	51	9.6
	A	0.48	26.5	1.4

[a]The data were determined by gel exclusion chromatography on Superose 12 using the experimental protocol and analysis of Corbett and Roche *(15)*. [b]The conditions for the native state (N) were pH 7.0, 0.15 M KCl; for the unfolded state (U), pH 2.0, 5 M Gdn.HCl; and for the molten globule state (A), pH 2.0, 0.5 M KCl.

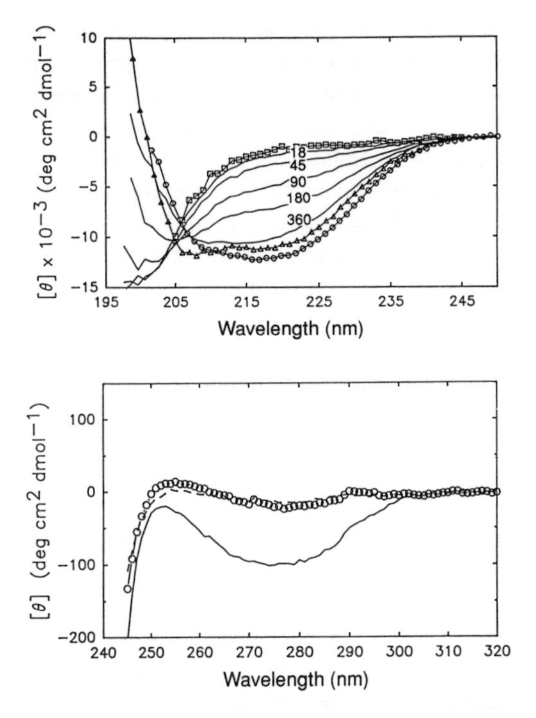

Fig. 2. The far-UV (**top**) and near-UV (**bottom**) circular dichroism spectra of beta-lactamase in the native (solid line), acid-unfolded state (broken line) and acid-induced A state (o).

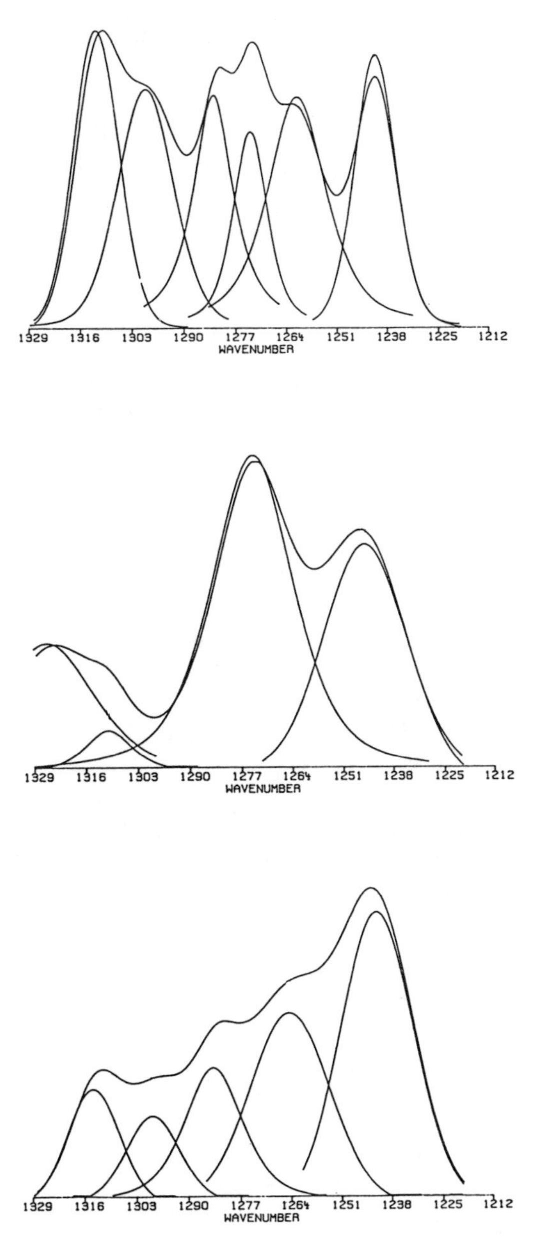

Fig. 3. FTIR spectra in the amide III region for cytochrome c in the native (**top**), A state (**middle**), and unfolded (**bottom**) states. The spectra were deconvoluted using the Nicolet FOCAS program.

also is distinct for the A state, namely, an emission maximum similar to that of the native state, i.e., reflecting a hydrophobic environment, but quenched – presumably due to dynamic quenching (Fig. 1A).

We return to the question of what is responsible for the transition from the acid-unfolded state of apomyoglobin at pH 2, U_A, to the A state. Once the protein is fully protonated, as at pH 2, the further addition of protons is not expected to cause any conformational effects. This points to the chloride as being responsible for the transition, which was confirmed in experiments in which we find that by maintaining the pH at 2 and increasing the chloride concentration by the addition of KCl a similar transition to the molten globule state is observed (Fig. 4) *(6)*.

It is also easy to show that it is the anion in general and not a specific chloride effect that is responsible. In fact, most other anions are more effective than chloride in bringing about the U_A to A transition *(2)*. Comparing the effectiveness of different acids or salts indicates that the order of effectiveness (Eq. 1)

$$Fe(CN)_6^{3-} > SO_4^{2-} >> CCl_3COO^- > SCN^- > ClO_4^- > I^- > NO_3^- > CF_3COO^- > Br^- > Cl^- \quad (1)$$

corresponds to what is known as the electro-

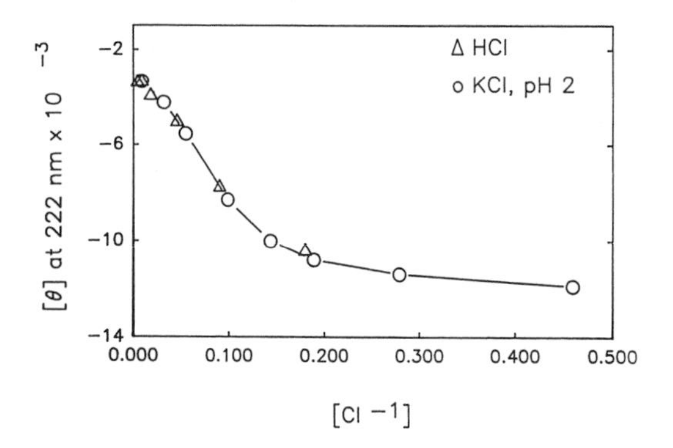

Fig. 4. The coincidence of the acid-unfolded to A state transitions as induced by HCl (triangles) and KCl at pH 2 (circles), for apomyoglobin.

selective series, which reflects the affinity of quaternary ammonium ion exchange resins for anions (7). Polyvalent anions are the most effective in bringing about the U_A to A transition.

Using Tanford's model *(8)* for the preferential binding of ligand to the A state compared to the U_A state reveals that, for apomyoglobin, only between 2 and 3 anions preferentially bind to the A state *(2)*. The association constants are in the order of 10^2 to 10^5 M^{-1}. Thus, it is the binding of anions to the ammonium groups of the unfolded state that masks the repulsive positive charge, allowing the intrinsic hydrophobic forces to cause the collapse to a compact state. The presence of secondary structure can be explained by the intrinsic properties of the polypeptide as a consequence of basic polymer physics and conformational entropy [as recently shown in models by Dill and coworkers *(9, 10)*].

Under commonly used conditions for acid denaturation, most proteins will yield a molten globule, even at pH 2 or 3. For example, consider the effect on apomyoglobin of repeating the above-mentioned HCl titration, but now in the presence of salt. Instead of the single broad unfolding transition, biphasic unfolding is observed at low ionic strength, followed by the transition to the MG state at lower pH (Fig. 5). At higher salt concentrations, a single "unfolding" transition is observed, leading directly from the native to the MG state. In fact, if other acids are used, e.g., sulfuric, the native state goes directly to the A state, even in the absence of salt. This is due to the effectiveness of the anion in binding to the ammonium groups and thus causing the U to A transition.

Using the data for apomyoglobin as a function of chloride concentration, we constructed a phase diagram (Fig. 6) *(11)*. The region corresponding to the unfolded state is

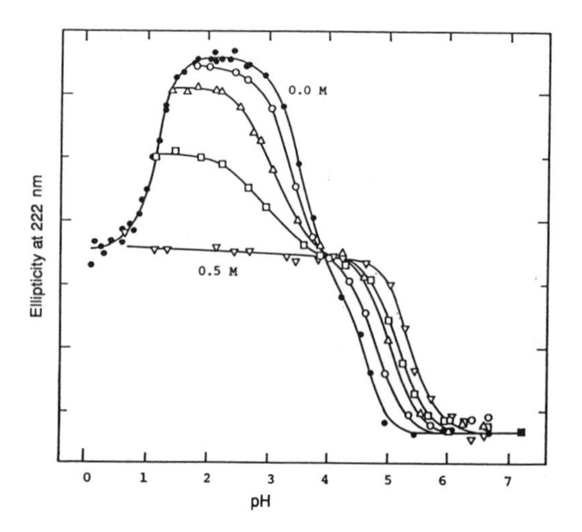

Fig. 5. The HCl-induced denaturation of apomyoglobin as a function of chloride concentration, monitored by ellipticity at 222 nm. The anion concentration was increased by the addition of KCl. Filled circles correspond to the salt-free HCl titration. The open symbols correspond to increasing KCl concentrations. The inverted triangles corresponding to 0.5 M KCl.

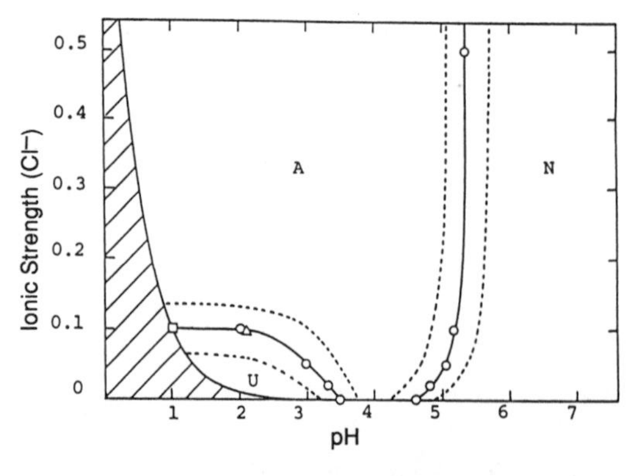

Fig. 6. Phase diagram for apomyoglobin as a function of pH and chloride ion concentration. The phase boundaries, indicated by solid lines, correspond to the midpoints of the transition curves monitored by ellipticity at 222 nm. The broken lines correspond to 20% or 80% completion of these transitions.

small. Comparison with other proteins indicates that both the position of the N/U_A boundary and the U_A/A boundary change for different proteins. Thus, the observed behavior on acid denaturation can be explained in terms of the relative size of the U_A region of the phase diagram. (The region to the left, shaded in the figure, reflects the minimum chloride concentration at that pH with HCl alone.) Interestingly, a simple, general model for charge effects in protein folding, developed by Stigter and Dill (see chapter 3), predicts the phase boundary between the native and A states very well, and qualitatively also indicates the boundary between the A state and U_A state, in terms of the shape and location of the boundary between low (U_A) and (A state) high density states.

Moving now to the generality of this model, we find that the behavior of other proteins falls into one of three categories. class I: N to U_A to A; class II: N to A; and class III: N to N′, where N′ represents a native-like conformation. For Class I proteins, we observe that the degree of unfolding in the U_A state varies considerably, from essentially completely unfolded, to only a rather small fraction. Similarly, for these proteins, the degree of compactness and amount of secondary structure in the A state vary from very similar to that of the native state to approximately half as much.

Examples of class I include beta-lactamase, apomyoglobin, cytochrome c *(6)*, papain, parvalbumin, and staphylococcal nuclease. Alpha-lactalbumin and carbonic anhydrase are good examples of the second class (Fig. 7) *(12)*, and staphylococcal protein A, ubiquitin, chicken lysozyme, and T4 lysozyme are examples of the third class (Fig. 8). Interestingly, the third class can be converted to the first by the addition of low concentrations of urea (and, as noted, the behavior of the first class is converted to that of the second by the addition of salts). For example, T4 lysozyme, in the presence of 2M urea shows an HCl titration very similar to that of apomyoglobin, and a similar salt titration at pH 2 (Fig. 9).

We conclude, therefore, that the conformational states induced by acid denaturation are governed by a fine balance among the hydrophobic interaction (i.e., the drive to minimize the exposed hydrophobic surface area, which leads to compact molten globule states), the offsetting chain conformational entropy and intramolecular charge-charge

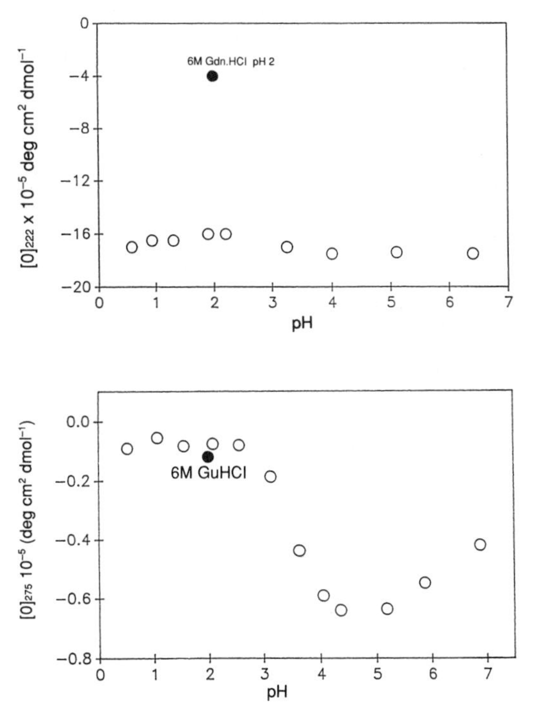

Fig. 7. The HCl-induced transitions of alpha-lactalbumin monitored by near- and far-UV circular dichroism. The solid circles show the values for protein unfolded with 6 M Gdn.HCl at pH 2.

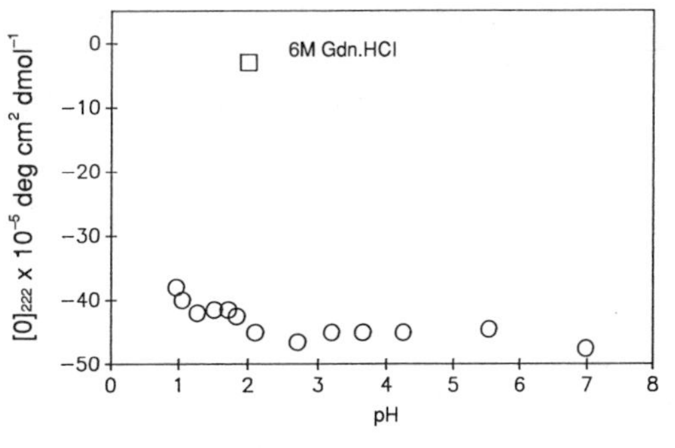

Fig. 8. The effect of HCl on salt-free T4 lysozyme, as monitored by far-UV circular dichroism. The solid circle shows the value for protein unfolded in 6 M Gdn.HCl, pH 2.

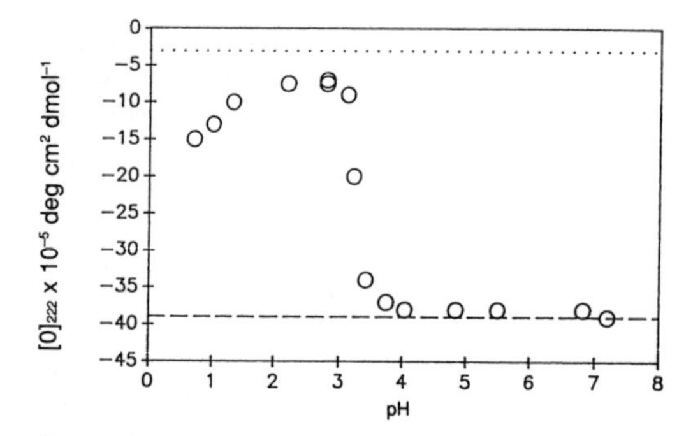

Fig. 9. The titration of T4 lysozyme by HCl in the presence of 2 M urea, as monitored by ellipticity at 222 nm.

repulsion, and the affinity of anions for the positively charged ammonium groups. The unique structure of a given protein, particularly the distribution and number of ammonium groups, and the presence of disulfide cross links, salt bridges, or metal-binding sites, determines how readily the molecule will unfold to a random coil-like structure or MG-like structure. This, in turn, determines the conformational state diagram and, hence, the observed behavior on acid titration.

For many proteins, we have found optimum conditions to make stable molten globules, in order to study their structures in more detail. One reason for this interest is their presence in the early stages of folding *(3, 4, 13, 14)*. We have studied the kinetics of the transformations between the unfolded, molten globule, and native states and find that the transition from U to MG is very fast – in many cases, less than the dead time of fluorescence or CD stopped-flow, i.e., lifetimes of less than a few msecs. Thus our picture of folding is a rapid collapse to MG state, followed by slower transformation to a native-like intermediate, followed in turn by final rearrangements to yield the N state *(4)*.

Acknowledgement

This research was supported by a grant from the National Science Foundation (DMB-8716292).

References

1. Aune, K. C., A. Salahuddin, M. H. Zarlengo, C. Tanford, *J. Biol. Chem.* **242,** 4486 (1970).
2. Goto, Y., N. Takahashi, A. L. Fink, *Biochemistry,* **29,** 3480 (1990).
3. Ptitsyn, O. B., *J. Prot. Chem.* **6,** 273 (1987).
4. Kuwajima, K., *Proteins* **6,** 87 (1989).
5. Goto, Y., and A. L. Fink, *Biochemistry* **28,** 945 (1989).
6. Fink, A. L., L. J. Calciano, Y. Goto, D. R. Palleros, in *Current Research in Protein Chemistry,* J. Villafranca, Ed. (Academic Press, NY, 1990).
7. Gjerde, D. T., G. Schmuchler, J. S. Fritz, *J. Chromatog.* **187,** 33 (1980).
8. Tanford, C., *Adv. Prot. Chem.* **24,** 1 (1970).
9. Chan, H. S., K. A. Dill, *Macromolecules* **22,** 4559 (1989).
10. Chan, S. H., and K. A. Dill, in press (1990).
11. Goto, Y., and A. L. Fink, *J. Mol. Biol.* **214,** 803 (1990).
12. Kuwajima, K., *J. Mol. Biol.* **206,** 547 (1989).
13. Semisotnov, G. V., N. A. Rodionova, V. P. Kutyshenko, B. Ebert, J. Blanck, O. B. Ptitsyn, *FEBS Letts.* **224,** 9 (1987).
14. Ikeguchi, M., K. Kuwajima, M. Mitani, S. Sugai, *Biochemistry* **25,** 6965 (1986).
15. Corbett, R. J. T., and R. S. Roche, *Biochemistry* **23,** 1888 (1984).

12

Characterization of Unfolded and Partially Folded States of Proteins by NMR Spectroscopy

Christopher M. Dobson, Claire Hanley, Sheena E. Radford
Jean Baum, Philip A. Evans

Introduction

We have become familiar with the nature of the compact globular states of proteins during the last 20 years, initially from studies of protein crystals by diffraction methods *(18)* and more recently from structure determination in solution by nuclear magnetic resonance (NMR) techniques *(23)*. Additionally, many details of the dynamical properties of these folded states have been revealed from both these experimental techniques and from theoretical studies *(12)*.

By contrast, very little is known about the unfolded or partially folded states of proteins, primarily because of the intrinsic difficulties inherent in their study. Crystallization of proteins in such states has not been achieved, and indeed seems unlikely. NMR spectroscopy, however, is ideally suited to their characterization, as it provides both structural and dynamical information about molecules in solution. The methods required for such studies may be very different from those used for studies of proteins in their globular states. Furthermore, the nature of any conformational description may need to be significantly different from that used for the globular state, because the conformational freedom is likely to be much greater in the unfolded or partially folded states.

We have been increasingly interested in the development and exploitation of suitable NMR methods for studying these states of proteins, as well as the more familiar globular states. The reasons for this interest are as follows: (i) Establishing the degree of stability of particular elements of protein structure will be of substantial value in understanding the nature of the factors important in stabilizing folded proteins and in the design and modification of proteins for specific purposes. (ii) Investigation of non-native states of proteins provides the basis for experimental studies concerned with the pathways of protein folding. Even the characterization of partly folded states of proteins that are not directly implicated in folding pathways will help to establish some fundamental understanding of such states. The prospects are, however, particularly exciting in those cases where direct comparison can be made with transient folding intermediates such as those detected in hydrogen exchange trapping methods *(19, 20, 22)*. Then, the study of the stable partially folded states offers the chance to provide a level of detail about such intermediates that is unlikely to be available from studies of transiently stable species.

In the pursuit of these objectives, we have been using NMR to study the c-type lysozymes and the closely related α-lactalbumins. Although the former class of proteins exhibit classical two-state folding and unfolding behavior, the α-lactalbumins can, by appropriate choice of conditions, be obtained in a partly folded state, called a molten globule, which has already been implicated as an intermediate on the pathway of the folding of these proteins *(11)*. In addition to the opportunity to begin the characterization of such a state, the comparison of the two proteins should promote exploration of the underlying features that stabilize such states.

Lysozyme

Lysozyme from hen egg white is a small, extremely well-characterized protein containing 129 residues. The ^{1}H NMR spectrum of lysozyme in its native state is very well resolved and, using two-dimensional (2-D) methods, has been analyzed in detail *(17, 21)*. Resonances have been assigned virtually completely to individual residues, permitting extensive study of the structure, dynamics, and binding properties of the protein in solution.

The situation with respect to the unfolded state of lysozyme is, however, very different. The ^{1}H NMR spectrum is poorly resolved in this case, whether unfolding is induced by temperature, chemical denaturants, or other means. Like the NMR spectra of other proteins in their unfolded states, it displays none of the characteristics of the spectra of proteins in the native state such as large chemical shift dispersion, extensive inter-residue nuclear Overhauser effects, or very slowly exchanging amide hydrogens, which have been crucial to their assignment and interpretation. Despite this, it is clear that the spectra of unfolded states are not identical to those of simple unstructured peptides and that they also differ with the nature of the denaturing conditions *(6, 9)*. The question is, how do we attempt to analyze such spectra?

One key to this comes from an early observation that, under certain conditions, lysozyme can be taken reversibly through the thermal unfolding transition and that the NMR spectrum reflects dramatically the change in the conformational state *(13)*. Furthermore, the rates of unfolding and refolding are such that at conditions close to the midpoint of the transition, where both folded and unfolded states are present in significant concentrations, the spectra of the folded and unfolded states of the protein are superimposed, rather than averaged as would be the case if the kinetics were fast. Under these conditions, the relative intensities of the resonances of individual protons in both the folded and unfolded states provide a direct measure of the equilibrium constant, reflecting the unfolding process for individual residues in the protein. By studying the temperature dependence of the equilibrium constant, the enthalpy and entropy changes associated with the transition can be determined. The data obtained in this way for lysozyme are in excellent agreement with calorimetric and other studies, and the measurements for individual residues show directly the high cooperativity (two-state nature) of the unfolding transition *(5)*.

Although the kinetics of the unfolding and folding transition are sufficiently slow for the resonances in the two conformational states to be observed separately, the equilibrium is, of course, a dynamic one; protein molecules do interconvert between the folded and unfolded states. Under these conditions, provided that the rates of interconversion are not slow compared with the nuclear relaxation rates, they can be probed by magnetization transfer methods. If, for example, saturation of the resonance of a residue in the folded state is carried out, this saturation may be transferred to the resonance of the same residue in its unfolded state. Experiments of this type have proved to be of great value in probing the kinetics of folding and unfolding of this and other proteins *(5, 10)*. In the context of the objectives described in this paper, however, they have an additional and rather different significance. This arises because transfer of magnetization between the folded and unfolded states enables the correlation of resonances of the same residue in the different states. Hence assignments in the spectrum of the folded state may be transferred to that of the unfolded state.

The exchange of magnetization between the folded and unfolded states can be detected by 2-D as well as 1-D experiments; here it is manifest in cross peaks linking the resonances of exchanging nuclei (Fig. 1). The resolution of the 2-D experiments permits the study of the exchange process for many resonances not resolvable in 1-D experiments. In the case of lysozyme, many resonances have been identified in this way, allowing the resonances of many residues to be identified in the unfolded state. Moreover, because of the very limited resolution of resonances in this state, assignment would be virtually impossible by direct methods such as those applied to the folded states of proteins.

We have analyzed the spectra of unfolded states of lysozyme, using this and other experimental strategies, in some detail. The conclusions of these studies are that local conformational preferences exist in the protein in its unfolded state, and there is evidence for clustering of hydrophobic residues *(4, 9)*. Although the description of the unfolded state may differ significantly from that of a classical "random coil," there is no evidence for persist-

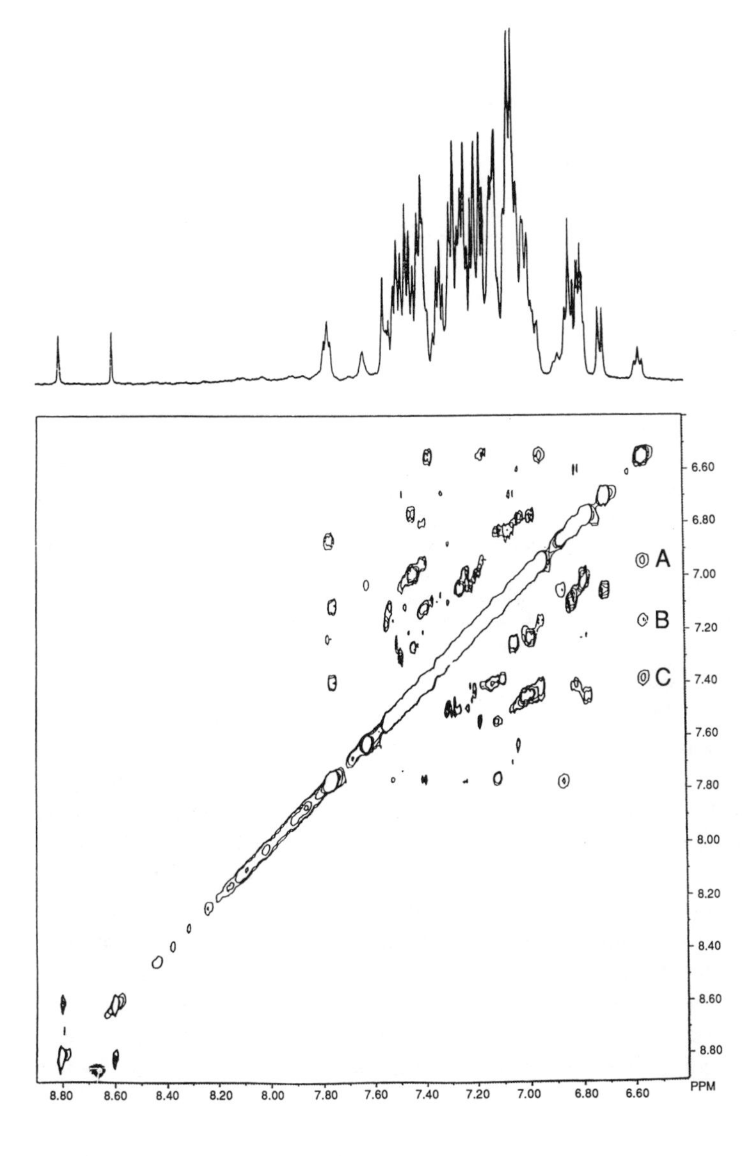

Fig. 1. Aromatic region of a 500 MHz 2-D exchange nuclear Overhauser effect (NOESY) spectrum of hen lysozyme in $^{2}H_2O$, pH 3.8, at the mid-point of the thermal unfolding transition (77°C). The two resonances at 8.6 and 8.8 ppm are from His 15 in its folded and unfolded states; the cross peaks arising from exchange between the two states are clearly visible. In more crowded regions of the spectrum, such peaks enable resonances of the same residue in the different states to be correlated. The cross peak labeled **(A)** is an example of this for the residue Trp 108 *(7)*. [The cross peaks labeled **(B)** and **(C)** correspond to NOE effects in the folded protein.]

ent secondary structure or, indeed, for any significant population of native-like structural features.

α-Lactalbumin

A very different situation from that of denatured lysozyme is, however, found for the molten globule state of α-lactalbumin. The α-lactalbumins and lysozymes share high sequence homology and their folded structures are closely similar. Even so, their unfolding behavior is quite different. In particular under certain circumstances such as exposure to low pH, cooperative unfolding of the native structure of α-lactalbumin leads to a so-called molten globule state in which significant helical secondary structure persists, and which retains a degree of compactness little different from that of the native state *(8)*. Only in the presence of high concentrations of chemical denaturants do these characteristics disappear, and the state then resembles a "normal" unfolded one.

The NMR spectrum of α-lactalbumin in the molten globule state is clearly very different from that of both the native and the fully unfolded protein (Fig. 2). The spectrum is broad and poorly dispersed, and hence almost as difficult to study directly by NMR as the unfolded state of lysozyme described above. The 2-D exchange experiment has, however, again proved to be extremely valuable, and it has enabled a direct correlation of resonances in the molten globule state with those in the native state. The identity of many of the resonances most perturbed from their position in the unfolded protein has been established *(3)*. These resonances are amongst those most perturbed in the native state, implying that native-like structural features could be present in the molten globule state.

Considerably more insight into the detailed nature of this state arises from a different experiment, which shares the philosophy of the correlation of resonances in the molten globule state with those of the fully folded state, and with the experiments developed to study transient folding inter-

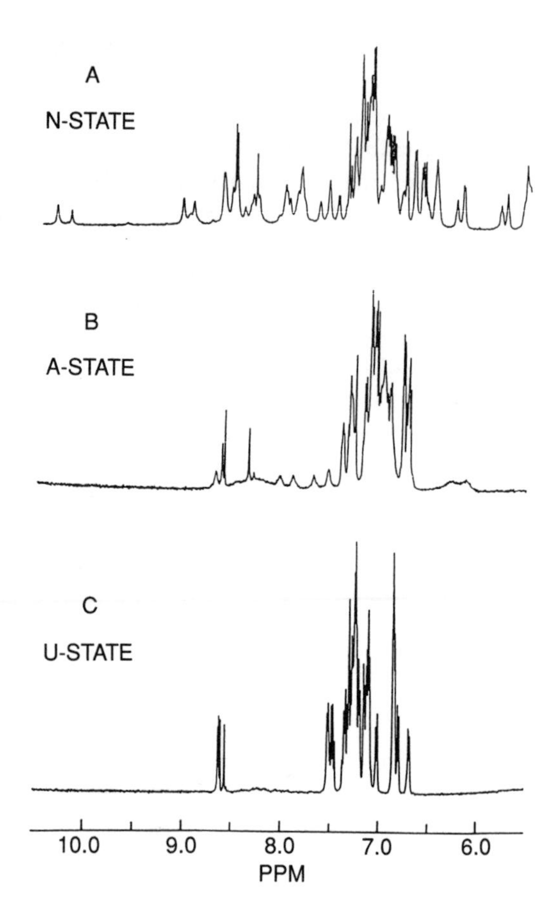

Fig. 2. Low-field regions of the 500 MHz ^{1}H NMR spectra recorded at 52°C of guinea pig α-lactalbumin at **(A)** pH 5.4 (the native state), **(B)** pH 2.0 (the molten globule or A-state) and **(C)** pH 2.0 in the presence of 9 M urea (the unfolded state). In each case, the protein was dissolved in ^2H_2O for 6 h prior to recording the spectrum. The resonances in the spectra shown correspond to protons from aromatic residues and from amide protons that have not exchanged with deuterons from the ^2H_2O solvent. [Reprinted from *(3)* with permission. Copyright 1989 by the American Chemical Society.]

mediates. One of the most striking observations in the spectrum of α-lactalbumin in its molten globule state is that there are several amide protons whose resonances persist in the spectrum hours or even days after the protein has been dissolved in ^2H_2O. Such resonances are completely absent in the spectra of unfolded lysozyme, or indeed of α-lactalbumin in the presence of high concentrations of denaturants. Slowly exchanging amides are, however, common in the spectra of folded proteins where they usually imply the existence of persistent secondary structure in an environment protected from solvent.

The 2-D exchange experiment described above proved difficult to utilize for the identification of these resonances, not least because hydrogen exchange was relatively fast under the conditions necessary for the efficient performance of the experiment. To identify the resonances, the following procedure was adopted *(3)*. After dissolution of the protein in ^2H_2O in the molten globule state at pH 2 for several hours, the pH was changed to pH 5.5. At this pH, the protein refolds rapidly to its native state. Observation of the NMR spectrum reveals the presence of several amide proton resonances. As the whole operation was carried out in ^2H_2O, the origin of these protons can only have been that they were present in the molten globule state. Conventional 2-D NMR experiments allow clear identification of the resonances (Fig. 3). They arise predominantly from one of the α-helical regions of the folded structure, the "C helix," which includes residues 89 to 97, and, therefore, indicates that this helix is stable in the molten globule state. It is clear from the spectra, however, that other regions of secondary structure, with somewhat less well-protected amide hydrogens, also exist. Indeed, recent experiments have resulted in the identification of several resonances in another helix, the "B helix," which in the native structure includes residues 23 to 34. This observation is of particular importance because it identifies helical regions in the molten globule state in both the C- and N-terminal regions of the structure that in the native state are close

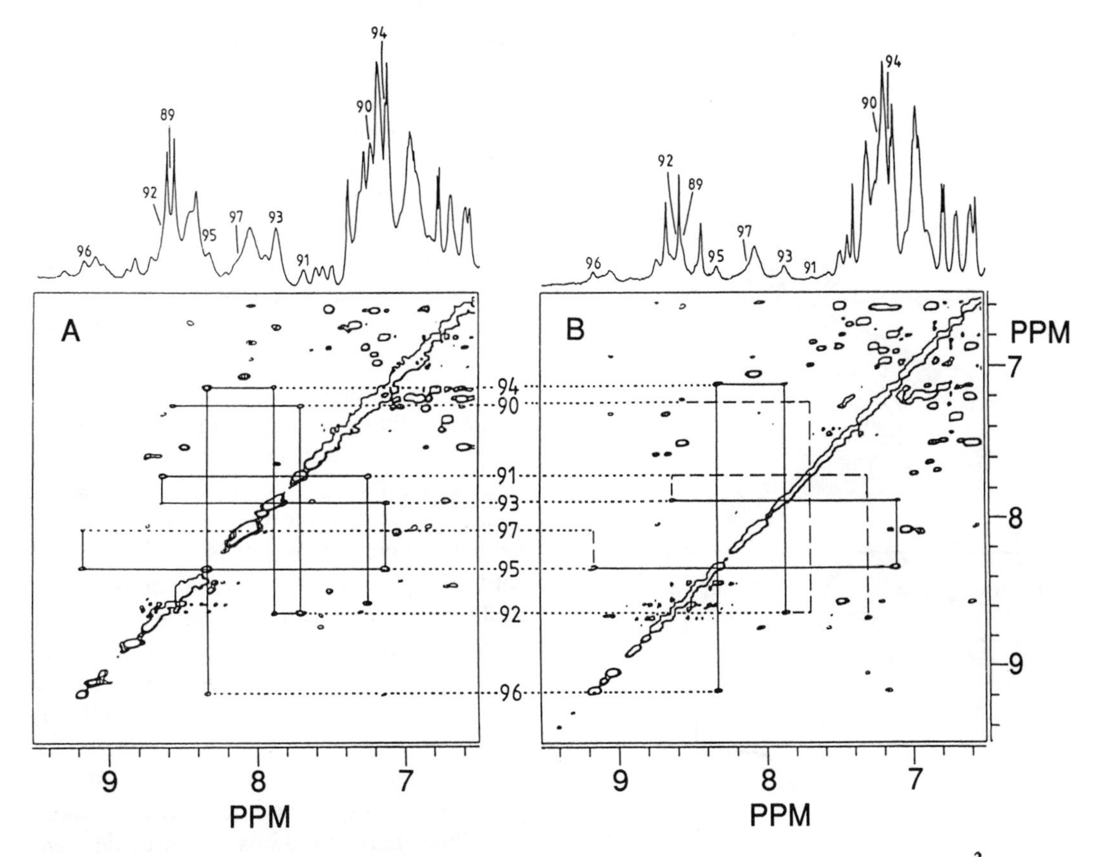

Fig. 3. Low-field regions of the 500 MHz NOESY spectra of α-lactalbumin at 35°C, freshly dissolved **(A)** in ^2H_2O at pH 5.4 and **(B)** in ^2H_2O at pH 5.4 following a pH jump from pH 2, where the protein had been allowed to remain for 10 h. The assignments for the resonances of residues 90–97 are indicated. More recent studies have allowed the identification of additional resonances in **(B)**, including some resonances arising from the helical region between residues 23 and 34. [Reprinted from *(3)* with permission. Copyright 1989 by the American Chemical Society.]

together in one of the "subdomains" of the molecular structure (Fig. 4). This bears similarities to the observations of the nature of the transient intermediate in cytochrome c *(20)*.

By contrast, no evidence for persistent native-like secondary structure in the molten globule state has been found in the segment of the protein that forms the other subdomain and includes a significant region of β-sheet structure. This is of particular interest in the light of recent NMR studies of the refolding of lysozyme from its denatured state *(14)*. These studies indicate that native-like structure able to protect amides against hydrogen exchange forms significantly more rapidly in the helical "subdomain" of lysozyme than in the remainder of the molecule. This would be consistent with at least a degree of similarity between the stable molten globule state of α-lactalbumin and the transient intermediate formed during the refolding of lysozyme.

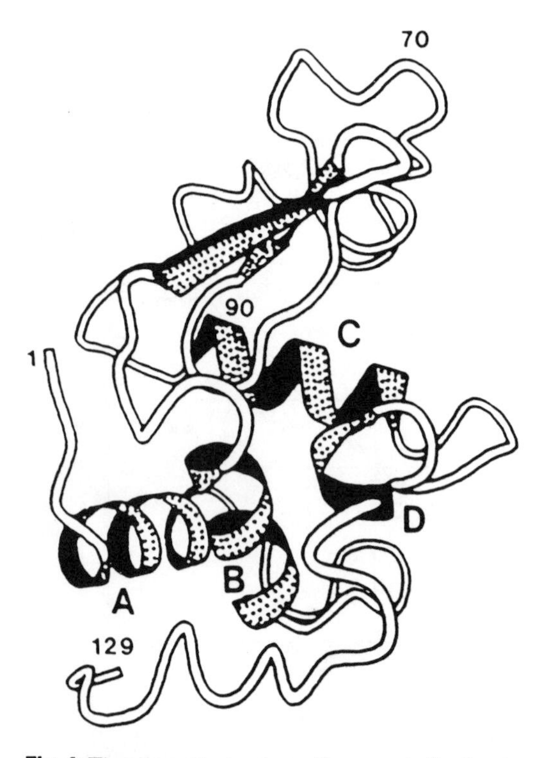

Fig. 4. The schematic structure of lysozyme indicating the locations of the major helical and sheet regions. [Reprinted from *(1)* with permission. Copyright 1989 by Academic Press Ltd.] The α-lactalbumin structure is closely similar.

These experiments reveal, for the first time, regions of native-like structure that persist in the molten globule state. There are, however, major differences between the spectra of the molten globule and that of the native state. These show, in particular, that the molten globule state has much greater conformational flexibility than the native state. Indeed, there is little evidence for any high degree of ordering of side-chain conformations within the molten globule state.

Conclusions

The NMR experiments described here illustrate that it is now becoming possible to study proteins in detail in states other than the fully folded native state. Continued study using the approaches described here and others that are being developed will undoubtedly produce a detailed description of the lysozyme/α-lactalbumin system of states with differing degrees of order and native-like structural features. An intriguing and exciting prospect is to establish the features that determine the degree of order in particular cases.

Why should α-lactalbumin form a stable molten globule state and not lysozyme? The answer to this does not appear to be a simple consequence of the more stable native state for lysozyme as compared with α-lactalbumin *(11)*. Thus, studies of lysozyme derivatives of reduced stability have as yet failed to generate a molten globule state comparable with that of α-lactalbumin *(16)*. One possible explanation could be related to the intrinsic stability of the helical regions of α-lactalbumin compared with lysozyme; we are examining this possibility through the study of synthetic peptide fragments. Such studies have provided great insight into the stability of individual regions of bovine pancreatic trypsin inhibitor (BPTI) *(15)*.

Characterization of non-native states of proteins by NMR methods such as those described here promises to provide fundamentally new information about the stability of individual features of protein structures and the nature of interactions between them. In conjunction with studies of transient species, such information should enable considerable advances to be made in our understanding of the means by which proteins fold from disordered denatured states to their compact and highly ordered native states.

Acknowledgment

This is a contribution from the Oxford Centre for Molecular Sciences which is supported by the Science and Engineering Research Council and Medical Research Council.

References

1. Acharya, K. R., D. I. Stuart, N. P. C. Walker, M. Lewis, D. C. Phillips, *J. Mol. Biol.* **208,** 99 (1989).
2. Archer, D. B., *et al., Biotechnology* **8,** 741 (1990).
3. Baum, J., C. M. Dobson, P. A. Evans, C. Hanley, *Biochemistry* **28,** 7 (1989).
4. Broadhurst, R. W., C. M. Dobson, P. J. Hore, S. E. Radford, M. L. Rees, *Biochemistry* **30,** 405 (1991).
5. Dobson, C. M., and P. A. Evans, *Biochemistry* **23,** 4267 (1984).
6. Dobson, C. M., P. A. Evans, K. L. Williamson, *FEBS Lett.* **168,** 331 (1984).
7. Dobson, C. M., P. A. Evans, R. O. Fox, C. Redfield, K. D. Topping, *Prot. Biol. Fluids* **35,** 433 (1987).
8. Dolgikh, D. A., *et al., Eur. Biophys. J.* **13,** 109 (1985).
9. Evans P. A., C. M. Dobson, K. D. Topping, D. N. Woolfson, *Proteins,* in press.
10. Evans, P. A., R. A. Kautz, R. O. Fox, C. M. Dobson, *Biochemistry* **28,** 362 (1989).
11. Kuwajima, K., *Proteins* **6,** 87 (1989).
12. McCammon, J. A., and M. Karplus, *Accts. of Chem. Res.* **16,** 187 (1983).
13. McDonald, C. C., W. D. Phillips, J. D. Glickson, *J. Am. Chem. Soc.* **93,** 235 (1971).
14. Miranker, A., S. E. Radford, M. Karplus, C. M. Dobson, *Nature* **349,** 633 (1991).
15. Oas, T. G., and P. S. Kim, *Nature* **336,** 42 (1988).
16. Radford, S. E., D. Woolfson, G. Lowe, S. R. Martin, C. M. Dobson, *Biochem. J.(272, 211 (1991).*
17. Redfield, C., and C. M. Dobson, *Biochemistry* **27,** 122 (1988).
18. Richardson, J. S., *Adv. Protein Chem.* **29,** 433 (1981).
19. Roder, H., and K. Wuthrich, *Proteins* **1,** 34 (1986).
20. Roder, H., G. A. Elove, S. W. Englander, *Nature* **335,** 700 (1988).
21. Smith, L. J., M. J. Sutcliffe, C. Redfield, C. M. Dobson, *Biochemistry* **30,** 986 (1991).
22. Udgaonkar, J. B., and R. L. Baldwin, *Nature* **335,** 694 (1988).
23. Wuthrich, K., *Science* **243,** 45 (1989).

Part IV

Auxiliary Factors and Folding: Membranes and Catalysis

Barry T. Nall

A basic tenet of protein folding is that the amino acid sequence encodes the information needed by a polypeptide to "self-assemble" into a unique three-dimensional structure. Is this *all* that is required? Perhaps not. The principle that amino acid sequence determines structure is accepted by a majority of those studying folding, but is often one of the first concerns of the newly initiated. The novice has a point. If sequence is all that is needed, why is it that so many proteins are irreversibly inactivated when unfolded? Clearly something else is required, at least some of the time.

Of course there are many excuses for why unassisted folding of some proteins fails and is only partly successful in other cases. Nonfolding reactions, particularly intermolecular associations and aggregation, get in the way and divert the polypeptide chain from the true path. Catalysts are required to facilitate folding and reduce the kinetic barriers for disruption of nonproductive associations during folding.

What about membrane associated or intrinsic membrane proteins? For membrane proteins structure may well be determined by sequence, but only in conjunction with the membrane. In the absence of membrane components, structure may be defective or wholly lacking. That membranes can modify the structure-determining role of sequence is not much of a surprise. Chemically the membrane is much like a protein interior or even a protein denaturant, and thus is expected to have a drastic effect on protein structure.

The importance of these "other players" in the game of protein conformation and folding is much neglected. It is important to learn how biological systems deal with the "other players," either by direct exploitation as with membrane proteins, or by catalyzing desired reactions so that undesired interactions are kinetically excluded.

Rhodanese provides Professor Horowitz with a classic example of a protein that folds in strict accordance with Murphy's laws. At first, everything is amiss, and folding is completely unsuccessful. Dissection of the process reveals that even the most obstinate of proteins adheres to chemical common sense. With strict avoidance of certain unsavory hydrophobic and oxidizing characters, rhodanase successfully attains its structural destiny writ in amino acid sequence.

Professor Wallace shows us that membrane proteins require relaxation of the sequence-one structure principle, and that perhaps membrane components should be included as adjunct members of the sequence. For gramicidin in particular, the rule is that sequence (plus a membrane-like environment) yields a family of structures. Structural preferences depend not only on hydrophobic solvent associations but also on cation binding.

Professor Schmid introduces prolyl isomerase, an in vitro folding catalyst. This enzyme has a well-defined function, the catalysis of prolyl imide bond cis-trans isomerization. Slow folding kinetic phases are often coupled to proline isomerization, so prolyl isomerase is also a catalyst of these slow folding reactions. Less clear is whether the intracellular function of prolyl isomerase is to catalyze folding and conformational changes. Even more intriguing, but completely lacking an explanation is the observation that potent inhibitors of prolyl isomerase activity also act as immunosuppressants.

Professor Freedman introduces us to a better known and more thoroughly characterized catalyst of folding within and outside cells, protein-disulfide isomerase. This enzyme catalyzes disulfide interchange reactions, removing a potent kinetic trap for folding of disulfide bond-containing proteins. Thus, protein-disulfide isomerase facilitates folding by allowing a protein to attain the lowest free energy state encoded by amino acid sequence.

13

Teaching Proteins to Fold

Paul M. Horowitz

Introduction

We recognize a functional protein as having a specific three-dimensional (3-D) structure, and there is great interest in determining how the protein acquires that structure. It is a generally accepted hypothesis that the amino acid sequence, exerted through side-chain and backbone interactions, determines the 3-D structure of a protein. This is basically a thermodynamic statement that says a protein's structure is the conformation with the lowest free energy. To most of us, though, successful refolding means acquiring the "native" 3-D structure within a reasonable time so that kinetic considerations are important. Many natural or recombinantly expressed proteins do not fold properly or are found not to refold efficiently after being unfolded. Our laboratory has become interested in these "orphan proteins" that appear to be impossible to refold *(1, 2)*. The enzyme rhodanese has proven a useful system for understanding some of the issues involved *(3)*.

Every isolated protein has an in vivo history. The protein was synthesized, subjected to processing, folding, transport, and compartmentation *(4, 5)*. Even if the conformation of the isolated protein is a global minimum, the protein may have had different folding and functional requirements at different times of its life. Thus, interactions with the cellular machinery and environments may represent thermodynamic constraints, and processing may change the potentials inherent in the sequence. For example, translation of the genetic information yields a polypeptide chain that sequentially appears from the synthetic apparatus, the ribosome. The folding itself, though it may follow a pathway, is influenced by processing that may include proteolysis, side-chain modification, or cofactor interactions.

In the cell, folding may occur in an environment where the protein concentration is estimated to be as much as 300 mg/ml *(6)*. Furthermore, interactions with accessory components such as chaperonins or heat shock proteins may modulate the protein folding to permit the functional compartmentation of the protein *(6–8)*. Further processing and conformational adjustments, which may involve energy-dependent steps, would then result in the active structure. Finally, every protein is subjected to aging, degradation, and turnover, which may follow paths involving further processing steps or conformational changes.

The proteins we want to fold are isolated either as a mature protein that has been subjected to the stresses of normal life or as one that has been subjected to synthesis and processing in a foreign cell, e.g., *Escherichia coli*. Because of the different life histories of isolated proteins, we expect a range of behavior in vitro, from proteins that would readily refold to those that have been extensively processed and will not refold.

Folding the "Unfoldable"

Until recently, the protein rhodanese (thiosulfate sulfurtransferase, EC 2.8.1.1) was considered unfoldable *(9, 10)*. We now know that the apparent irreversibility was the result of kinetic traps that lead to conformational cul de sacs that slow the folding process *(11)*. Less stable but more rapidly formed states compete with the desired conformation and slow its formation to the point where folding appears irreversible. Also, chemical reactions such as oxidation change the folding potential of the isolated protein *(12)*.

Rhodanese is synthesized in the cytoplasm of eukaryotic cells and transported into the matrix of the mitochondrion *(13)*. This intramitochondrial protein consists of a single polypeptide chain with a molecular weight of 33,000. Comparisons with the cDNA indicate that there is little processing of the synthesized chain, so all folding information with which the protein was born is present in the mature, isolated enzyme *(14, 15)*. The x-ray structure of rhodanese shows why we expect this protein to refold easily *(16)*. No cofactors or side-chain modifications are required for the observed structure. The protein is folded into two equal-sized domains, and the active site of this enzyme, whose job it is to transfer sulfur atoms, is at a cysteine residue that is in the interdomain region. This interdomain region is strongly hydrophobic. Active rhodanese has all four of its sulfhydryl groups reduced; therefore, there is no requirement for the correct formation of disulfides.

Renaturation problems

Renaturation attempts fail whether rhodanese is denatured by heat, guanidinium hydrochloride (GdmHCl), urea, or acid. There is no regain of either activity or native structure, and the lack of refolding does not depend on the interval between unfolding and the start of refolding. The basic reasons for this failure are that unfolding/refolding produces intermediate forms that are sensitive to oxidation at sulfhydryl groups and have exposed hydrophobic surfaces *(17)*. These effects are coupled, in that protein oxidation leads to apolar exposure, and perturbations that produce apolar exposure lead to increased sensitivity to oxidation *(18)*. What we see is that attempted refolding, even at low protein concentrations, leads to aggregation that competes with renaturation.

These considerations can be expanded to show how they led to the ability to refold rhodanese. Rhodanese oxidation leads to a loss of activity and exposure of hydrophobic surfaces as assessed by the increased fluorescence of the hydrophobic probe, 1-anilinonaphthalene, 8-sulfonate (ANS) *(18)*. The exposure begins immediately, and there is very little lag in the appearance of hydrophobic surfaces. Reducing agents such as dithiothreitol (DTT) lead to rhodanese oxidation as a consequence of their ability to form hydrogen peroxide as they auto-oxidize. Therefore, it is the partial reduction of oxygen to species like hydrogen peroxide that gives rise to many of the irreversible effects.

This inactive enzyme can be reactivated if reductants are added soon after inactivation. Oxidation continues in the inactive enzyme. The protein loses its ability to be reactivated by reduction, and after a considerable lag period, there is protein aggregation as evidenced by formation of turbidity. The aggregates that form are stabilized by both covalent disulfide formation and hydrophobic interactions, as can be demonstrated by comparisons of sodium dodecyl sulfate (SDS) gel patterns produced by running aggregated protein in the presence and absence of reductants. To reactivate oxidized rhodanese, it is necessary to choose the correct reducing agents in the correct amounts. Mercaptoethanol at 200 mM is effective.

Even without oxidation, structural perturbations expose hydrophobic surfaces and lead to aggregation. The precipitation, which occurs maximally at a sharply defined critical concentration of GdmHCl, removes almost all of the protein from solution. The presence of a nondenaturing detergent such as lauryl maltoside completely prevents precipitation under conditions where the detergent has no effect on the activity of the native enzyme *(17)*.

It thus became clear that there were three experimental requirements for successful refolding of rhodanese. First, oxidation had to be prevented by the inclusion of an appropriate reductant at concentrations that would avoid effects of reduced oxygen species formed during auto-oxidation. Second, a detergent needed to be included to prevent aggregation due to the extensive hydrophobic exposure during unfolding/refolding. Third, it was important to arrange conditions to stabilize the protein after refolding. Thus, useful "renaturation" of a functional protein requires simultaneous refolding, reactivation, and "long-term" stabilization, three requirements that are not necessarily equivalent. Rhodanese is particularly sensitive to these difficulties, and efficient renaturation has been hampered because even under conditions favorable for global folding, the protein is sensitive to inactivating events in localized regions.

Experimental results

A variety of experimental results can be summarized *(3)*. After denaturation in 6M GdmHCL, there was no reactivation when rhodanese was diluted to 50 μg/ml in buffer supplemented singly, with substrate, reductant, or detergent alone. Although many enzymes refold significantly in the presence of their substrates, denatured rhodanese could not be reactivated in the presence of its substrate, thiosulfate. The presence of the pairs of reagents – detergent and thiosulfate or β-mercaptoethanol (BME) and thiosulfate – were not very effective. However, the combination of detergent and reductant was somewhat effective, and the best results under these conditions led to the recovery of about 30% of the activity when thioglycolate was used as a reductant. The combined effects of the three components was much greater than the sums of their individual effects, and GdmHCl denatured rhodanese regained more than 90% of its activity when allowed to refold and reactivate in the simultaneous presence of detergent, BME, and thiosulfate.

Enzyme activity was measured as follows: the time dependence of the appearance of the product, thiocyanate, was determined after diluting rhodanese, preincubated in various GdmHCl concentrations, into assay mixes. The activity at any instant is the slope of these progress curves taken at the appropriate time. The initial slopes can be used to estimate the activities at any GdmHCl concentration. Above 2M GdmHCl, the initial slope was close to zero, and the enzyme always regained activity after an induction period. There was no induction period at concentrations below 1.5M GdmHCl, and the progress curves were linear as expected for steady state behavior. The nonlinear curves might be the consequence of the formation of an inactive intermediate that forms on the refolding path.

The reversibility of unfolding was studied using various parameters such as loss of enzyme activity, exposure of tryptophan residues as assessed from fluorescence spectroscopy, and the loss of secondary structure from the far-ultraviolet circular dichroism (UV CD). The fractional denaturation was obtained by interpolating between the properties of the fully native and the fully denatured protein. All three properties gave rise to transition curves that indicate reversibility, and all three properties gave different midpoints of transition. Noncoincidence of reversible transition curves is often taken as clear proof of the presence of intermediates during refolding. The variation of dissymmetry around the tryptophan residues was assessed from the near-UV CD and followed the transition curve for activity. The transition curve for the fluorescence response was markedly asymmetric.

Urea denaturation

Similar approaches were tried with urea denaturation to determine the extent to which ionic interactions were critical to the observations. The results were very similar, except for the expected requirements for higher urea concentrations to produce structural perturbation *(19)*. Importantly, the reversible unfolding with urea gave an asymmetric transition by fluorescence that could not be described by a two-state transition. However, the observed transition could be described as the sum of two individual two-state transitions. The fitting procedure assumed an intermediate with a wavelength maximum at 345 nm. One transition had midpoint at 3.6M urea and a $\Delta G = 9$ kcal/mole. This transition takes the enzyme from a wavelength maximum of about 335 nm characteristic of the native enzyme to an intermediate with a wavelength maximum at 345 nm. A second transition had a midpoint at .0M urea and a $\Delta G =$ 3 kcal/mole. This transition takes the enzyme from the assumed intermediate state to the denatured state (355 nm).

It is possible to trap a species with properties expected for an intermediate. If the denatured enzyme is diluted into buffers with all the required components except thiosulfate, a species is formed within the time of manual mixing (seconds) that has a fluorescence maximum at 345 nm. If tested quickly (within 30 min), there is a single, reversible unfolding transition that follows one of the two theoretical transitions for the equilibrium unfolding discussed above. This species is enzymatically inactive.

If samples are preincubated for long times after dilution from high urea concentrations, the fluorescence properties return toward those expected of the native enzyme. The samples, though, only regain a very small amount of enzyme activity. If this refolded but inactive enzyme is supplemented with the thiosulfate, the activity returns in a first order process. Under typical conditions, this return has a $t_{1/2} = 80$ min.

The ability to produce an intermediate-like state permits us to measure a number of its properties. This intermediate is monomeric in the refolding mixture. The fluorescence maximum is consistent with the intermediate having its tryptophan residues only partially exposed to the solvent, but more exposed than the native enzyme. This conclusion is strengthened by results showing that iodide can quench the fluorescence of the tryptophan residues. Both the intermediate and native enzyme have greatly restricted iodide access compared with the denatured protein. The intermediate has lost those tertiary interactions that account for the induced CD signal at 280 nm that is observed in the native enzyme. The far-UV CD shows, though, that the intermediate retains a great deal of secondary structure. Rotational relaxation times assessed from fluorescence measurements show that the motion of tryptophan residues in the intermediate is between those for the denatured and native enzyme.

X-ray structure

The results can be understood in terms of the x-ray structure. The main elements are that the active site SH group is able to oxidize to form monosulfhydryl adducts such as sulfenic and sulfinic acids. The oxidation can go further, and it has been shown that the active site can form disulfides with other SH groups *(20)*. However, the x-ray indicates that these states can not form from the native structure because of steric constraints.

A major source of accessible hydrophobic surfaces is at the interdomain interface. These surfaces can be made accessible with very little change in the structure of the individual domains. The tryptophan residues are distributed through the structure in a way that makes them responsive to conformational changes. The structure makes it very natural to understand the relationship between folding, hydrophobic exposure and sensitivity to oxidation.

As an aside, it is interesting that the active site SH must be ionized to act as a nucleophile for catalysis. Two helices in the interdomain region point with the positive ends of their dipoles oriented toward this SH group. This has the effect of lowering the pKa of the SH *(21)*. This effect is strong only when the dipoles are shielded from the solvent *(22)*. Weakening the interdomain interactions would increase access of these helices to the solvent and reduce the nucleophilicity of the active site sulfhydryl group. This would account for the correlated exposure of hydrophobic surfaces, exposure of Trp residues, and loss of enzyme activity.

Folding Model for Rhodanese

A folding model can be formulated for rhodanese which, although provisional, incorporates what is known so far about this system. On dilution from high concentrations of denaturants, unfolded rhodanese rapidly collapses to a relatively compact state, driven to a large extent by hydrophobic interactions with formation of a large amount of secondary structure. This could correspond to nucleation of the individual domains. This state has the characteristics associated with the so-called molten globule state *(23)* in that it is compact, yet it has a high degree of flexibility – a structure that is relatively open to access by solute components, has exposed hydrophobic surfaces, and contains sulfhydryl groups in a state in which they can be easily oxidized. This structure is enzymatically inactive. A second step that is rate limiting would allow the enzyme to regain its final global structure to give a native-like conformer that is still enzymatically inactive. The addition of thiosulfate would then permit final local readjustments at the active site and lead to reactivation. In addition to solubilizing sticky intermediates, the flexibility induced by the presence of lauryl maltoside might assist the conformational search.

For rhodanese, then, the major conformational impediment to successful refolding is kinetic competition of aggregation due to association of monomers that follow the initial formation of exposed, organized surfaces. This effect is minimized as reported here by the use of nondenaturing detergents. It appears, though, that many methods that limit protein association will have a salutary effect. Thus, we have now been able to achieve high degrees of refolding by (i) renaturation at high dilution, (ii) immobilizing rhodanese on Sepharose, and (iii) using the chaperonins cpn60 and cpn10 from *E. coli.* The latter procedure is particularly interesting since it replaces the need for detergents and makes the refolding process adenosine triphosphate (ATP) dependent. These considerations have allowed us to design a protocol for isolation of the active enzyme that forms inclusion bodies during expression in *E. coli.*

In conclusion, successful refolding of a "problem" protein requires knowledge about that specific protein and a recognition that there can be both chemical and conformational restrictions on folding. A major objective is to prevent kinetic traps that represent conformational cul de sacs that limit refolding on a useful time scale.

References

1. Kane, J. F., and D. L. Hartly, *Trends Biotechnol.* **6,** 95 (1988).
2. Marston, F. A. O., *Biochem. J.* **240,** 1 (1986).
3. Tandon, S., and P. M. Horowitz, *J. Biol. Chem.* **264,** 9859 (1989)
4. King, J., *Chem. Eng. News* **67,** 32 (1989).
5. Jaenicke, R., *Prog. Biophys. Mol. Biol.* **49,** 117 (1987).
6. Rothman, J., *Cell* **59,** 591 (1989).
7. Randall, L. L., S. J. S. Hardy, J. R. Thom, *Ann. Rev. Microbiol.* **41,** 507 (1987).
8. Goloubinoff, P., A. A. Gatenby, G. H. Lorimer, *Nature* (London) **342,** 884 (1989).
9. Horowitz, P. M., and D. Simon, *J. Biol. Chem.* **261,** 13887 (1986).
10. Stellwagen, E., *J. Mol. Biol.* **135,** 217 (1979).
11. Tandon, S., and P. M. Horowitz, *J. Biol. Chem.* **265,** 5967 (1990).
12. Tandon, S., and P. M. Horowitz, *J. Biol. Chem.* **264,** 3311 (1989).
13. Horowitz, P. M., and M. G. Douglas, *Fed. Proc. Abstr.* **1836,** 1859 (1981).
14. Boggaram, M., P. M. Horowitz, M. R. Waterman, *Biochem. Biophys. Res. Commun.* **130,** 407 (1985).
15. Miller, D. M., R. Delgado, J. M. Chirgwin, S. J. Hardies, P. M. Horowitz, *J. Biol. Chem.* **266,** 4686 (1991).
16. Ploegman, J. H., *et al., Nature* **273,** 124 (1978).
17. Horowitz, P. M., and N. L. Criscimagna, *J. Biol. Chem.* **261,** 15652 (1986).
18. Horowitz, P. M., and S. Bowman, *J. Biol. Chem.* **262,** 8728 (1987).
19. Horowitz, P. M., and N. L. Criscimagna, *J. Biol. Chem.* **265,** 2576 (1990).
20. Horowitz, P. M., and S. Bowman, *J. Biol. Chem.* **264,** 3311 (1989).
21. Hol, W. G. J., *Adv. Biophys.* **19,** 133 (1985).
22. Gilson, M. K., and B. H. Honig, *Proteins* **3,** 32 (1988).
23. Ptitsyn, O. B. J., *Protein Chem.* **6,** 272 (1987).

14

Alternate Folding Motifs for Gramicidin

Crystallographic and Spectroscopic Analyses of Polymorphism

B. A. Wallace

Gramicidin is a hydrophobic linear polypeptide produced by *Bacillus brevis* at the onset of sporulation. It consists of 16 hydrophobic residues (15 amino acids with alternating L- and D-configurations, plus a C-terminal ethanolamine) with the following sequence *(1)*:

formyl-L-val-gly-L-ala-D-leu-L- ala-D-val-L-val-D-val-L-trp-D-leu- L-trp- D-leu-L-trp-D-leu-L-trp- ethanolamine.

The biological functions of this molecule are as yet unknown, but in vitro it has been shown to have a number of activities, including those of an ion channel *(2)*, an RNA polymerase inhibitor *(3)*, and a regulator of differentiation and development *(4)*. Its functional properties as a monovalent cation channel have been the best explored because its regular and well-behaved channels have provided an excellent model system for conductance studies. Its polysemous functional roles are reflected in the polymorphism of its structure, and, as such, this molecule is an excellent system for examining protein folding.

This chapter will compare and contrast the structures adopted by gramicidin in a variety of different environments, including membranes, solutions, and crystals. It will further examine the profound effects that ion binding has on the different structures of this molecule. Lastly, it will describe emerging studies on the interconversion and refolding of this molecule and the energetics involved in these processes.

Gramicidin affords us an excellent example for studying environmental effects on protein folding. This molecule has been the object of study for more than 40 years by a wide variety of spectroscopic, diffraction, chemical modifications and modeling techniques. Not only has its structure in the solid state been determined at high resolution by x-ray crystallography *(5, 6)*, but detailed information on its solution structure is also available from two-dimensional nuclear magnetic resonance (NMR) spectroscopy *(7, 8, 9)*. Furthermore, two different crystal structures, one with *(5)* and one without *(6)* ions present, have been solved, and there is a wealth of other spectroscopic data relating these structures to structures in other solvent systems.

Polymorphism in Solution

Because gramicidin has no hydrophilic side chains nor free amino- or carboxy-terminal groups that would permit formation of zwitterionic structures, this molecule is insoluble in water but soluble in a wide range of organic solvents as well as in the hydrophobic interiors of lipid bilayers and detergents. Circular di-

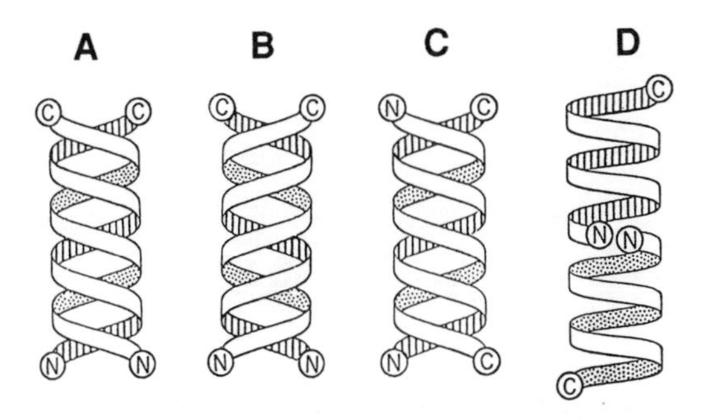

Fig. 1. Schematic drawings of the types of motifs adopted by gramicidin in different environments, illustrating the polymorphic nature of this molecule. **(A)** Left-handed, parallel double helix; **(B)** Right-handed, parallel double helix; **(C)** Left-handed, antiparallel double helix; **(D)** Helical dimer.

chroism spectroscopic studies have examined the overall folding of the molecule in a range of organic solvents, including alcohols of increasing chain length as well as dioxane and ethyl acetate *(10, 11, 12)*.

In all of these solvents, the net spectra obtained are very similar and suggest that the molecule may adopt similar conformations, regardless of the hydrophobicity or net dipole moment of the solvent. However, the situation is not as simple as that observation would suggest. In each of these solvents, gramicidin adopts, not a single conformation, but a family of conformations. The spectra obtained are actually a linear combination of four different spectra resulting from four different conformations that are present in different proportions, depending on the solution conditions. These four structures interconvert slowly on the NMR time scale and can be separately identified *(9, 13)*.

Two of the structures are left-handed, parallel double helices, which differ from each other in the stagger between the ends of their polypeptide chains (Fig. 1A). One is a right-handed, parallel double helix (Fig. 1B) and one is a left-handed, antiparallel double helix (Fig. 1C). In these solvents, as under most conditions, gramicidin is dimeric in nature *(10, 14)*. Only in very dilute solutions or in dimethyl sulfoxide *(10, 15)* has the molecule been found to form predominantly monomeric structures. While the four solution structures are very different topologically and might, at first glance, appear to differ considerably in energies of stabilization, all possess a similar number of intermolecular hydrogen bonds and are folded in such a way that their hydrophobic side chains come in contact with the hydrophobic solvent, and their hydrophilic polypeptide backbones are sequestered from extensive contact with solvent molecules. While all have roughly similar stabilities (as evidenced by their nearly equimolar abundances at equilibrium) only one of these structures has thus far been found to be stabilized in the environment of an ordered crystal: that of the left-handed, antiparallel double helix *(6)*.

Polymorphism in Crystals

In crystals, as well as in solution, there is a polymorphism seen for gramicidin (Fig. 2). In this case, it is manifest as crystallization in a variety of different space groups with correspondingly different intermolecular contacts (Table 1). Despite the differences in unit cell dimensions and number of molecules per assymetric unit, however, the volume per gramicidin dimer is essentially constant for all crystal types *(16)*. Only one of the ion-free crystal forms (the orthorhombic form from ethanol) has been solved at atomic resolution *(6)*; whether the ion-free monoclinic form from methanol will have a similar fold to that of the orthorhombic crystals is yet to be seen *(17)*. For the ion-containing crystals, again only one crystal form (the gramicidin-CsCl complex) has been solved *(5)*, but there is evidence that even such subtle changes as anion type may cause differences in the crystal structure *(18)*. Crystals formed under identical conditions, with the same cations but different anions, exhibit differences in at least the positions and nature of their ion-binding sites.

Similarity of Solution and Crystal Structures

Given the evidence for polymorphism of gramicidin in different environments, it is interesting to see that one of the forms present

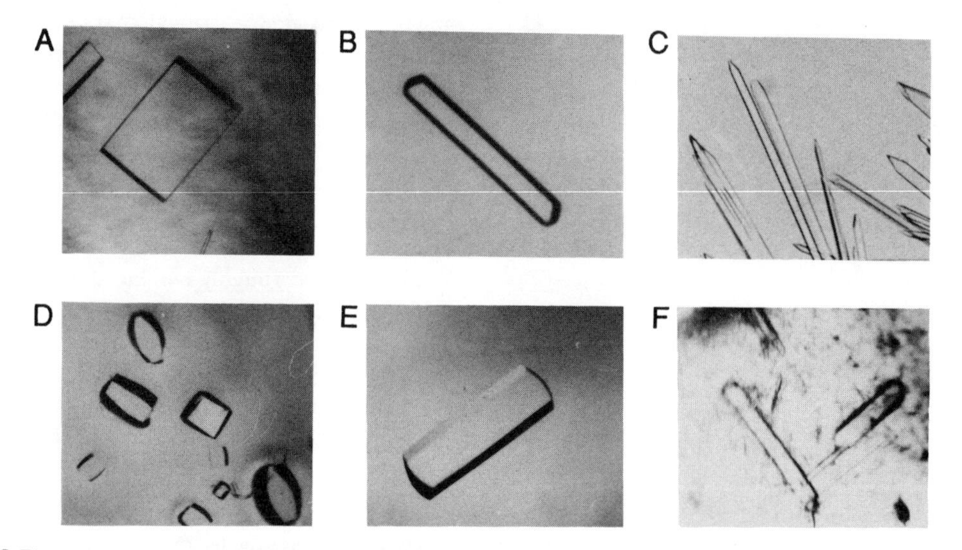

Fig. 2. Photographs of gramicidin crystals showing their polymorphism. (A) Ion free (from methanol); (B) Ion free (from ethanol); (C) CsCl-I complex; (D) CsSCN complex; (E) KSCN complex; (F) Lipid complex.

in solution (albeit a minor component on a molar basis) is very similar to that found in the ion-free crystals *(6)*. Both in solution and in crystals, in the absence of ions, gramicidin appears to form a left-handed, antiparallel double helix; in both states, the pitch and stagger of the helices are similar.

In the crystal structure, the backbone of the molecule is quite irregular and thus differs substantially from a number of the models built for this form of the polypeptide *(19)*. It also differs in this way from the regular backbone structure proposed from the NMR studies. However, it is not clear that the spectroscopic studies could have picked up this structural detail, so this may not represent a true difference between solution and crystals. Thus, while gramicidin exhibits a polymorphism within a particular type of environment that is a reflection of the relative stability of alternate forms for the molecule in any given solvent, environmental effects such as those associated with the isotropic environment of a solution versus the anisotropic and highly concentrated environment of a crystal do not appear to be particularly notable.

That the polypeptide backbone structures in the solid state and in solution are similar may not be so surprising because those are regions of the molecule that in the crystal are sequestered away from the intermolecular contacts. What might be more expected to

Table 1. Unit cell parameters for different gramicidin crystals.

Components	Space group	Unit cell dimensions (Å)	Dimers unit/cell	Volume/ dimer ($Å^3$)
Gramicidin/MeOH	$P2_1$	15.2 × 26.7 × 31.7	2	6433
Gramicidin/EtOH	$P2_12_12_1$	24.8 × 32.4 × 32.7	4	6569
Gramicidin/CsCl-I	$C222_1$	33.9 × 35.9 × 91.4	16	6952
Gramicidin/CsCl-II	$P2_12_12_1$	32.1 × 52.1 × 31.2	8	6522
Gramicidin/CsCl-III	$P2_12_12_1$	32.1 × 26.1 × 31.2	4	6535
Gramicidin/CsSCN	$P2_12_12_1$	32.3 × 53.2 × 31.8	8	6830
Gramicidin/KSCN	$P2_12_12$	31.1 × 52.2 × 31.1	8	6311
Gramicidin/Lipid	$P222_1$	26.8 × 27.5 × 32.8	2	6714

vary between solution and crystals is those parts of the molecule that form the interfaces between dimers as they pack in the crystal. In the case of gramicidin, in which the periphery of the molecule is formed from the amino acid side chains, one might expect, therefore, to see a difference in the relative orientations of the side chains and/or stacking of aromatic side chains.

While the orientations of the side chains are clearly seen in the crystal, the NMR studies did not provide any information on this feature, so direct comparisons cannot be made. However, other spectroscopic techniques such as fluorescence and optical spectroscopy measure the relative orientations of the tryptophan side chains *(20, 12)*. From those studies, it appears that the tryptophans are not stacked relative to each other in solution, in agreement with the observations in the crystals. Thus, although differences in the details of side-chain orientations will undoubtedly be found in solution and in crystals, to date there is no evidence of polymorphism between solution and crystal structures.

Polymorphism in Membranes

In membranes, as opposed to solutions, a single conformational type, "the channel," seems to predominate, although there is evidence for a small fraction of the gramicidin molecules being in one of two alternate conformations. Such alternate conformers are reflected functionally in the existence of (i) channels with much longer mean channel lifetimes and slightly lower conductances than those of the predominant form or (ii) channels with similar mean channel lifetimes but lower conductances. The former structures may be represented by the "pore" form found in solution *(16)*, while the latter, designated "minichannels," have been suggested to have overall conformations that are slight variants on the predominant "channel" structure *(21)*. The factors influencing the equilibria between these various functional states are currently being examined in a number of laboratories.

Differences Between Solution and Membrane-Bound Structures

In contrast to the lack of environmental effects between solutions and crystals, a dramatic environmental effect is seen between the structures of gramicidin dissolved in the isotropic environments of organic solvents and embedded in the more ordered, hydrocarbon chains of membrane lipid fatty acids. Circular dichroism spectroscopy has shown that the backbone folds of gramicidin in solution and in membranes are entirely different *(22)*. NMR studies *(23)* have indicated that in vesicles the predominant form of the gramicidin molecule is not any of the double helical-type structures found in solution, but rather a unique, helical dimer topology in which two monomers are held together at a small interface consisting of roughly six intermolecular hydrogen bonds (Fig. 1D).

While details of the structure in membranes are not yet available from either NMR or x-ray crystallographic analysis, it is clear that, not only are the backbone folds different, but the relative orientations of the tryptophan side chains differ between the two environments. Circular dichroism and fluorescence spectroscopic studies *(12, 20, 24)* indicate stacking of tryptophans relative to each other. NMR studies of gramicidin in detergent (sodium dodecyl sulfate) indicates the molecule folds as a helical dimer *(8)* very similar to that originally proposed from modeling studies *(25)*. Again, this is entirely different from any of the structures in solution.

Optical spectroscopic studies indicate that the structure of this molecule in the detergent is similar, although not identical, to that in membranes *(8, 26)*. More detailed information on the structure of the molecule in complex with lipids will be forthcoming from cocrystals prepared containing both gramicidin and lipid molecules (Fig. 2F) *(27)*. Preliminary work on characterizing those crystals (Table 1) again shows a distinct difference between the lipid-containing crystals and those formed in the absence of lipid *(16, 28)*.

In summary, gramicidin exhibits a strong

environmental influence in solutions and in membranes. Those effects are not merely solvent-hydrophobicity effects, as evidenced by the fact that solutions of gramicidin and lipids in alcohols have the same circular dichroism spectra *(29)* as solutions without the lipid in alcohol, but differ from the spectra of vesicle samples prepared from those same lipid molecules.

This then suggests that the anisotropy of the environment and/or very specific geometric interactions between the gramicidin and lipid molecules produce the alternate folding forms. With respect to the latter, preliminary studies using lipid molecules in which the carbonyl oxygen has been eliminated (thereby producing an ether lipid) indicate that the helical dimer structure is destabilized relative to the double helix structure *(29, 35)*. Perhaps this destabilzation is due to the absence of a hydrogen bond between the lipid carbonyl oxygen and the ethanolamine hydroxyl hydrogen of the polypeptide. Such subtleties in conditional differences suggest that the energetic differences between the various membrane conformations of gramicidin may be small.

Polymorphism as a Result of Cation Binding

Just as the presence of membranes has a profound effect on the conformation of gramicidin, so too does the presence of ions have an influence on the folding of this molecule. This was first seen in solution by circular dichroism experiments in which ion-free gramicidin and gramicidin that had been saturated with ions were compared *(22)*. Gramicidin has two or more binding sites per dimer for monovalent cations, and these studies suggested that the effect of ion binding might be a change in the pitch of the helices in order to accommodate ions in the central pore of the molecule. This suggestion was reinforced by crystallization studies that compared unit cell dimensions of crystals with and without the ions *(30, 31)*. Those studies suggested that ion-containing dimers were shorter and fatter structures than ion-free dimers.

Finally, from the recent solutions of the crystal structures of gramicidin with and without ions present *(5, 6)*, it can clearly be seen that the major effect of ion binding on the structure of the molecule is to change the pitch of the helix from one in which there are 4.8 residues per turn to one in which there are 6.4 residues per turn, resulting in a foreshortening of the molecule and an increase in the pore diameter (Fig. 3). There are other more subtle changes that include an increase in the regularity of the backbone and a shift in the stagger of the two monomers. In solution, another dramatic effect of ion binding was seen: while in the absence of ions there is an equilibrium mixture of four conformers, in the presence of ions, only a single stable conformer is seen *(7)* – that of an antiparallel double helix.

The relationship between the crystal structure and the solution structure of the ion-bound form is not nearly as straightforward as that seen for the ion-free form. The crystal structure *(5)* indicates that the molecule is a left-handed, antiparallel double helix with a stagger of the chains such that it results in nearly flat ends of the molecule. The molecule has a superhelical twist that results in backbone hydrogen bonds lying nearly parallel to the helix axis. In contrast, the NMR studies *(7)*, while also suggesting an antiparallel double helix motif, indicate that the structure is right-handed in nature and with a stagger of the chains such that the ends of the molecule are not flat. Those studies reported no evidence of a superhelical twist. However, it is not clear that they would have been able to detect such a structural feature. Whether the polymorphism between the crystal and solution forms of ion-bound gramicidin is real or is due to interpretation of the data has yet to be determined.

There are a number of experimental differences between the conditions used that might have influenced the results seen: (i) different solvents (while the crystals are from methanol, the solution study is from

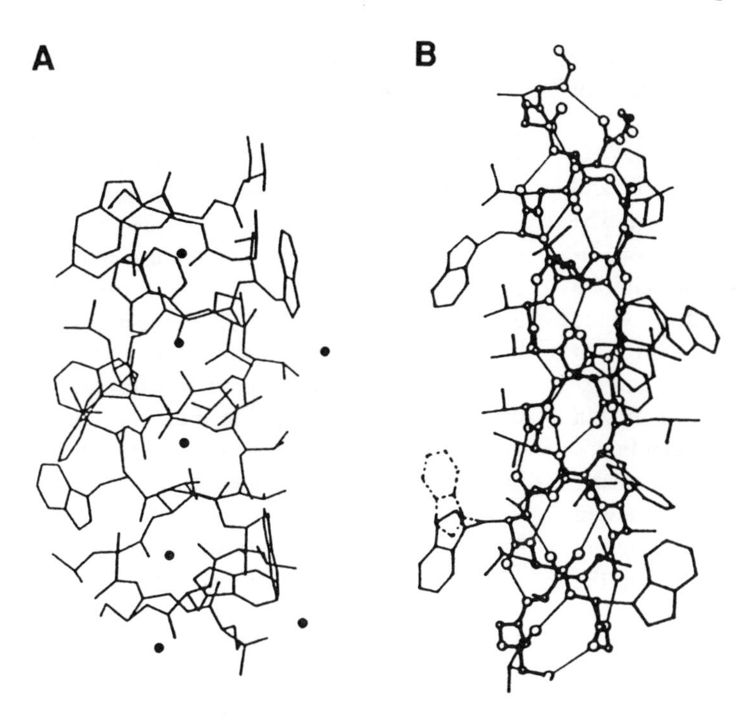

Fig. 3. Crystal structures of gramicidin (A) with *(5)* and (B) without ions *(6)*, showing the influence of ion binding on the conformation. [Reprinted, with permission, from the *Annual Review of Biophysics and Biophysical Chemistry,* vol. 18, copyright 1989 by Annual Reviews, Inc.]

chloroform/methanol), (ii) different counterions [the crystals are formed with cesium chloride, the solutions with cesium thiocyanate, which recent results suggest may have an influence on the crystals *(18)*], and (iii) the cesium concentrations differ considerably. Alternatively, as in the case of the ion-free form, the crystals may represent a form that is relatively sparsely populated on a molar basis in solution.

While there appears to be a single major folding motif present in the ion-containing forms, gramicidin exhibits a polymorphism (albeit less dramatic) even under these conditions. In the crystals, the asymmetric unit is two dimers rather than a single dimer – the result of subtle differences between two different dimer conformations present in this ordered and anisotropic environment. The two conformations have common topologies and very similar backbone conformations, but they differ (by < 0.1 Å root-mean-square deviation) in details of their folding, mostly in the side-chain orientations *(32)*. Thus, while the binding of ions tends to have a dramatic effect on the overall conformation of the molecule and stabilizes one conformation relative to other structures, there is still some polymorphism, even in the ion-bound structures.

Refolding and Interconversion Between Gramicidin Conformers

As described above, gramicidin exhibits a number of different conformational states, depending on its environment, and the similar relative molar abundances of the various conformations suggest a very small difference in energy between the different motifs. Some studies on refolding have shed light on the processes of interconversion between the different forms and the energetics involved. Early circular dichroism *(10)* studies of solution forms of gramicidin indicated, not surprisingly, that when crystals of the left-handed, antiparallel double helix were first solubilized, they produced a single conformation (the left-handed, antiparallel double helix). But, in time, the molecule transformed to an equilibrium mixture of four different conformers. Depending on the solvent type, the half time for that interconversion could range from minutes to days. Furthermore, chromatographic studies indicated that if any one of those four conformers were isolated, each had the potential to undergo the interconversion and again form the equilibrium mixture of four conformers. Their relative ratios at equilibrium differed by about a factor of three, suggesting differences in energies of stabilization of less than 1 kcal.

As the various forms of double helices can interconvert, so, too, can the double helix and helical dimers present in membranes. The topology and energetics of this process are interesting: the refolding from a double helix to a helical dimer requires an unwinding of the molecules and, more importantly, a breaking of 28 intermolecular hydrogen bonds and reformation of a different set of 28 hydrogen bonds, although only 6 of the hydrogen bonds are intermolecular in the helical dimer case. This refolding procedure can be followed spectroscopically *(33, 20)*. The double helical form binds to membranes; upon heating, the structure transforms to the helical dimer. The rate and temperature at which this occurs is dependent on the nature of the lipids as well as the ionic strength of the surrounding medium and the lipid-to-gramicidin mole ratio *(33, 34)*. Depending on the conditions chosen, it is a process that can be observed over a time scale ranging from less than one minute to many hours and provides a means for examination of the energetics as well as the intermediates in the process *(33)*.

As conductance studies of gramicidin in membranes suggest, there is one dominant (helical dimer) conformation present, with a small amount of a second conformation that has been attributed to antiparallel double helix. It is clear that the energetics of stabilization between the two forms in membranes are not very different from one another: the relative abundance on a molar basis of ~ 10,000:1 *(36)* suggests an energy difference of around six kcals, or roughly the contribution of one hydrogen bond. It is not surprising that the energies of the two forms do not differ that much, as each dimer is stabilized by approximately the same number of hydrogen bonds, although each has a different number of inter- and intramolecular hydrogen bonds. It is also not surprising that the activation energy for breaking and reforming all 28 hydrogen bonds may be substantial and, hence, the reason for the large energy input necessary to effect this transformation.

Summary: Gramicidin, A Polymorphic Molecule

In summary, gramicidin is a relatively small polypeptide that is hydrophobic in nature and partitions strongly into solvent environments with low dielectric constants. While it tends to adopt very regular structures that are stabilized by a large number of backbone hydrogen bonds, it is clearly polymorphic in nature as evidenced by studies in both solution and in crystals. This polymorphism is likely a consequence of the relatively small size of the molecule, which results in a regular secondary structure but does not give rise to a globular molecule with extensive tertiary interactions. That these different conformational motifs can interconvert on time scales that are compatible with spectroscopic measurements means that this molecule is an excellent candidate for investigating the folding and refolding of polypeptides in hydrophobic environments.

Acknowledgements

I thank the students, postdoctoral associates, colleagues, and collaborators with whom I have worked with on gramicidin: Bob Bittman, E. R. Blout, David Busath, Nancy Buswell, Jon Callahan, Tamie DiNolfo, Keith Dunker, Michelle Fausel, Wayne Hendrickson, Bob Jones, Martha Kimball, Roger Koeppe, K. Ravikumar, Will Veatch, Shula Weinstein, and Sean Wormuth. This work was supported by the Biophysics Program of the National Science Foundation (currently grant DMB88-16981).

References

1. Sarges, R., and B. Witkop, *J. Am. Chem. Soc.* **86**, 1862 (1964); Sarges, R., and B. Witkop, *J. Am. Chem. Soc.* **87**, 2011 (1965).
2. Andersen, O. S., *Ann. Rev. Physiol.* **46**, 531 (1984).
3. Sarker, N., D. Langley, H. Paulus, *Proc. Nat. Acad. Sci. U.S.A.* **74**, 1478 (1977).
4. Katz, E., and A. L. Demain, *Bact. Rev.* **41**, 449 (1977).

5. Wallace, B. A., and K. Ravikumar, *Science* **241,** 182 (1988).
6. Langs, D. A., *Science* **241,** 188 (1988).
7. Arseniev, A. S., I. L. Barsukov, V. F. Bystrov, *FEBS Lett.* **180,** 33 (1985).
8. Arseniev, A. S., I. L. Barsukov, V. F. Bystrov, A. L. Lomize, Y. A. Ovchinnikov, *FEBS Lett.* **186,** 168 (1985).
9. Bystrov, V. F., and A. S. Arseniev, *Tetrahedron* **44,** 925 (1988).
10. Veatch, W. R., and E. R. Blout, *Biochemistry* **13,** 5257 (1974).
11. Veatch, W. R., E. T. Fossel, E. R. Blout, *Biochemistry* **13,** 5249 (1974).
12. Wallace, B. A., *Biophys. J.* **49,** 295 (1986).
13. Fossel, E. T., W. R. Veatch, Y. A. Ovchinnikov, E. R. Blout, *Biochemistry* **13,** 5264 (1974).
14. Veatch, W. R., R. Mathies, M. Eisenberg, L. Stryer, *J. Mol. Biol.* **99,** 75 (1975); Veatch, W., and L. Stryer, *J. Mol. Biol.* **113,** 89 (1977).
15. Braco, L., C. Abad, A. Campos, J. E. Figueruelo, *J. Chromatogr.* **353,** 18 (1986).
16. Wallace, B. A., *Annu. Rev. Biophys. Biophys. Chem.,* **19,** 127 (1990).
17. Koeppe, R. E., II, K. O. Hodgson, L. Stryer, *J. Mol. Biol.* **121,** 41 (1978).
18. Ravikumar, K., and B. A. Wallace, *ACA Proceedings* **17,** 116 (1989).
19. Chandrasekaran, R., and B. V. V. Prasad, *CRC Crit. Rev. Biochem.,* p. 125 (1978); Colonna-Cesari, F., S. Premilar, F. Heitz, G. Spach, B. Lotz, *Macromolecules* **10,** 1284 (1977); Koeppe, R. E., II, and M. Kimura, *Biopolymers* **23,** 23 (1984).
20. Masotti, L., P. Cavatorta, G. Sartor, E. Casali, A. G. Szabo, *Biochim. Biophys. Acta* **862,** 265 (1986).
21. Busath, D., and G. Szabo, *Biophys. J.* **53,** 689 (1988).
22. Wallace, B. A., *Biopolymers* **22,** 397 (1983).
23. Weinstein, S., B. A. Wallace, E. R. Blout, J. Morrow, J., W. R. Veatch, *Proc. Natl. Acad. Sci., U.S.A.* **76,** 4230 (1979); Weinstein, S., B. A. Wallace, J. Morrow, W. R. Veatch, *J. Mol. Biol.* **143,** 1 (1980).
24. Scarlata, S. F., *Biophysical J.* **54,** 1149 (1988).
25. Urry, D. W., M. C. Goodall, J. D. Glickson, D. F. Mayers, *Proc. Nat. Acad. Sci.* **68,** 1907 (1971).
26. Masotti, L., A. Spisni, D. W. Urry, *Cell Biophys.* **2,** 241 (1980).
27. Wallace, B. A., and R. W. Jones, *J. Mol. Biol.,* **217,** 625 (1991).
28. Short, K. W., B. A. Wallace, R. A. Myers, S. P. A. Fodor, A. K. Dunker, *Biochemistry* **26,** 557 (1987).
29. Wallace, B. A., W. R. Veatch, E. R. Blout, *Biochemistry* **20,** 5754 (1981).
30. Koeppe, R. E., II, J. M. Berg, K. O. Hodgson, L. Stryer, *Nature* **279,** 723 (1979).
31. Kimball, M. R., and B. A. Wallace, *Ann. N.Y. Acad. Sci.* **435,** 551 (1984).
32. Wallace, B. A., W. A. Hendrickson, K. Ravikumar, *Acta Crystallog.,* **B46,** 440 (1990).
33. Wallace, B. A., *Biophys. J.* **45,** 114 (1985).
34. LoGrasso, P. V., F. Moll III, T. A. Cross, *Biochemistry* **26,** 6621 (1987); Killian, J. A., K. U. Prasad, D. Hains, D. W. Urry, *Biochemistry* **27,** 4848 (1988).
35. Callahan, J. D., R. Bittman, B. A. Wallace, unpublished results.
36. Durkin, J., and O. Andersen, unpublished results.

15

Prolyl Isomerase

Its Role in Protein Folding and Speculations on Its Function in the Cell

Franz X. Schmid, Kurt Lang, Thomas Kiefhaber Sabine Mayer, E. Ralf Schönbrunner

Introduction

In 1973, Garel and Baldwin *(1)* discovered that unfolded ribonuclease (RNase) A is a heterogeneous mixture of fast- (U_F) and slow-folding (U_S) species, which coexist in a slow equilibrium. These U_F and U_S species give rise to parallel fast (in the milliseconds time range) and slow (in the minutes range) phases in the refolding of RNase A. Since then, similar heterogeneities have been detected in the folding of many other proteins. In 1975, Brandts *et al.* *(2)* proposed the proline hypothesis as a molecular explanation for these experimental results.

The basic assumption of this hypothesis is that the fast- and slow-folding molecules, U_F and U_S, differ in the conformational state (cis or trans) of one or more Xaa-Pro peptide bonds. In the U_F molecules, the prolyl peptide bonds are in the same isomeric state as in the native protein, N. U_S molecules have at least one incorrect, non-native proline isomer. The cis ⇋ trans isomerizations of Xaa-Pro peptide bonds are intrinsically slow processes with a high activation energy of about 20 kcal/mol. Therefore, it was postulated that folding of the U_S molecules is limited in rate by the initial, slow re-isomerizations of incorrect prolines, which is then followed by rapid structure formation. This model explained some aspects of slow folding, although it became evident that, depending on the folding conditions, extensive formation of ordered secondary and tertiary structure is possible prior to the reversal of incorrect proline isomers *(3–5)*. Consequently, proline re-isomerization usually occurs late in the folding process, and it can be accelerated in extensively folded protein molecules, i.e., an "intramolecular catalysis" of re-isomerization appeared to be possible, mediated by the prior formation of ordered structure around the incorrect prolyl peptide bonds *(6)*. It should be pointed out, however, that the folding of polypeptide chains with incorrect proline isomers is much slower than the rapid folding of the U_F molecules, which have correct proline isomers.

Soon after the proline hypothesis was suggested, the search began for a putative enzymatic activity that catalyzes the isomerization of Xaa-Pro peptide bonds. In 1984, Fischer and his co-workers *(7)* were successful in detecting a prolyl isomerase activity in porcine kidney. Their assay is based on the conformational specificity of chymotrypsin, which cleaves the 4-nitroanilide moiety from the assay peptide glutaryl-Ala-Ala-Pro-Phe-4-

nitroanilide only when the Ala-Pro peptide bond is in the trans conformation.

The assay peptide consists of a mixture of 90% trans and 10% cis form in solution. Therefore, in the presence of a high concentration of chymotrypsin, 90% of the hydrolysis reaction occurs within the dead time of manual mixing. Hydrolysis of the remaining 10% is slow, limited in rate by the cis → trans isomerization of the Ala-Pro bond of the peptide. By using this assay, they were able to detect and purify an enzyme from porcine kidney that accelerates this isomerization very efficiently. Accordingly, the enzyme was named peptidyl-prolyl cis/trans isomerase or PPIase. It is a monomeric protein with a molecular weight of 17000. The K_M value for the assay peptide lies in the range of 0.5 mM, the k_{cat}/K_M value is estimated to be about $7 \times 10^6\ M^{-1}\ S^{-1}$ *(25)*.

A surprising result emerged from the sequencing of PPIase *(8, 9)*. It was found to be identical with cyclophilin, the major binding protein for the immunosuppressive drug cyclosporin A (CsA) in the cell. This protein was discovered in the same year as PPIase *(10)*. No sequence homologies with other known proteins were found. Apparently the cyclophilins/PPIases represent a new class of ubiquitous, highly conserved proteins. They are present in all organisms and tissues that were examined so far, and occurrence is not at all restricted to cells of the immmune system.

Binding of CsA to mammalian 17k-PPIase is very tight, with a dissociation constant in the nanomolar range *(10)*. CsA inhibits PPIase activity efficiently with an inhibition constant of the same order of magnitude *(9)*. The simplest explanation for this finding is that the binding site for CsA and the site of PPIase activity are identical. Recently other prolyl isomerases have been discovered, which differ from the 17k-PPIase in molecular weight and in sequence, and which do not bind CsA *(11)*. They are not considered in this chapter and the term PPIase is used for the prolyl isomerases that bind to CsA with high affinity.

PPIase Catalyzes Some but Not All Slow Steps in Protein Folding

Bovine RNase A was used initially to investigate whether prolyl isomerase is able to accelerate slow, rate-limiting reactions in protein folding. This protein was selected since good, but indirect, evidence existed for the involvement of proline isomerization in slow steps of its unfolding and refolding. The experiments carried out in our laboratory as well as in others, however, gave essentially negative results.

The slow phases of RNase A refolding are basically insensitive to the presence of PPIase *(12–14)*. This negative result could be explained in different ways. Slow folding of RNase A might not involve proline isomerization at all, or else PPIase might not recognize folding polypeptide chains as substrates. Another possibility is that rapid formation of ordered structure during folding renders the prolyl peptide bonds inaccessible for the isomerase. This explanation is supported by results on the S-protein fragment of RNase A (residues 21–124), which contains all proline residues. The S-protein is less stable and folds much more slowly than the parent protein. Folding of this fragment is indeed accelerated by PPIase, even though the efficiency of catalysis is not very good *(13)*.

Subsequently, the action of prolyl isomerase in folding was examined for a number of proteins *(12–15)*. Small proteins were selected for these experiments, where kinetic data suggested that proline isomerization was involved in slow folding steps. A clearcut answer was not obtained. Good catalysis was observed for some molecules such as the immunoglobulin light chain or RNase T1. For other proteins such as porcine RNase, cytochrome c, or pepsinogen the folding rates were only moderately enhanced. Similar to bovine RNase A, slow refolding of thioredoxin was not catalyzed by PPIase at all *(15)*.

There is evidence from experiments with an engineered mutant that a specific proline residue, Pro76, is involved in a late, slow step of the folding of thioredoxin *(16)*. The simplest explanation for the lack of catalysis is that, as in RNase A, rapid formation of ordered structure renders Pro76 inaccessible to PPIase.

Clearly, not all slow steps in protein folding are related with proline isomerization. In particular, the very slow folding reactions of large proteins can be determined by other events such as correct domain pairing *(17)*. The in vitro unfolding and refolding of many large proteins is only partially reversible. This is caused by aggregation reactions that compete with correct folding, presumably early on the folding pathway, when hydrophobic regions are still exposed *(18)*. We examined whether prolyl isomerase can suppress aggregation by increasing the rate of critical early folding steps. Attempts to find such an effect of prolyl isomerase were not successful so far. The presence of PPIase affected neither the rate, nor the yield of reactivation of two large oligomeric proteins – lactate and malate dehydrogenase *(26)*.

Proline-Containing Sequences Are Targets of PPlase in Its Catalysis of Slow Folding

The name given to PPIase suggests that it acts by catalyzing the isomerization of prolyl peptide bonds. This has been shown conclusively for the short synthetic peptide glutaryl-Ala-Ala-Pro-Phe-4-nitroanilide. This artificial assay was selected because it allowed the exploitation of the conformational specificity of chymotrypsin for an efficient screening procedure *(7)*. The acceleration of slow protein folding reactions by PPIase might occur by another, yet unknown, mechanism. Such an alternative mechanism should be considered, since the correlation between the kinetic evidence for the involvement of proline isomerization into the folding of several proteins and the respective catalysis by PPIase is not very good. An alternative would be that PPIase is a general "polypeptide-binding protein" that interacts with exposed chain segments and thereby facilitates folding.

To examine the mechanism of action of PPIase in protein folding, we used RNase T1 as a model system. The refolding of this protein occurs in three phases: in a minor fast phase in the milliseconds time range and in two slow phases (the "intermediate" and the "very slow" phase), both of which show some properties characteristic of proline-controlled processes *(5)*. These two phases are catalyzed by prolyl isomerase with very different efficiencies *(9)*.

The intermediate phase is strongly accelerated by PPIase, whereas catalysis of the very slow phase is fairly poor. The presence of 0.6 μM PPIase leads to a 30-fold increase in rate of the intermediate phase, but only to a 2.5-fold increase of the very slow phase. Native RNase T1 contains two cis prolines at positions 39 and 55 *(19)*. After unfolding, cis prolines isomerize largely to the more favorable trans conformer; therefore, we considered these two cis prolines to be involved in the observed two slow phases of the refolding of RNase T1.

To test this assumption, the cis Ser54-Pro55 prolyl peptide bond was replaced by a trans Gly54-Asn55 peptide bond. These substitutions were chosen since they occur naturally in RNase C2, a close relative of RNase T1. The folding results obtained with this mutant were clearcut *(27)*. The intermediate phase of folding was almost completely absent, and the good catalysis of folding by PPIase was no longer observed.

This result bears two implications. First, isomerization of Pro55 is a rate-limiting step in the folding of RNase T1, and second, the good catalysis of refolding of the wild-type protein by PPIase depends on the presence of Pro55. This suggests that Pro55 is, indeed, the target of PPIase. The efficient catalysis of this intermediate phase of RNase T1 folding supports the original assumption that PPIase accelerates slow steps in protein folding by catalyzing the isomerization of incorrect prolyl peptide bonds.

The investigation of the slow folding of RNase T1 in the presence of prolyl isomerase led to another important result, i.e., its catalytic efficiency in Pro39 isomerization decreases strongly with increasing formation of stable structure during refolding. In refolding molecules where both Pro39 and Pro55 are still in the incorrect trans state, initially formed ordered structure is not very stable. At this stage, Pro39 is still accessible for PPIase, and catalysis of its isomerization is good. However, as soon as Pro55 reverts to the native cis conformation, this ordered structure gains additional stability, thus leading to a strong decrease in the accessibility to prolyl isomerase of the remaining incorrect trans Pro39. This proline residue becomes "trapped" in a folding intermediate.

The simplest explanation for these results is that the presence of folded structure restricts the accessibility of prolines for PPIase and hence decreases or even abolishes its capacity as a catalyst for folding. They also suggest that PPIase can act much more efficiently at early steps of folding. This important role of accessibility nicely explains the lack of catalysis observed in the in vitro folding of bovine RNase A and of thioredoxin. In both cases, native-like folded intermediates were found to accumulate rapidly in refolding prior to the rate-limiting, proline-controlled steps *(6, 16)*.

Since the evidence is good now that PPIase does, in fact, accelerate proline-limited steps in protein folding, this enzyme can be used as a diagnostic tool to identify proline isomerization steps in complex folding reactions. However, only a "positive" answer is meaningful. When catalysis is not observed, this could mean that proline isomerization is not a rate-limiting step in folding, or alternatively results from a shielding of prolyl peptide bonds by early formation of ordered structure.

PPIase as a Catalyst

In the initial experiments to detect rate enhancements of protein folding reactions in the presence of PPIase, fairly high concentrations were employed, which were similar to the concentration of the refolding protein. As an example, 1.6 μM PPIase were required for a sevenfold acceleration of the folding of 2 μM immunoglobulin light chains *(13)*. Therefore, it appeared possible that PPIase does not act as an enzyme in protein folding but rather as a stoichiometric binding protein.

Such an alternative is clearly ruled out for the action of PPIase on the intermediate phase of RNase T1 folding. Good catalysis is observed when the PPIase concentration is much smaller than RNase T1 concentration. Catalysis depends on both PPIase and RNase T1 (substrate) concentration in a nearly linear fashion. These results suggest that, under the conditions employed, the saturation of PPIase with the substrate RNase T1 is very low. This is not surprising with respect to the small substrate concentrations that are used in refolding experiments and in view of the K_M value of 0.5 mM obtained for the assay peptide. K_M values for protein substrates cannot be measured because of the prohibitively high protein concentrations that would be required in the folding experiments. If, in a first approximation, we assume a K_M value of 0.5 mM for RNase T1 as well, then the catalytic efficiency of PPIase as a catalyst of folding would be fairly high with a turnover number of about 100 s^{-1} (at 10°C). Under the same conditions, the respective value for the assay peptide is 3000 s^{-1}.

Other Cyclophilins Catalyze Protein Folding

A number of proteins have been detected that are homologous with the mammalian 17k-PPIases/cyclophilins. Notable examples are the product of the ninaA gene of *Drosophila*, which is involved in visual signal transduction *(20)*, and cyclophilins from the cytoplasm and the mitochondria of *Neurospora crassa* and of yeast, whose in vivo functions are unknown *(21)*. These molecules are 40–60% identical in amino acid sequence with the mammalian 17k-PPIases. Prolyl isomerase activity was also detected in archaebacteria.

The cyclophilins from *N. crassa* and from yeast are very similar in their catalytic properties to PPIase from porcine kidney. They are active in the coupled assay with the test peptide and chymotrypsin, and they are inhibited by $HgCl_2$ and by CsA in the same manner as porcine PPIase. Folding experiments with RNase T1 as a model system indicate that the cytoplasmic cyclophilins from yeast and *N. crassa* catalyze the proline-controlled slow steps in the folding of RNase T1 with an efficiency similar to that of porcine PPIase. Taken together, these results show that both sequence and function are conserved in different species. Prolyl isomerase activity and catalysis of slow protein folding reactions seem to be correlated with each other.

In Vivo Role of PPIase: Speculations

Before speculating on the role of PPIases in the cell, let us first summarize some of the properties of this new class of enzymes that might be important for the allocation of physiological functions. PPIases are ubiqitous proteins, which have been found in practically all organisms and tissues that have been examined to date. Most of them are soluble proteins, localized in different cell compartments such as the cytoplasm and the mitochondria. Sequence homologies with members of other protein families can not yet be detected. A putative nucleotide binding motif in the sequence remains ambiguous *(22)*.

Isoforms of PPIase appear to exist in mammals. Up to twenty different genes for PPIases have been detected in the rat *(23)*, but it is not yet clear whether these are authentic genes or pseudogenes. This is probably not the case for the lower eukaryotes. Only one gene was found in *N. crassa* that codes for both the cytoplasmic and the mitochondrial cyclophilin *(21)*. The presumed function of cyclophilin/PPIase in mediating the immunosuppressive effect of CsA is still not understood at a molecular level. The evidence is good that CsA affects the control of transcription of the genes for interleukin 2 and interferon gamma in T cells. Whether this role is mediated by binding of CsA to the major cytoplasmic PPIase or by another mechanism remains an open question *(24)*.

We propose that, in their in vivo function, PPIases also act on proline-containing chain regions (β turns?) of their target proteins. Whether these targets are folding nascent polypeptide chains and/or folded proteins with accessible prolyl peptide bonds is presently unknown. An effect of PPIase on folded proteins has not yet been demonstrated.

When speculating about the functions of PPIases, it is important to consider that proline isomerization is an intrinsically slow process, in unfolded polypeptides as well as in folding intermediates and in native proteins. PPIases might participate directly in signal transduction pathways as specific transmitters. Alternatively, they might work in concert with other, yet unidentified, effector proteins that carry the specificity for particular signal transduction steps. By simultaneous binding of the two proteins – effector and PPIase – to their target, the effector could shift the equilibrium at a particular prolyl peptide bond of the target protein to the other isomer at the expense of energy. PPIase could catalyze this isomerization. After dissociation of the effector and the isomerase, the target protein would slowly revert back to its original isomeric state. By such a speculative mechanism, it would be possible to activate or deactivate a protein and at the same time set a molecular timer, which would keep this protein in an alternative conformational state for a period of time.

Such a model would imply that PPIases are not specific factors in a signal transduction chain by themselves, but rather that they work together with other factors that determine specificity and supply energy. The novel function of proline cis/trans isomerization would be to provide a time standard for the return of the target protein to its original functional state. Proline-containing chain regions would be very well suited as "molecular timers" because they frequently occur in solvent-exposed turns and because the intrinsically low isomerization rates can be modified to some extent by the structural environments of the prolylpeptide bonds.

We admit that this is a highly speculative model, that is not backed by any experimental evidence. It might, however, stimulate alternative ways of thinking and lead to new efforts to elucidate the role of this class of proteins. It is, of course, not our intent to dismiss the possibility that PPIases are involved in the de novo folding of nascent polypeptide chains in the cell. With regard to the activity in in vitro protein folding, such a function appears to be "straightforward," but positive experimental evidence for it is still lacking.

References

1. Garel, J.-R., and R. L. Baldwin, *Proc. Natl. Acad. Sci. U.S.A.* **70**, 3347 (1973).
2. Brandts, J. F., H. R. Halvorson, M. Brennan, *Biochemistry* **14**, 4953 (1975).
3. Kim, P. S., and R. L. Baldwin, *Ann. Rev. Biochem.* **51**, 459 (1982).
4. Schmid, F. X., R. Grafl, A. Wrba, J. J. Beintema, *Proc. Nat. Acad. Sci. U.S.A.* **83**, 872 (1986).
5. Kiefhaber, T., R. Quaas, U. Hahn, F. X. Schmid, *Biochemistry* **29**, 3053 (1990); Kiefhaber, T., R. Quaas, U. Hahn, F. X. Schmid, *Biochemistry* **29**, 3061 (1990).
6. Schmid, F. X., and H. Blaschek, *Eur. J. Biochem.* **114**, 11 (1981).
7. Fischer, G., H. Bang, C. Mech, *Biochmed. Biochim. Acta* **43**, 1101 (1984).
8. Takahashi, N., T. Hayano, M. Suzuki, *Nature* **337**, 473 (1989).
9. Fischer, G., B. Wittmann-Liebold, K. Lang, T. Kiefhaber, F. X. Schmid, *Nature* **337**, 268 (1989).
10. Handschumacher, R. E., M. W. Harding, J. Rice, R. J. Drugge, D. W. Speicher, *Science* **226**, 544 (1984); Harding, M. W., R. E. Handschumacher, D. W. Speicher, *J. Biol. Chem.* **261**, 8547 (1986).
11. Harding, M. W., A. Galat, D. E. Uehling, S. L. Schreiber, *Nature* **341**, 758 (1989); Siekierka, J. J., S. H. J. Hung, M. Poe, C. S. Lin, N. H. Sigal, *Nature* **341**, 755 (1989).
12. Fischer, G., and H. Bang, *Biochim. Biophys. Acta* **828**, 39 (1985).
13. Lang, K., F. X. Schmid, G. Fischer, *Nature* **329**, 268 (1987).
14. Lin, L.-N., H. Hasumi, J. F. Brandts, *Biochim. Biophys. Acta* **956**, 256 (1988).
15. Lang, K., Thesis, Regensburg, W. Germany (1988).
16. Kelley, R. F., and F. M. Richards, *Biochemistry* **26**, 6765 (1987).
17. Vaucheret, H., L. Signon, G. Le Bras, J.-R. Garel, *Biochemistry* **26**, 2785 (1987).
18. Jaenicke, R., *Prog. Biophys. Molec. Biol.* **49**, 117 (1987).
19. Heinemann, U., and W. Saenger, *Nature* **299**, 27 (1982).
20. Schneuwly, S., *et al.*, *Proc. Natl. Acad. Sci. U.S.A.* **86**, 5390 (1989); Shieh, B.-H., M. A. Stamnes, S. Seavello, G. L. Harris, C. S. Zuker, *Nature* **338**, 67 (1989).
21. Tropschug, M., *et al.*, *J. Biol. Chem.* **263**, 14433 (1988); Tropschug, M., I. B. Bartelmess, W. Neupert, *Nature* **342**, 953 (1990).
22. Gschwendt, M., W. Kittstein, F. Marks, *Biochem. J.* **256**, 1061 (1988).
23. Danielson, P. E., *et al.*, **DNA 7**, 261 (1988).
24. Drugge, R. J., and R. E. Handschumacher, *Transplantation Proceedings* **20**, 301 (1988).
25. Schönbrunner, E. R., unpublished data.
26. Kongsbak-Reim, M., and F. X. Schmid, unpublished results.
27. Kiefhaber, T., H.-P. Grunert, U. Hahn, F. X. Schmid, *Biochemistry* **29**, 6475 (1990).

16

Protein-Disulfide Isomerase

An Enzyme That Catalyzes Protein Folding in the Test Tube and in the Cell

Robert B. Freedman

Introduction

In the context of this volume, protein folding has been approached primarily as a problem in macromolecular physical chemistry. Analysis of protein stability and structure, of the significance of electrostatic interactions for protein conformation, dynamics, and energetics, and of the relation of amino acid sequence to folded structure, all emphasize the classical statement of the "protein folding problem" as a problem of *information*. The sequence of amino acid residues represented by an unfolded polypeptide is seen as encoding the biologically significant, three-dimensional folded structure of the protein. The problem is to understand the nature of this code, which allows both decoding – in order to predict the structure of a protein of known sequence – and encoding – in order to design a sequence that will generate specified structure.

It is characteristic of this *informationist* approach to protein folding that questions of mechanism or pathway of folding are regarded as secondary. However, it has long been known that protein cannot explore all possible conformations in the time scale available during folding, so that folding occurs *via* intermediate states along kinetically favoured pathways. While it is still generally assumed that the ultimate state corresponds to the global free energy minimum, this cannot be demonstrated rigorously; the state achieved can only be guaranteed to be the lowest in energy of the kinetically accessible states.

Even if this qualification is seen as insignificant, it is not the most serious limitation of the single-mindedly informationist approach to protein folding. The observation that proteins can be refolded in the test tube has been interpreted as establishing that proteins, in general, contain in their sequences all the information to specify their folded state and do not require the input of exogenous information. In short, they do not need to be *told* how to fold up. But this correct assertion has had serious misapplications. It has been taken as implying that the folding process in the test tube is an adequate description of the process as it occurs in the cell. It has been taken as implying that since no *information* is necessary for the process, no cellular *machinery* is involved. It has led to considerable initial disappointment in attempts to generate functional, folded recombinant proteins by cloning and expressing in bacteria or lower eukaryotes, genes for valuable mammalian proteins. The mistake has been to infer that because proteins do not need to be *instructed* how to fold, they therefore do not need to be *helped* to fold.

Catalysis of Protein Folding: Cellular Context

In the past few years, cell biologists interested in the biosynthesis of proteins, the assembly of complex multicomponent cellular structures, and the translocation of newly synthesized proteins to their cellular site of action, have developed powerful model systems for analyzing cotranslational and posttranslational events in the life of a protein. This work has highlighted new "informational" aspects of protein structure, namely, the presence within translation products of amino acid sequences that function as "targeting signals," which are recognized by cellular machinery and ensure the transfer of the protein to the specified subcellular compartment *(1)*.

The work in this field has had two major conceptual consequences, however. Firstly, it led to the realization that in the cell, protein folding occurs as part of and concomitant with a host of other processing events. A polypeptide that forms part of an oligomeric cytosolic enzyme (a glycolytic enzyme, for example) will fold, following biosynthesis in the cytosol; associate with other polypeptides to form the oligomer; and then undergo further folding to generate the final tertiary and quarternary structure *(2)*. A polypeptide forming part of one of the respiratory complexes of the inner mitochondrial membrane may be synthesized in the cytoplasm, associate posttranslationally with the mitochondrion, transfer across both mitochondrial membranes, be proteolytically processed to remove its mitochondrial targeting signal, and then integrate with many other polypeptides into the functional membrane-associated complex *(3)*. A secretory polypeptide such as an immunoglobulin heavy chain will be recognized cotranslationally by signal recognition particle so that it is docked with the translocation machinery of the endoplasmic reticulum (ER) membrane; passed across the membrane; proteolytically processed; subjected within the ER lumen to posttranslational modifications, including disulfide bond formation and glycosylation; and associated with other chains to form the intact immunoglobulin *(4)*.

All this work emphasizes that within a cell, many processes intervene between the initial decoding of genetic information – the production of a polypeptide of specified sequence – and the ultimate formation of a functional protein. These key cotranslational and posttranslational processes include assembly, translocation, and covalent modification in addition to folding. Most significantly, these processes are concurrent or overlapping rather than simply consecutive. The corollary is that understanding of protein folding in a cellular context cannot neglect these parallel processes.

The second major conceptual change to emerge from this cell biological tradition is that the cellular processes of protein folding, assembly, translocation, and modification are facilitated by proteins, just as other cellular processes are. There already appear to be several ubiquitous families of proteins whose role is not fully defined at a mechanistic level, but which function as facilitators of folding *(5, 6)*. Proteins do, indeed, need help in folding when this process is seen in a cellular context and on a cellular timescale.

Catalysis of Protein Folding: Rate-Determining Steps

The recognition that protein folding from the unfolded state in the test tube proceeds via kinetically favoured pathways has led to extensive study of the kinetics and mechanism of folding by spectroscopic and other physical techniques and by chemical trapping *(2, 7)*. A mass of data, apparently specific for each protein, has slowly yielded unifying principles, but recent technical and theoretical advances reviewed in Ptitsyn's chapter 10 in this volume and elsewhere *(8)* suggest that more rapid progress can now be expected.

Emergence of these general principles has been obscured, to some extent, by the existence of two specific processes found in most, but not all, protein folding pathways, which are generally slow in the conditions used for protein refolding and hence dominate the kinetics. The first – cis/trans isomerization of peptide bonds to prolyl residues – is discussed fully by Schmid in chapter 15. The existence of an equilibrium mixture of cis and trans isomers of prolyl peptide bonds in unfolded proteins, while the same bonds in the folded protein must be in one or other conformation, makes prolyl-peptidyl cis/trans isomerization a significant process in protein folding in the test tube. However, it is not clear whether this process is a significant factor in protein folding in the cell. An enzyme capable of catalyzing this reaction, prolyl-peptidyl cis/trans isomerase (PPI), has been identified and isolated and is clearly a protein with an important role, since it is the target for agents that act to suppress immune function *(9)*. But, as presently understood, the properties of this enzyme suggest that it is more likely to function in signal transduction, by acting on key membrane proteins, than to function in initial protein folding *(10)*.

The second distinct process that is rate-determining in the refolding of some proteins is the formation and isomerization of disulfide bonds. The oxidation and refolding in vitro of reduced, unfolded ribonuclease to produce active enzyme with the correct set of four disulfide bonds was the classical example of spontaneous refolding *(11)*. Because the disulfide bond is covalent, unlike the other intramolecular interactions that stabilize native protein structure, it can be readily trapped. The study of the formation of disulfide bonds during refolding of reduced proteins in vitro has been especially instructive and influential *(7)*.

Table 1 summarizes some general features of the process of disulfide formation in protein refolding, drawn from the chemical studies of Creighton, Scheraga, Wetlaufer, and others. The slowest steps in such pathways are disulfide isomerizations at a late stage in the overall process, which lead to formation of the native conformation; the isomerizations themselves involve thiol:disulfide interchange reactions, and since these reactions require attack of the thiolate (-S) anion, the reaction rates increase with pH. The slowness of protein thiol:disulfide interchange at physiological pH suggests the involvement of an enzymic catalyst in the cellular process, and such an enzyme protein-disulfide isomerase (PDI) – has been identifed. This enzyme and its properties are the major theme of this chapter.

Table 1. The formation of native disulfide bonds in the refolding of reduced, denatured mature proteins in vitro.

(i)	Proceeds by kinetically determined pathways
(ii)	Involves intermediates with non-native disulfides
(iii)	Involves nonrandom subset of such intermediates
(iv)	Proceeds via thiol:disulfide interchanges
(v)	Has $t_{1/2}$ of 10^3–10^5 sec at pH > 8
(vi)	Even slower at physiological pH

See *(7)* for review of original data.

Formation of Disulfides as a Model for Posttranslational Processes

Analysis of the cellular mechanism responsible for the formation of native disulfide bonds is particularly valuable since the formation of disulfide bonds is broadly representative of the range of posttranslational events leading to the formation of a functional protein. As the previous section demonstrates, formation of disulfide bonds is an intrinsic and essential aspect of the *folding* of some proteins and has been widely used, both as a marker for following folding and as a probe for exploring the folding process. Secondly, the formation of disulfide bonds between different polypeptides, as in the generation of immunoglobulins or procollagens, represents the process of *assembly* whereby different gene products come together to form a single functional protein. Finally, the formation of disulfide bonds is a chemical reaction, a genuine modification of amino acid side chains, and hence is representative of many *posttranslational modifications* characteristically undergone by proteins proceeding through the secretory pathway *(4)*. Disulfide bonds are found in practically every class of secretory or cell-surface protein and, indeed, are essentially absent from cytosolic proteins.

Evidence for the Cellular Role of Protein-Disulfide Isomerase

The proposal that PDI functioned as a cellular catalyst of the formation of native disulfide bonds in the biosynthesis and folding of secretory proteins was first made over 20 years ago *(12)* but was not immediately supported by extensive evidence. Even in 1977, while the formation of disulfide bonds could be identified as an important cellular process, and PDI was recognised as an enzyme with interesting catalytic properties, it was not possible to demonstrate a convincing connection between them *(13)*. However, this is no longer the case.

The distribution of PDI, its developmental properties, and its subcellular location all strongly suggest a role in secretory protein biosynthesis. The data, which have been reviewed previously *(14–16)*, show that PDI is a major resident protein of the ER lumen and hence is abundant in the cellular compartment where protein disulfide formation occurs in the cell. PDI is abundant in cells with well-developed rough ER and active in the biosynthesis of disulfide-bonded secretory proteins. Thus, in a collection of lymphoma cell lines that varied widely in rate of immunoglobulin secretion, there was a 40-fold variation in PDI activity, correlating closely with the level of Ig synthesis and secretion. In a number of developmental situations (chick embryonic bone development, wheat endosperm ripening, B lymphocyte differentiation and maturation), levels of PDI vary with developmental state, with high levels of PDI always correlating with a high level of biosynthesis of disulfide-bonded proteins.

Thus PDI is present at the right place, in the right cells at the right time, and is now generally accepted as the cellular catalyst of native disulfide formation.

Catalytic Properties of Protein-Disulfide Isomerase

Depending on the redox state of the protein substrate and of low relative molecular mass thiol and disulfide compounds present, PDI can catalyze net oxidation, reduction, or isomerization of protein-disulfide bonds. Essentially it can catalyze breakage and formation of protein disulfides by thiol:disulfide interchange, and the consequence is the accumulation of the set of disulfides consistent with the most stable state of the system as a whole. The enzyme shows little specificity in its protein substrates and acts on a wide variety of proteins ranging from small single domain proteins to large oligomers with multiple domains *(15)*.

The most relevant properties of PDI as a catalyst can be demonstrated by its actions on bovine pancreatic trypsin inhibitor (BPTI), procollagen, and ribonuclease S. BPTI is the protein whose folding pathway in vitro is best understood. Refolding of reduced BPTI in presence of a disulfide oxidant involves a well-defined set of intermediates containing one or two disulfide bonds, and the kinetics and thermodynamics of the folding process are well-characterized *(7)*. In the presence of PDI, the refolding of BPTI is catalyzed *(17)*; the same intermediates are detected as in the uncatalyzed process and in the same kinetic sequence, but the rate-determining steps that involve both thiol:disulfide interchange and conformational change, are all accelerated (Fig. 1). PDI is active in this reaction as a true catalyst, turning over many molecules of protein substrate per molecule of enzyme.

The refolding of reduced procollagen is more complex in that procollagen comprises three polypeptides, the N- and C-termini of which form globular domains, while the central main portion of the molecule contains the three chains extended in the characteristic triple-helical conformation. Assembly of this structure from the unfolded chains (both in vivo and in vitro) involves chain association, formation of interchain disulfide bonds within the C-terminal globular domain, and then the folding of the central part of the chains into the triple helix. Data on the time-course of refolding of procollagen types I and II in vitro are shown in Table 2. Assembly here refers to the formation of interchain disulfide bonds, whereas folding refers to the formation of the

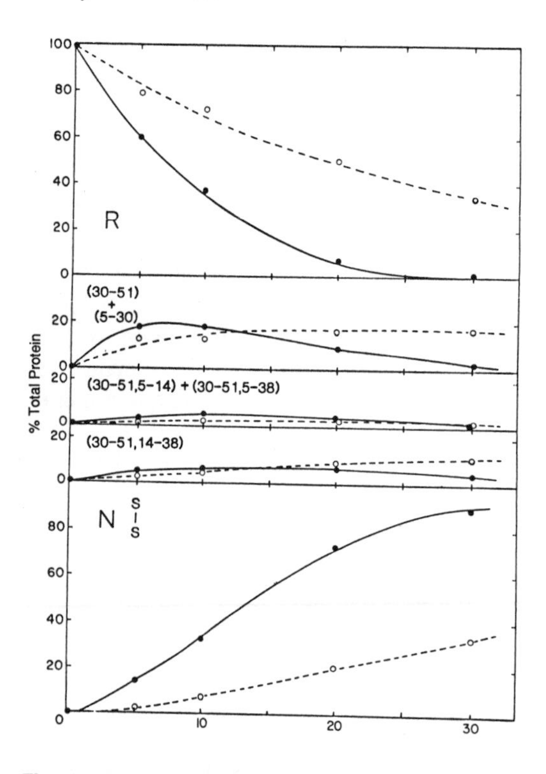

Fig. 1. Catalysis by PDI of the refolding of bovine pancreatic trypsin inhibitor (BPTI). Reduced BPTI (30 μM) was reoxidized by oxidized dithiothreitol (20 mM) in presence — and absence (- - - -) of PDI (0.3 μM). See *(17)* for full discussion. (Reproduced with permission of Academic Press Inc. (London) Ltd.)

triple helix. The former process can be detected by the presence of oligomers on nonreducing SDS-PAGE gels, while the latter can be monitored by the protease resistance provided by the triple-helical folded conformation.

The data on type I procollagen show that assembly is significantly accelerated by the presence of PDI but that the interval between assembly and folding remains approximately six minutes and is not reduced by the presence of PDI. Thus, PDI only catalyzes the assembly process, which depends on disulfide formation, not the later folding step. The half times for assembly and folding observed here in vitro in the presence of PDI are close to those observed in the cell, and the discrepancy in rates of assembly and folding between types I and II procollagen found in these in vitro experiments is also observed in the cell *(18)*.

Table 2. Procollagen assembly and folding in vitro ($t_{1/2}$ (min)).

	GSH/GSSG		+PDI	
	Assembly	Folding	Assembly	Folding
Type I	11.2	17.1	3.7	9.4
Type II	89.1	—	11.6	17.3

Data from *(18)*.

Ribonuclease S is the product of limited action of subtilisin on ribonuclease A. A single peptide bond (between residues 20 and 21) is broken, and the resultant N-terminal peptide (S peptide) remains noncovalently associated with the remainder of the molecule (S protein) to give a complex that retains ribonuclease activity. All four disulfide bonds of the original ribonuclease A are present in the S protein. PDI has no action on ribonuclease S; that is, it leaves the four disulfides within the S protein portion intact. However, when S protein is isolated from S peptide, PDI acts on the S protein to "shuffle" the disulfides; if S peptide is added back, PDI now acts to reinstate the original disulfide bonds (19).

These findings demonstrate that the "correct" disulfides are not dictated by local sequence clues recognised by PDI but by the energetics of the system as a whole; the disulfides that are characteristic of native ribonuclease A and that are stable in ribonuclease S, the noncovalent complex between S peptide and S protein, are acted on and removed in the isolated S protein. Thus, the correct set of disulfides within the S protein is dependent on the presence or absence of S peptide.

The conclusion from these three specific instances of the catalytic action of PDI is that PDI is a genuine catalyst, which can turn over many molecules of substrate, does not determine the product of the reaction, and can bring about disulfide formation at rates comparable to those observed in the cell.

PDI Action on Nascent and Newly Synthesized Proteins

All the work summarized above refers to the action of PDI on mature proteins that have been reduced and unfolded. Are the conclusions relevant to the cellular situation in which a nascent polypeptide is transferred across the ER membrane and where disulfide formation and folding occur in the context of other cotranslational and posttranslational processing events? We have approached this question by attempting to analyze disulfide formation and folding in products of translation in standard in vitro systems for translation and processing.

Work on cotranslational and posttranslational disulfide formation in the wheat storage protein γ-gliadin showed that when translation took place in presence of conventional dog pancreas microsomes, intramolecular disulfide bonds formed rapidly and in good yield, but that this cotranslational disulfide formation was defective in microsomes from which PDI had been selectively removed by an alkaline wash *(20, 21)*. Reintroduction of homogeneous PDI into the interior volume of these PDI-deficient microsomes reversed the defect *(21)*, implying that it is the action of PDI that ensures rapid cotranslational disulfide formation. We have subsequently confirmed these observations with other protein substrates and with microsomes depleted of PDI by other methods *(22)*. The result confirms that the work on the catalytic properties of PDI on mature, unfolded proteins is relevant to its action on newly synthesized proteins in the cell.

Molecular Properties of PDI

PDI is an abundant protein in secretory tissues such as liver and pancreas, and this, together with its solubility in aqueous buffers, has facilitated its purification and characterization. It has been purified to homogeneity from mammalian liver and shown to be a homodimer with a polypeptide relative molecular mass of 57,000 *(16)*. Early work suggested that a vicinal dithiol group was involved in its mechanism of action *(23)* but that this group was found oxidized as a disulfide in the enzyme as isolated. The enzyme could, therefore, be pictured as undergoing cycles of change in redox state, involving the dithiol, the disulfide, and mixed disulfides with its substrate proteins *(17)*.

The major insight into the enzyme's structure and activity came from the cloning and sequencing of the rat liver enzyme by Rutter and co-workers in 1985 *(24)*. Their data (Fig. 2) showed that the PDI polypeptide comprised a number of distinct regions and that two of these (a and a′) were highly homologous, both to each other and to the well-characterized small bacterial protein thioredoxin. Thioredoxin effects oxidoreductions in a number of cellular processes through its pair of Cys residues (Cys 32 and Cys 35), which can interconvert between the dithiol and disulfide states; x-ray diffraction data on thioredoxin crystals show these residues to be located on a prominent loop, extending out from the body of the thioredoxin molecule *(25)*.

Both chemical reactivity studies and structural modeling show that the a and a′ domains of PDI each closely resemble thioredoxin *(26)*. Since PDI is a dimer, a single PDI molecule contains four such domains. We have shown that a molecule of PDI, with four thioredoxin-like domains, is 80-fold better as a catalyst of disulfide isomerization in a standard substrate than is a molecule of thioredoxin. It is tempting, therefore, to picture PDI as an enzyme evolved to bring multiple

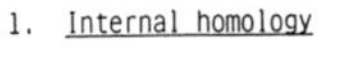

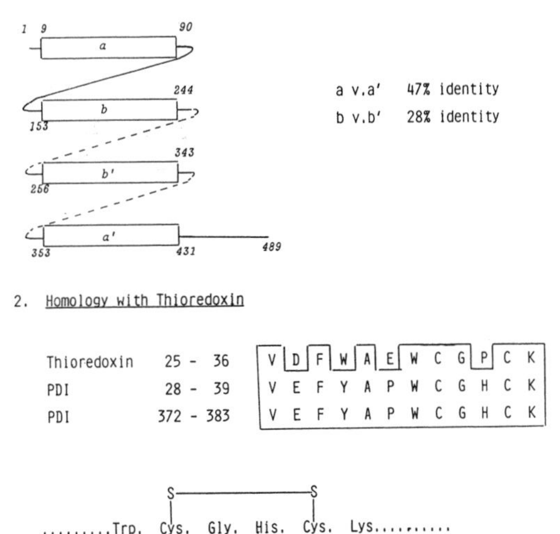

Fig. 2. Analysis of the polypeptide sequence of rat PDI. Based on *(24)*.

thioredoxin-like domains to bear simultaneously on the various disulfide bonds in a protein substrate. At present there is no direct evidence for this.

Proposed Additional Functions of PDI

Although PDI was first recognized, purified, and cloned in the context of a function in protein-disulfide formation, the molecule has been linked to a variety of other functions, on the basis of biochemical or molecular biological evidence *(28)*. Some of these functions and the status of the proposals are summarized in Table 3.

The evidence appears incontrovertible that, in collagen-synthesizing tissues, some PDI molecules combine with the α chains of the enzyme prolyl-4-hydroxylase to form the tetrameric holoenzyme. Thus, PDI is identical to the β chains of prolyl-4-hydroxylase *(29)*. The function of the PDI chains in this enzyme is obscure. The α and β subunits cannot be isolated from prolyl-4-hydroxylase with retention of activity, so that it is difficult to establish their respective functions, but it is unlikely that the subunits (PDI) play a major role since prolyl-4-hydroxylases from lower eukaryote sources appear to consist only of chains homologous to the vertebrate α chain *(30)*.

A third proposed function for PDI was as a component of the enzyme oligosaccharyl transferase, which transfers an oligosaccharyl group from dolichol to an asparagine group of a nascent protein acceptor in the first step of synthesis of N-linked carbohydrate side chains. This proposal arose from intriguing evidence of the specific labelling of a PDI-like protein by a photoaffinity analogue of the peptide acceptor sequence for glycosylation *(31)*. The proposal has some plausibility in that glycosylation, like prolyl-hydroxylation and disulfide-bond formation, is an early posttranslational modification for which the substrate would be an unfolded nascent polypeptide. It was reasonable to think that PDI might play an analogous role in each modification.

However, the system of PDI-depleted microsomal membranes that was developed to study the role of PDI in cotranslational disulfide formation *(20, 21)* has now also been applied to glycosylation. The first result (Fig. 3), showed that PDI-depleted microsomes are as effective as control microsomes in generating a high relative molecular-mass translation product of complement component C9 *(20)*. Similar results have now been obtained with

Table 3. Proposed multiple functions of PDI polypeptide.[a]

Name	Function	Status
Protein disulfide-isomerase (PDI) (2 × 57,000 homodimer)	Catalysis of native disulfide bond formation	
Prolyl-4-hydroxylase (β-subunits of $\alpha_2\beta_2$ tetramer)	Undefined role in cotranslational hydroxylation of prolyl residues	Established as identical to PDI
Dolichol: acceptor protein oligosaccharyltransferase (glycosylation-site binding protein (GSBP))	Soluble, luminal component of oligosaccharyltransferase involved in recognition of ..Asn.Xaa.Thr/Ser.. consensus sequence	See text
Thyroid hormone binding protein (THBP)	Physiological function not established	Probably identical to PDI
Iodothyronine 5′-monodeiodinase	Thyroxine (T4) deiodination to T3	Claimed identity to PDI inconsistent with a considerable body of data

[a]For a full discussion, see *(28)*.

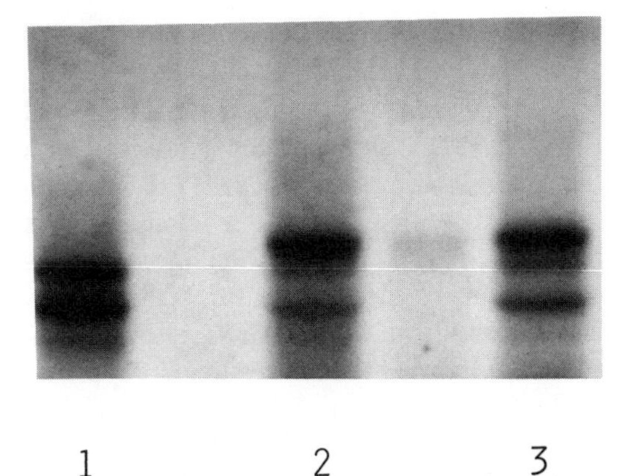

Fig. 3. Cotranslational glycosylation in PDI-deficient microsomes. mRNA coding for complement component C9 was translated in a rabbit reticulocyte lysate in absence of microsomes (track 1), in presence of conventional dog pancreas microsomes (track 2), and in presence of pH 9-washed microsomes that were depleted of PDI (track 3). From *(42)*. We are grateful to Dr. K. Stanley, EMBL, for his support of this work.

mRNAs for a number of proteins and with microsomes depleted of PDI by various techniques *(22)*. Microsomes lacking detectable PDI, and defective in cotranslational disulfide formation, appear to glycosylate nascent proteins efficiently. Hence PDI cannot be an obligatory component of the machinery for core N-glycosylation.

PDI has also been identified as a protein that can bind thyroid hormone analogues, but the physiological significance of this binding is not clear *(32)*. The proposal was made that PDI is identical to iodothyronine-5′-monodeiodinase (5′-DI) *(33)*, a major enzyme of thyroid hormone metabolism, but there is a significant body of evidence against this *(34)*. For example, when rat liver microsomes are incubated with a radioactive affinity analogue of triiodothyronine (T3), two microsomal proteins are labeled, one of 57 kD and one of 27 kD (Fig. 4). Predigestion of the microsomes with trypsin has no effect on PDI activity but destroys 5′-DI activity; this treatment eliminates labeling of the 27 kD band but enhances labeling of the 57 kD band (Fig. 5). In untreated rat liver microsomes, labeling of the 27 kD band, but not of the 57 kD band, is inhibited by substrates and inhibitors of 5′-DI. Pancreas microsomes which lack 5′-DI activity show extensive labeling only of the 57 kD protein. All these lines of work suggest that the 57 kD protein is PDI, while the 27 kD protein is 5′-DI *(34)*.

Characterization of Yeast PDI

The most reliable definition of the cellular functions of a protein comes from genetic analysis of the phenotypes of organisms with mutations in the gene coding for that protein. To develop such an analysis, we aimed to characterize and clone PDI from yeast. Since PDI is present in the ER lumen in higher eukaryotes *(35)*, yeast microsomal fractions were prepared and shown to contain low levels of PDI activity (0.1–0.6 units/g) that could be activated by sonication (10–25-fold), implying that the enzyme was located within the interior of the microsomal vesicles *(36)*. This suggested the existence of yeast PDI homologous to that in vertebrates; hence it was inferred that the active site amino acid sequence of yeast PDI would be identical to the sequence that is conserved in vertebrate PDI (Fig. 5).

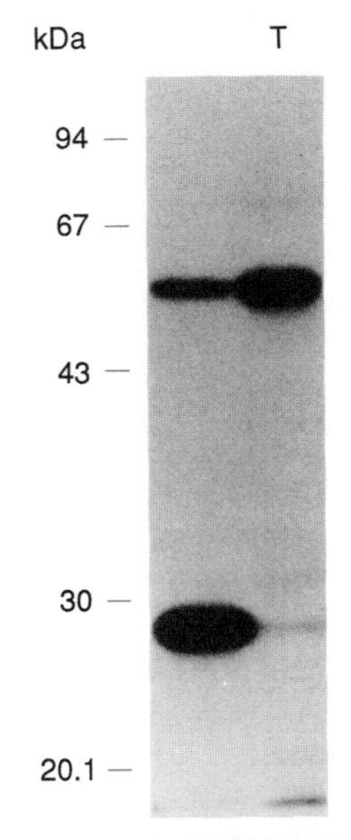

Fig. 4. Labeling of microsomal PDI by affinity analogue of triiodothyronine. Rat liver microsomes were incubated with trypsin (T) or control-treated (-) before being treated with ^{125}I-labeled bromoacetyl-3, 3′, 5-triiodothyronine. The higher M_r labelled band corresponds to PDI. See *(34)* for full discussion. (Reproduced with permission of Academic Press Inc. (London) Ltd.)

```
a)  Consensus active site sequence of vertebrate PDI

          F  Y  A  P  W  C  G  H  C  K

b)  Predicted sequences of homologous yeast PDI gene

          TTC TAC GCT CCA TGG TGT GGT CAC TGT AAG
          Phe Tyr Ala Pro Trp Cys Gly His Cys Lys

c)  Observed sequences in cloned yeast PDI DNA

          TTT TTT GCT CCA TGG TGT GGC CAC TGT AAG
      56  Phe Phe Ala Pro Trp Cys Gly His Cys Lys   65

          TAC TAT GCC CCA TGG TGT GGT CAC TGT AAG
     401  Tyr Tyr Ala Pro Trp Cys Gly His Cys Lys  410
```

Fig. 5. Sequences for cloning of yeast PDI. Nucleotides or amino acids underlined in (c) are those observed to differ from the predicted sequence.

On the basis of this prediction and the rules for yeast-preferred codons, the corresponding yeast DNA sequence was predicted and a unique 30-unit oligodeoxynucleotide was synthesized to probe for this sequence in a yeast genomic library. Two positive clones emerged from the screening *(36)*, one of which, when transformed into yeast on a multicopy plasmid, led to the appearance of two new strong bands on SDS-PAGE analysis of yeast extracts and to a 10-fold enhancement of PDI activity in the transformed yeast *(37)*. Sequencing of the insert in this plasmid established that it coded for an open reading frame with the properties expected of yeast PDI *(38)*. The overall length was similar to that of vertebrate PDIs and two putative active sites were found with sequences close to those of vertebrate PDI (Fig. 5).

The sequence of the C-terminal region of the yeast PDI is shown in Fig. 6, aligned with the corresponding region of rat PDI. The high level of homology in the a′ domains (rat residues 347–461) is clear; the alignment indicated shows 48 identities in this region, especially around the WCGHCK (residues 378–383) representing the active site. Comparison of the C-terminal domains is also interesting. In rat PDI, this domain is predominantly acidic (15 residues out of 28) with several adjacent acidics. The C-terminal sequence is -KDEL, which is characteristic of soluble proteins retained within the ER lumen in higher eukaryotes *(39)*. The acidic C-terminal domain has been proposed to confer Ca^{2+}-affinity on PDI and other resident proteins of the ER lumen, while the -KDEL tetrapeptide functions as a sorting signal for retention within (or recycling to) the ER. The corresponding region of the yeast PDI is similar but longer, mainly due to the presence of additional Ala residues, giving rise to runs of alternating alanine and acidic residues. Furthermore, the C-terminal tetrapeptide is -HDEL rather than -KDEL, a result previously observed in another yeast ER luminal protein *(40)*.

The cloning of yeast PDI has permitted some characterization of its cellular properties. Southern hybridization analysis has shown that the yeast PDI gene is present in a single copy; therefore, gene disruption experi-

```
yeast   ESKAIESLVKDFLKGDASPIVKSQEIFEN-|QDSSVFQLVGKNHDEIVNDPKKDVLVLYYA
        ... :. . ..::.:. .: ..:::. :. |.. .:  ::::: .:.. : ::.:.: .::
rat     TAEKITQFCHHFLEGKIKPHLMSQELPEDW|DKQPVKVLVGKNFEEVAFDEKKNVFVEFYA
          320                  340                  360

yeast   PWCGHCKRLAPTYQELADTYANATSDVLIAKLDHTENDVRGVVIEGYPTIVLYPGGKKSE
        :::::::.:::....:..:: .. ....:::.: :.:.: .: ....::. ..:.. ...
rat     PWCGHCKQLAPIWDKLGETY-KDHENIVIAKMDSTANEVEAVKVHSFPTLKFFPASADRT
          380                   400                  420

yeast   SVVYQGSRSLDSLFDFIKENGHFDVDG|KALYEEAQEKAAEEAEADAEAEADADAELADEE
         . :.:.:.::.. .:. :.:. : .:|..    . .:  . . :.:....:  :
rat     VIDYNGERTLDGFKKFL-ESGRQDGAG|DNDDLDLEEALEPDMEEDDDQKAVKDEL.
           440                  460                  480

yeast   DAIHDEL
```

Fig. 6. Alignment of the C-terminal regions of rat and yeast PDI sequences. The vertical bars define the region homologous to thioredoxin and designate the a′ domain.

ments were attempted. Haploid yeast carrying the disrupted PDI gene were not viable, and when viable diploids carrying the disrupted PDI gene were allowed to sporulate, only two of the four spores from each ascospore were viable *(41)*. This experiment establishes that PDI is an essential gene for yeast viability but does not directly indicate what is the essential role that PDI plays in this organism.

Conclusion

Studies on protein folding in vitro have implicated two specific processes, disulfide isomerization and cis/trans isomerization of prolyl peptide bonds, as rate-determining steps. Enzymes catalyzing these two processes have been identified. In parallel, cellular studies have implicated several families of proteins as facilitators of the process of protein folding in the cell. PDI is the only protein in common between these groups. On the one hand, it is an enzyme of known sequence catalyzing a relatively well defined reaction, and its mechanism of action is open to investigation. On the other hand, its biological properties clearly identify PDI as a facilitator of protein folding in the cell. Much will be learned from further analysis of the structure, mechanism and cellular function of protein-disulfide isomerase.

References

1. Alberts, B., *et al., Molecular Biology of the Cell* (Garland Publishing, Inc., ed. 2, 1989), chapter 8.
2. Jaenicke, R., *Prog. Biophys. Mol. Biol.* **49,** 117 (1987).
3. Attardi, G., and G. Schatz, *Ann. Rev. Cell Biol.* **4,** 289 (1988).
4. Freedman, R. B., *Biochem. Soc. Trans.* **17,** 331 (1989).
5. Ellis, R. J., and S. M. Hemmingsen, *TIBS* **14,** 339 (1989).
6. Rothman, J. E., *Cell* **59,** 591 (1989).
7. Creighton, T. E., *Prog. Biophys. Mol. Biol.* **33,** 231 (1978).
8. Kim. P. S., and R. L. Baldwin, *Ann. Rev. Biochem.* **59,** 631 (1990).
9. Fischer, G., B. Wittman-Liebold, K. Lang, T. Kiefhaber, F. X. Schmid, *Nature* **337,** 476 (1989).
10. Brandl, C. J., and C. M. Deber, *Proc. Natl. Acad. Sci. U.S.A.* **83,** 917 (1986).
11. Anfinsen, C. B., *Science* **181,** 223 (1973).
12. Goldberger, R. F., C. J. Epstein, C. B. Anfinsen, *J. Biol. Chem.* **238,** 628 (1963).
13. Freedman, R. B., and H. C. Hawkins, *Biochem. Soc. Trans.* **5,** 348 (1977).
14. Freedman, R. B., Trends Biochem. Sci. 9, 438 (1984).
15. Freedman, R. B., B. E. Brockway, N. Lambert, *Biochem. Soc. Trans.* **12,** 929 (1984).
16. Freedman, R. B., N. J. Bulleid, H. C. Hawkins, J. L. Paver, in *Gene Expression. Regulation at the RNA and Protein Levels, Biochem. Soc. Symp.* **55,** 167 (1989).
17. Creighton, T. E., D. A. Hillson, R. B. Freedman, *J. Mol. Biol.* **142,** 43 (1980).
18. Koivu, J., and R. Myllyla, *J. Biol. Chem.* **262,** 6159 (1987).
19. Kato, I., and C. B. Anfinsen, *J. Biol. Chem.* **224,** 1004 (1969).
20. Paver, J. L., H. C. Hawkins, R. B. Freedman, *Biochem. J.* **257,** 657 (1989).
21. Bulleid, N. J., and R. B. Freedman, *Nature* **335,** 649 (1980).
22. Bulleid, N. J., and R. B. Freedman, *EMBO J.* **9,** 3527 (1990).
23. Ramakrishna Kurup, C. K., T. S. Raman, T. Ramasarma, *Biochim. Biophys. Acta* **113,** 255 (1986).
24. Edman, J. C., L. Ellis, R. W. Blacher, R. A. Roth, W. J. Rutter, *Nature* **317,** 265 (1985).
25. Holmgren, A. *Annu. Rev. Biochem.* **54,** 237 (1985).
26. Freedman, R. B., H. C. Hawkins, S. J. Murant, L. Reid, *Biochem. Soc. Trans.* **16,** 96 (1988).
27. Hawkins, H. C., E. C. Blackburn, R. B. Freedman, unpublished observations.
28. Freedman, R. B. *Cell* **57,** 1069 (1989).
29. Pihlajaniemi, T., *et al., EMBO J.* **6,** 643 (1987).
30. Myllyla, R., D. D. Kaska, K. I. Kivirikko, *Biochem. J.* **263,** 609 (1989).
31. Geetha-Habib, M., R. Noiva, H. A. Kaplan, W. J. Lennarz, *Cell* **54,** 1053 (1988).
32. Horiuchi, R., *et al., Europ. J. Biochem.* **183,** 529 (1989).
33. Boado, R. J., D. A. Campbell, I. J. Chopra, *Biochem. Biophys. Res. Commun.* **155,** 1297 (1988).
34. Schoenmakers, C. H. H., I. G. A. J. Pigmans, H. C. Hawkins, R. B. Freedman, T. J. Visser,

Biochem. Biophys. Res. Commun. **162**, 857 (1989).
35. Lambert, N., and R. B. Freedman, *Biochem. J.* **228** 635 (1985).
36. Murant, S. J., Thesis, University of Kent at Canterbury (1989).
37. Farquhar, R., R. B. Freedman, M. F. Tuite, unpublished observations.
38. Farquhar, R., P. Bossier, R. B. Freedman, M. F. Tuite, unpublished observations.
39. Munro, S., and H. R. B. Pelham, *Cell* **48,** 899 (1987).
40. Normington, K., K. Kohno, Y. Kozutsumi, M-J. Gething, J. Sambrook, *Cell* **57,** 1223 (1989).
41. Honey, N., R. B. Freedman, M. F. Tuite, unpublished observations.
42. Paver, J., Thesis, University of Kent at Canterbury (1987).

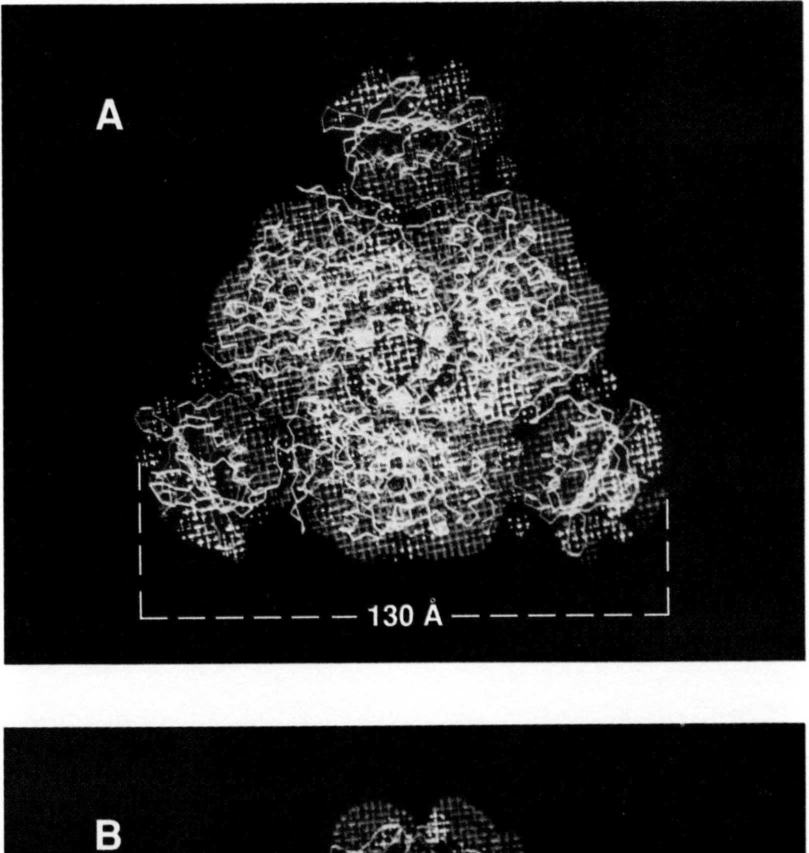

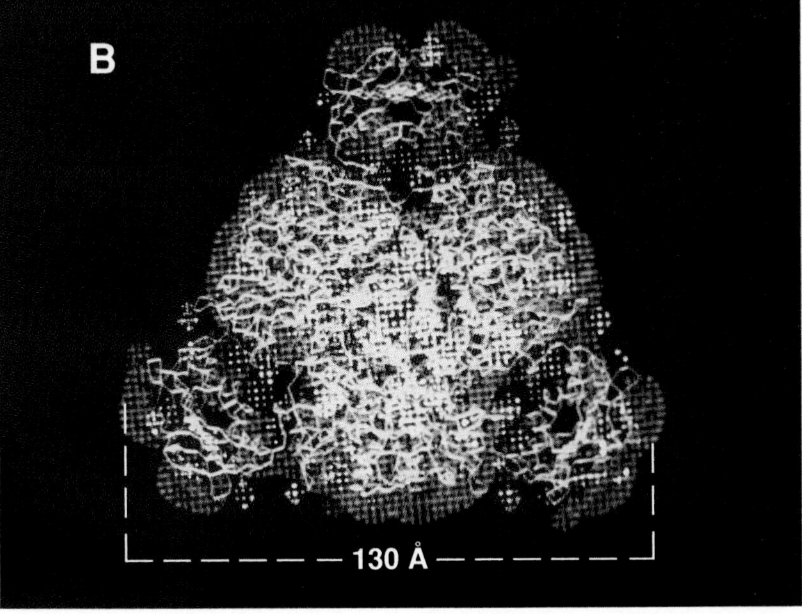

Plate I. Potential surfaces around **(A)** PALA-liganded ATCase and **(B)** CTP-liganded ATCase as viewed down the three-fold axis. Regions of positive potential > 2kT are blue; regions of negative potential < –2kT are red. Regions of positive potential are associated with active and nucleotide binding sites.

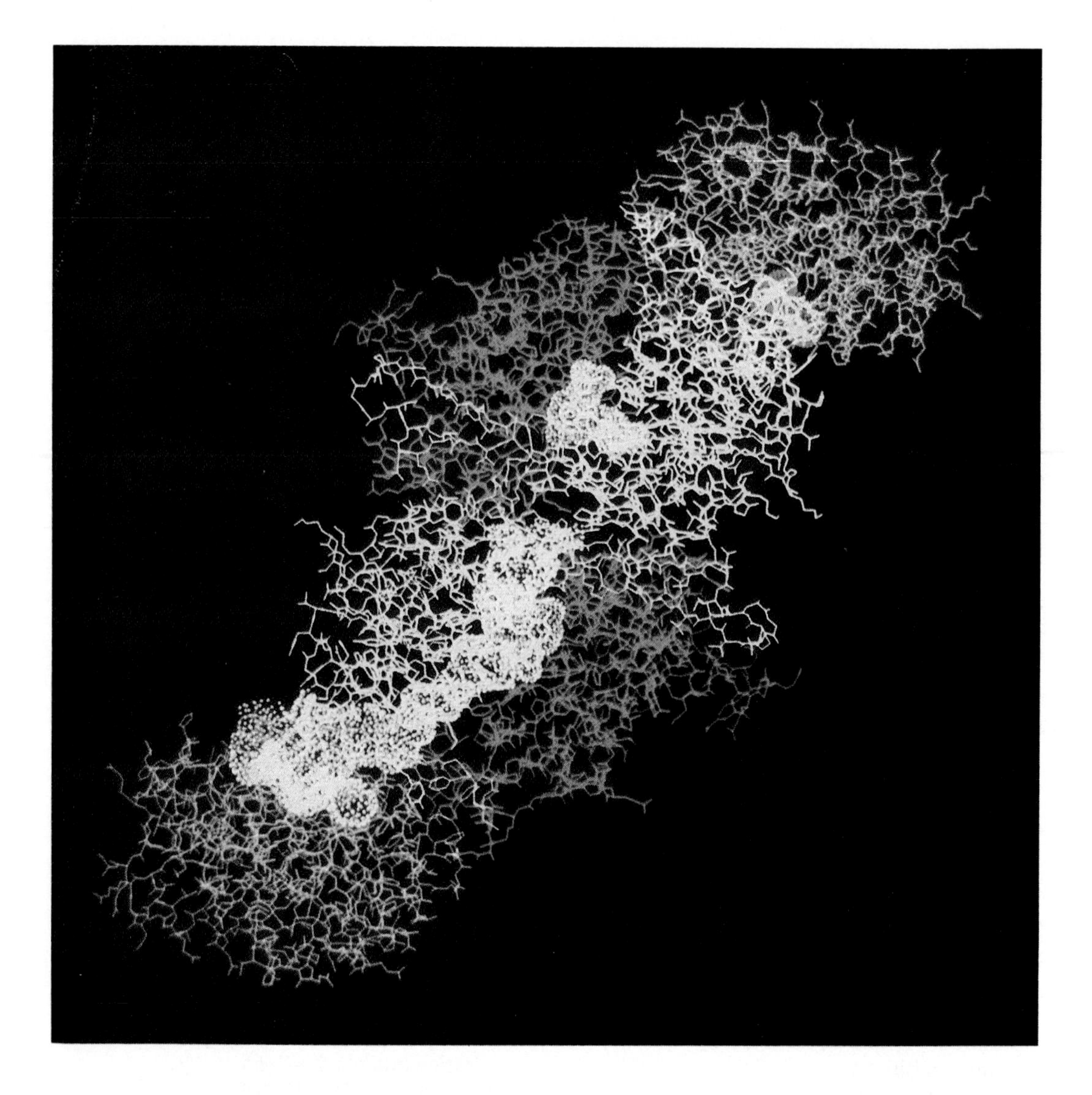

Plate II. View of the *S. typhimurium* $\alpha_2\beta_2$ tryptophan synthase complex, looking approximately down the two-fold axis of symmetry between ββ-subunit pairs. The smaller β subunits (blue) are distant from each other on opposite ends of the β-subunit dimer. The β-subunit N-terminal residues (1–204) and C-terminal residues (205–397) are shown in yellow and red, respectively. The dot surfaces highlight the positions of bound indole propanol phosphate (red) in the active sites of the α subunits and the coenzyme pyridoxal phosphate (dark blue) in the active sites of the α/β barrels. A tunnel that connects the two active sites (light blue) is shown in one αβ-subunit pair. Reprinted from the *Journal of Biological Chemistry* **263,** 17857 (1988).

Index